Aptamers in Biotechnology

Aptamers in Biotechnology

Edited by Fleur Lewis

www.statesacademicpress.com

States Academic Press,
109 South 5th Street,
Brooklyn, NY 11249, USA

Visit us on the World Wide Web at:
www.statesacademicpress.com

ISBN: 978-1-63989-732-2

Cataloging-in-Publication Data

Aptamers in biotechnology / edited by Fleur Lewis.
    p. cm.
Includes bibliographical references and index.
ISBN 978-1-63989-732-2
1. Oligonucleotides--Biotechnology. 2. Biotechnology. I. Lewis, Fleur.
QP625.O47 A68 2023
572.85--dc23

Table of Contents

Preface

Over the recent decade, advancements and applications have progressed exponentially. This has led to the increased interest in this field and projects are being conducted to enhance knowledge. The main objective of this book is to present some of the critical challenges and provide insights into possible solutions. This book will answer the varied questions that arise in the field and also provide an increased scope for furthering studies.

Aptamers are short sequences of artificial DNA or RNA that bind to specific target molecules such as peptides, proteins, carbohydrates, toxins, small molecules or live cells. In molecular biology, systematic evolution of ligands by exponential enrichment (SELEX) is a combinational chemistry technique used for generating oligonucleotides of either single-stranded DNA or RNA that specifically bind to a target ligand or ligands. Aptamers can be classified into two types, namely, peptide aptamers and DNA/RNA aptamers. There are several applications of aptamers, which can be categorized into therapeutic, diagnostic, reagent production, and engineering. Aptamers are suitable in diagnostic applications since they exhibit properties of tailored specificity and affinity, high thermal stability, and scalability of production. In therapeutics, aptamers can replace antibodies and facilitate the controlled release of therapeutic biomolecules such as growth factors. This book outlines the biotechnological applications of aptamers in detail. It aims to equip students and experts with the advanced topics and upcoming concepts in this area of study.

I hope that this book, with its visionary approach, will be a valuable addition and will promote interest among readers. Each of the authors has provided their extraordinary competence in their specific fields by providing different perspectives as they come from diverse nations and regions. I thank them for their contributions.

Editor

Rational Design of Aptamer-Tagged tRNAs

Takahito Mukai⬤

Department of Life Science, College of Science, Rikkyo University, 3-34-1 Nishi-Ikebukuro, Toshima-ku, Tokyo 171-8501, Japan; takahito.mukai@rikkyo.ac.jp

Abstract: Reprogramming of the genetic code system is limited by the difficulty in creating new tRNA structures. Here, I developed translationally active tRNA variants tagged with a small hairpin RNA aptamer, using *Escherichia coli* reporter assay systems. As the tRNA chassis for engineering, I employed amber suppressor variants of allo-tRNAs having the 9/3 composition of the 12-base pair amino-acid acceptor branch as well as a long variable arm (V-arm). Although their V-arm is a strong binding site for seryl-tRNA synthetase (SerRS), insertion of a bulge nucleotide in the V-arm stem region prevented allo-tRNA molecules from being charged by SerRS with serine. The SerRS-rejecting allo-tRNA chassis were engineered to have another amino-acid identity of either alanine, tyrosine, or histidine. The tip of the V-arms was replaced with diverse hairpin RNA aptamers, which were recognized by their cognate proteins expressed in *E. coli*. A high-affinity interaction led to the sequestration of allo-tRNA molecules, while a moderate-affinity aptamer moiety recruited histidyl-tRNA synthetase variants fused with the cognate protein domain. The new design principle for tRNA-aptamer fusions will enhance radical and dynamic manipulation of the genetic code.

Keywords: genetic code; anticodon; pseudo-anticodon; amber suppression

1. Introduction

RNA aptamers are small RNA molecules that specifically bind to their target molecules. Small hairpin RNA aptamers are naturally found in certain mRNAs [1–3] and other RNAs [4]. Diverse hairpin aptamers have been used for tagging many kinds of RNA molecules for recruiting their cognate target proteins or ligands on the fusion RNA molecules [4–7]. On the other hand, transfer RNAs (tRNAs) are small RNA molecules that transfer proteinogenic amino acids onto the growing polypeptide chain in the ribosome. tRNAs fold into the L-shape tertiary structure, while some tRNAs have an extra arm in varying size (variable arm or V-arm) (see Figure 1). Although several types of tRNA variants fused with hairpin RNA aptamers have been developed [7–13], these fusion RNAs usually lost the translational activity of tRNA. Recently, "V-Spinach tRNAs" were developed by inserting a Baby Spinach aptamer [14] into the V-loop of a few *E. coli* tRNA species [8], although the V-Spinach tRNAs showed only marginal tRNA activities [8,15]. More recently, the V-arm hairpin of a yeast serine tRNA species was replaced by the MS2 phage hairpin RNA (19 nucleotides in length) to recruit an enzyme fused with the MS2 coat protein on the premature form of these fusion RNA molecules in yeast cells [11]. These works encouraged us to design a new generation of tRNA-aptamer fusions in which a small hairpin aptamer constitutes the tip of the V-arm.

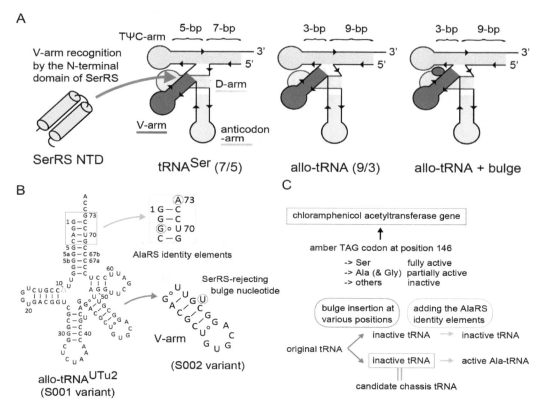

Figure 1. Strategy for tRNA chassis development. (**A**) Design of a new tRNA chassis derived from a few allo-tRNA species having the 9/3 composition. Compared to the *E. coli* tRNA^{Ser} having the canonical 7/5 composition, (9/3) allo-tRNA molecules may have a smaller space for the binding of the N-terminal domain (NTD) of SerRS. Thus, it was assumed that the interaction between SerRS and allo-tRNA molecules could be sterically hindered by the insertion of a bulge (flipped-out) nucleotide into the stem region of the V-arm. (**B**) A preliminary result suggested that the allo-tRNA^{UTu2} variant containing the bulge U was a poor substrate for SerRS, while its variant having the AlaRS identity elements was active for Ala-tRNA regardless of the bulge U insertion. (**C**) Experimental design for screening candidate chassis tRNAs by using a reporter gene in *E. coli*.

tRNA plays a central role in reprogramming the genetic code system and establishing new genetic codes [16–18]. In nature, the genetic code may have been expanded to acquire selenocysteine (Sec) and pyrrolysine (Pyl) as the 21st and 22nd amino acid, respectively. Interestingly, canonical tRNA^{Sec} has the largest tRNA structure, while tRNA^{Pyl} has the most compact tRNA structure [16], which may help them escape from non-cognate aminoacyl-tRNA synthetases (aaRSs). The genetic code was expanded in diverse organisms using synthetic biologists by developing various sets of orthogonal tRNA and aaRS pairs that do not cross-react with endogenous tRNAs and aaRSs and other orthogonal tRNA-aaRS pairs [17–22]. It has been a big challenge to artificially create orthogonal tRNA species with both high tRNA activity and high orthogonality in part due to saturation of recognition elements of tRNA [23]. One promising approach is "tRNA Extension (tREX)" [17] by which the tRNA activity and the orthogonality of a few orthogonal systems were significantly enhanced in a rapid and systematic manner towards their simultaneous use in *E. coli* [17]. Creation of a new tRNA identity may be achieved by introducing new tertiary structural elements [24,25] rather than by introducing small local modifications to tRNA. However, tRNA variants having an abnormal tertiary structure may be poorly compatible with the translation apparatus [8,16,26,27]. Thus, introduction of a small hairpin aptamer as the V-arm of orthogonal tRNAs may be a good solution for developing tRNA variants which are translationally active and have a characteristic tertiary structural element.

Recently, a new group of tRNAs, (9/3) allo-tRNAs, was identified in the raw short-read data of soil or marine metagenomic datasets [26,28]. These tRNAs have a non-canonical composition of the amino-acid acceptor branch (Figure 1A,B). While canonical tRNAs have a 7-bp acceptor stem and a 5-bp TψC stem, the (9/3) allo-tRNAs have a 9-bp acceptor stem and a 3-bp TψC stem (Figure 1A,B). In addition, the (9/3) allo-tRNAs have a long D-stem, like selenocysteine (Sec) tRNA species, and a long V-arm, like serine (Ser) and Sec tRNA species. Thus, the (9/3) allo-tRNAs intrinsically have identity elements for the Ser-tRNA synthetase (SerRS) and the Sec-tRNA synthesis apparatus [26,28,29]. Amber suppressor variants of two of the (9/3) allo-tRNA species were used for the site-specific Ser or Sec incorporation into proteins in response to the TAG amber stop codon in *E. coli* [29]. Two of the natural allo-tRNA species also have the essential identity elements for alanyl-tRNA synthetase (AlaRS) [26,29], and one of the two preferentially worked as tRNA^{Ala} rather than as tRNA^{Ser} in *E. coli* cells [26]. Despite the non-canonical secondary structure, many allo-tRNA species were highly compatible with the *E. coli* translation system and worked as exceedingly active nonsense or missense suppressor tRNAs in *E. coli* [26,29]. These findings prompted us to employ these (9/3) allo-tRNA scaffolds as tRNA chassis for making our tRNA-aptamer fusions. However, the strong intrinsic affinity of (9/3) allo-tRNA species for SerRS has limited their utility.

In the present study, a few kinds of SerRS-rejecting allo-tRNA chassis were developed. Several kinds of small hairpin aptamers inserted into these chassis were confirmed to be bound by their cognate proteins or protein domains expressed in *E. coli*. The new tRNA chassis, or "orthogonal allo-tRNAs", will help to greatly expand the design flexibility of new tRNA-aaRS pairs.

2. Results

2.1. Strategy for tRNA Chassis Development

In an attempt to develop a SerRS-rejecting (9/3) allo-tRNA^{Ser} variant, we realized that certain allo-tRNA species (allo-tRNA^{UTu2} series [29], see Figure 1B) having a G:U wobble base pair at the bottom of the V-stem worked as less active tRNA^{Ser} than allo-tRNA variants carrying Watson-Crick (WC) changes (data not shown). It is known that a G:U pair in an RNA double helix introduces a helical twist [30] and might cause steric hindrance against target proteins [31,32]. Thus, it was suggested that the original allo-tRNA^{UTu2} variant (or S001 variant) may have a narrow (but enough) space between the TψC arm and the V-arm for the insertion of the N-terminal domain (NTD) of SerRS (Figure 1A) [33]. To verify this idea, we inserted a bulged uracil (U) nucleotide into the V-stem of the S001 variant to make S002 (Figure 1A,B). In a preliminary experiment, the S002 variant hardly incorporated serine into proteins in *E. coli*, meaning that S002 is either SerRS-rejecting or simply unstable (Figure 1). On the other hand, the Ala-accepting variant of S002, or A002, which has the major identity elements for AlaRS (G3:U70 with A73) [26,29,34], incorporated alanine into proteins (data not shown here) (Figure 1C). Thus, the S002/A002 scaffold may retain the tertiary structural stability but may not have enough space for accommodating the NTD of SerRS (Figure 1). Encouraged by these preliminary results, I decided to examine seventy-nine kinds of bulge-carrying variants of nine kinds of amber suppressor (9/3) allo-tRNA species (Figure 1B and Figure S1) [26] in order to search for better chassis tRNAs than the S002 variant. Experiments were facilitated by use of *E. coli* cells with several reporter genes carrying an in-frame TAG codon (Figure 1C).

2.2. Searching for Candidate tRNA Chassis

In this study, all tRNA sequences were cloned in a high-copy-number plasmid vector under control of the arabinose promoter [26,29]. The starting amber suppressor allo-tRNA variants (named as S001, S012, S020, S026, S032, S043, S047, S051, and S055) were developed from nine allo-tRNA species (Figure 1B and Figure S1) chosen from ten allo-tRNA species found in nature [26]. Positional effects of single bulge nucleotide or two separate or successive bulge nucleotides on the V-arms of these allo-tRNA scaffolds were investigated by individually examining seventy-nine kinds of bulge-carrying variants (see supplementary Excel file). Their "Ala-acceptable" variants carrying the major identity elements for AlaRS (G3·U70 with A73) were also developed, although other structural elements contribute to the tRNA recognition by AlaRS [34,35]. A chloramphenicol acetyltransferase (CAT) reporter gene in which the Ser146 codon had been mutated to TAG, or *cat(Ser146TAG)* [26], was used for examining the TAG-translating activities of these tRNA variants in a suppressor-free strain of *E. coli* DH10B. While the replacement of Ser146 by Ala results in a partially active enzyme, replacement of Ser146 by bulkier residues than Ser resulted in loss of function [26,36]. Consequently, the *cat(Ser146TAG)* reporter is useful for detecting a strong Ala-inserting activity and even a weak Ser-inserting activity of allo-tRNAs in chloramphenicol (Cm)-sensitive *E. coli* (Figure 1C) [26]. Two conditions for a good candidate allo-tRNA chassis were that the Ala-acceptable variant inserts Ala, while the variant lacking the G3:U70 and A73 elements hardly inserts Ser (Figure 1C).

By comparing 78 variant pairs to the S002/A002 pair, S005/A005, S072/A072, and S073/A073 pairs (Figure 2A) passed the first screening using the *cat(Ser146TAG)* gene with Cm at a concentration of 34 µg/mL in the growth media (Figure 2B and Table S1). While S002 conferred marginal Cm resistance to cells, S005, S072, and S073 hardly conferred Cm resistance to cells (Figure 2B). A005 and A072 conferred up to 150 µg/mL Cm resistance, while A073 conferred only up to 67 µg/mL Cm resistance (Figure 2C). The S002/A002 and S005/A005 variant pairs are both derived from S001/A001 and have a bulged U or A, respectively, at different positions in the V-stem (Figure 2A). It is likely that S005 is more SerRS-rejective than S002. Because the S072/A072 and S073/A073 variant pairs differ only at the junction of the anticodon stem and the V-stem (Figure 2B), the A073 variant may be either less stable or less compatible to AlaRS than the A072 variant. Whereas the S079/A079 pair derived from S043 also passed the first screening (Table S1, Figure 2C), the other S043-derived variants retained their SerRS identity (Table S1) possibly due to the G20 residue (Figure S1). Thus, I assumed that the S005/A005 and S072/A072 chassis may be the most reliable ones.

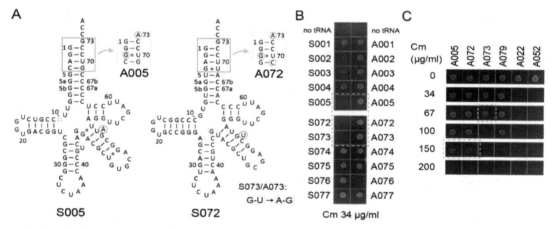

Figure 2. Candidate tRNA chassis. (**A**) The cloverleaf structures of the S005 and S072 tRNA variants and their Ala-accepting variants (A005 and A072, respectively). The S073/A073 variants have two indicated modifications compared to the S072/A072 variants, respectively. (**B**) The S005, S072, and S073 variants conferred lower chloramphenicol (Cm) resistance (less than Cm 34 µg/mL) to *E. coli* DH10B cells expressing the *cat(Ser146TAG)* reporter gene compared to the S002 variant. (**C**) The A005 and A072 variants conferred up to Cm 150 µg/mL resistance to cells, while the A073 variant conferred only Cm 67 µg/mL resistance.

2.3. Sequestration of Aptamer-tRNA Fusion RNAs

Tight protection of an aptamer moiety by its cognate protein will result in sequestration of the aptamer-tagged allo-tRNA molecules from the translation apparatus (Figure 3A,B). Previously, we observed that (9/3) allo-tRNA molecules were fully sequestered when selenocysteine synthase (SelA) molecules were overexpressed [29]. As shown in Figure 3A, (9/3) allo-tRNAs have a long D-stem which is specifically recognized by the *N*-terminal domain of SelA (SelA-N) [37] and the *C*-terminal domain of archaeal *O*-phosphoseryl-tRNA kinase (PSTK-C) [38]. As expected, the Ala-inserting activity of the A005 variant was almost fully removed when the *Aquifex aeolicus* (*Aa*) SelA-N or the

Methanopyrus kandleri (*Mk*) PSTK-C were expressed at a moderate level by using the *tac* promoter on a pGEX vector and IPTG at a concentration of 10 μM (Figure 3C). In the presence of these proteins, A005 hardly conferred 17 μg/mL Cm resistance to cells (Figure 3C), which demonstrated that this reporter assay system is valid. Interestingly, the *Methanocaldococcus jannaschii* (*Mj*) PSTK-C failed to sequester A005 tRNA molecules (Figure 3C), possibly because *Mj* tRNA[Sec] has a longer D-stem than those of *Mk* tRNA[Sec] and allo-tRNAs [38]. Similar results were obtained by using the A072 variant (data not shown). Next, the V-arm moiety was engineered to mediate the sequestration of tRNA molecules. Two kinds of small hairpin aptamers, the MS2 hairpin RNA [9,11] and the *E. coli fdhF* SECIS RNA [3], were transplanted to the A005 variant with varying lengths of adaptors (Figure 3A). The in vivo interaction of the transplanted MS2 RNA and SECIS RNA with a MS2 coat protein dimer [9,11] and the *C*-terminal WH3/4 domains of *E. coli* SelB [3], respectively, were examined. As expected, the A005-MS2 variants were sequestered by the coat protein, while the A005-SECIS variants were sequestered by the WH3/4 domains (Figure 3D). Among the three A005-MS2 variants, A005-MS2-1 with the medium-length adapter was most efficiently sequestered by MS2 coat protein (less than 17 μg/mL Cm resistance) (Figure 3D). Among the three A005-SECIS variants, A005-SECIS-2 with the longest adapter was most efficiently sequestered by the WH3/4 domains (up to 17 μg/mL Cm resistance) (Figure 3D). Thus, the optimal length of the adapter stem varied for each aptamer-protein pair. These experiments demonstrated that a small hairpin aptamer can be safely installed into an allo-tRNA chassis without significantly impairing the tRNA function in the absence of the target protein.

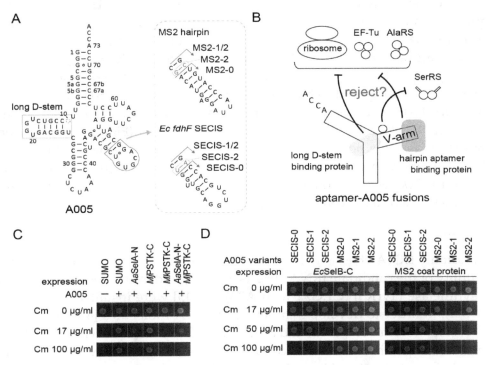

Figure 3. Sequestration of aptamer-tRNA fusions from the *E. coli* translation system via the expression of the cognate aptamer-binding proteins. (**A**) The cloverleaf structure of A005 tRNA variants having a hairpin RNA aptamer in the V-arm. The long D-stem of (9/3) allo-tRNAs and tRNA[Sec] species is a binding site for tRNA[Sec]-binding enzymes SelA and PSTK. (**B**) Strategy for the sequestration of the aptamer-A005 fusion variants. (**C**) The A005 tRNA molecules were almost fully sequestered by the *N*-terminal domain of *Aquifex aeolicus* (*Aa*) SelA and the *C*-terminal domain of *Methanopyrus kandleri* (*Mk*) PSTK but not by the *C*-terminal domain of *Methanocaldococcus jannaschii* (*Mj*) PSTK. (**D**) The SECIS-A005 fusions were sequestered by the *C*-terminal domain of *E. coli* SelB, while the MS2-A005 fusions were sequestered by the MS2 coat protein. Among the fusion tRNAs, the A005-MS2-1 variant was most effectively sequestered. The SUMO protein was used as a negative control and as a *N*-terminal solubilizing tag for the SelA, PSTK, and SelB domains.

Next, sequestration of fully active tRNA^{Ser} molecules was challenged. It is known that the allo-tRNA^{UTu1} variant (Figure 4A and Figure S2A) [29] (or S012 in this study) is as active a tRNA^{Ser} as one of the *E. coli* tRNA^{Ser} species in *E. coli*. S012 conferred at least 600 μg/mL Cm resistance to *E. coli* expressing the *cat(Ser146TAG)* gene (Figure S2B) and at least 34 μg/mL Cm resistance to cells additionally expressing the *Mk* PSTK-C (Figure S2C,D). Thus, full sequestration of these fully active tRNA molecules may require another tight interaction between an RNA structure and its target protein. Instead of the wildtype MS2 hairpin RNA and the wildtype MS2 coat protein, an MS2 hairpin RNA variant with a *C*-loop (U to C at the -5 position) [39,40] and the MS2 coat protein V29I variant [40,41] were used to enhance the RNA aptamer-coat protein interaction. To make a single-chain MS2 coat protein [42] together with an *Mk* PSTK-C domain, two V29I domains were fused with a long linker containing a SUMO domain, a 6xHis tag, and an *Mk* PSTK-C domain, in this order (Figure 4B). This chimeric fusion protein efficiently sequestered a S012 variant having an MS2 *C*-loop hairpin with a 1-bp adaptor, or S012-MS2c-1, in *E. coli* expressing the *cat(Ser146TAG)* gene (Figure 4A and Figure S3). However, it seems likely that the S012-MS2c-1 variant was less active than the S012 variant, as the original S012 variant was exceedingly toxic to cells (Figure 4C), probably due to its high suppression efficiency [26,29]. Thus, the V-arm adaptor length of the S012-MS2c-1 variant was optimized to restore its tRNA activity in the absence of the chimeric protein (Figure 4A). One variant that has the smallest V-arm (S012-MS2c-m2) showed a similar toxicity to the S012 variant (Figure 4A,C). Both the S012 and S012-MS2c-m2 variants still conferred at least 50 μg/mL Cm resistance to *E. coli* expressing the *cat(Ser146TAG)* gene and the *Mk* PSTK-C domain not fused with the MS2 coat protein (Figure 4D). Then, the sequestration efficiencies of S012 and S012-MS2c-m2 molecules by the chimeric fusion protein were examined (Figure 4E). Surprisingly, expression of the chimeric fusion protein completely repressed the Ser-inserting activities of S012-MS2c-m2 molecules (Figure 4E), while S012 molecules were not completely sequestered. To see the contribution of the *Mk* PSTK-C domain moiety and the single-chain coat protein moiety in the chimeric fusion protein, the *Mk* PSTK-C domain was removed or replaced by an *Mj* PSTK-C domain which did not sequester allo-tRNAs (Figure 4E). All fusion proteins sequestered S012-MS2c-m2 molecules in a complete manner (Figure 4E). Thus, the single-chain coat protein moiety may have bound to the MS2 hairpin moiety of the S012-MS2c-m2 molecules and inhibited their interaction with SerRS. In contrast, the partial sequestration of S012 molecules was dependent on the *Mk* PSTK-C domain (Figure 4E). These results demonstrated that the strong protein-RNA interaction led to the complete sequestration of allo-tRNA molecules.

2.4. Synthetic Tyrosine tRNAs

Can the candidate tRNA chassis gain any amino-acid identity other than alanine and serine? To confirm the utility of the candidate tRNA chassis, the major identity elements for the archaeal-type tyrosyl-tRNA synthetase (TyrRS) (C1:G72 with A73) [43–45] were installed in the candidate tRNA chassis (Figure 5A), in order to develop synthetic Tyr tRNA molecules [44]. Y005, Y072, and Y073 variants were developed from S005, S072, and S073, respectively. Two derivatives of Y072 (Y072-2 and Y072-3) were also developed because modifications of the second and third base pairs in the acceptor stem may affect aminoacylation (see Figure 5A). An amber suppressor variant of the tRNA^{Tyr} species of *Methanocaldococcus jannaschii* (*Mj*), or Nap3 [46], was used as a positive control. An archaeal TyrRS species from *Candidatus* Methanomethylophilus alvus Mx1201 (*Ma*) was used to be paired with these tRNAs. Three kinds of reporter genes were used for examining the orthogonality and the tRNA^{Tyr} activity of these tRNA variants (Figure 5B,C). One is an ampicillin/carbenicillin resistance gene (*bla*) having an in-frame TAG codon at a permissive site (Ala182) [47] (Figure 5B). The *cat(Ser146TAG)* reporter gene was cloned together with this *bla(Ala182TAG)* reporter gene to make a dual-reporter plasmid. The third one is a sfGFP gene in which the Tyr66 codon had been mutated to TAG, since Tyr66 is essential for the fluorophore formation [48] (Figure 5C). Using the *bla* reporter gene, it was revealed that the Y072 and Y073 variants conferred only marginal carbenicillin (Car) resistance to cells, while the Y005 variant and the *Mj* tRNA^{Tyr} Nap3 variant conferred up to 500 μg/mL Car resistance to cells

(Figure 5B). Note that the Nap3 variant has been used as a representative orthogonal tRNA^{Tyr} in *E. coli* [49]. This result indicated that the Y072 and Y073 variants were not activated by any endogenous aaRS species in *E. coli*. Although the Y072-2 variant conferred up to 500 µg/mL Car resistance to cells (Figure 5B), none of the Y072, Y072-2, and Y072-3 variants conferred Cm resistance to *E. coli* DH10B cells expressing the *cat(Ser146TAG)* reporter gene (Figure 5B). Thus, these variants showed no Ser identity. Next, by using the sfGFP(Tyr66TAG) reporter gene, it was confirmed that the Y072, Y072-2, and Y072-3 variants were charged with tyrosine by *Ma* TyrRS (Figure 5C).

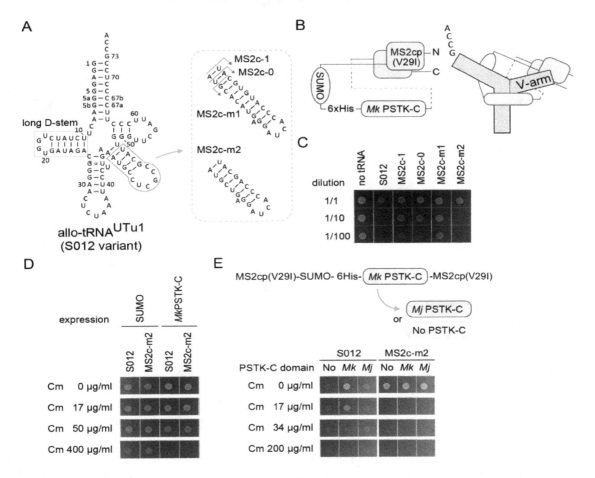

Figure 4. Sequestration of fully active allo-tRNA^{Ser} molecules. (**A**) The cloverleaf structure of S012 tRNA variants having a MS2 hairpin *C*-loop variant with varying lengths of adaptors. The long D-stem of S012 variants is tightly bound by the *C*-terminal domain of *Mk* PSTK. (**B**) A structural model of the fusion protein consisting of an MS2 coat protein V29I variant monomer, a SUMO tag, a 6xHis tag, a *Mk* PSTK-C, and another V29I monomer. The *N*-terminus and *C*-terminus are indicated. A binding model of a S012 variant and this fusion protein is shown, too. (**C**) The S012 and MS2c-m2 variants most efficiently impaired the growth of *E. coli* DH10B cells expressing the *cat(Ser146TAG)* reporter gene and the SUMO protein. Cell cultures were diluted (10 times and 100 times) or non-diluted and spotted on an agar plate. The arabinose concentration was raised to 0.1% from 0.01% to observe clear differences. (**D**) The *Mk* PSTK-C protein not fused with the MS2 coat protein failed to fully sequester both S012 and MS2c-m2 molecules in *E. coli* DH10B cells expressing the *cat(Ser146TAG)*. (**E**) The *Mk* PSTK-C domain of the chimeric fusion protein was replaced with an *Mj* PSTK-C domain or simply removed. All fusion protein variants perfectly sequestered MS2c-m2 molecules in *E. coli* DH10B cells expressing the *cat(Ser146TAG)*, indicating that the single-chain coat protein moiety tightly bound with the MS2 hairpin moiety of MS2c-m2. On the other hand, S012 molecules were sequestered only by the *Mk* PSTK-C-containing fusion protein. The other fusion proteins failed to repress the toxicity and the suppressor tRNA activity of the S012 molecules. Note that cells are unclear but were growing on the Cm free and Cm-containing agar plates under the S012 + No/Mj conditions.

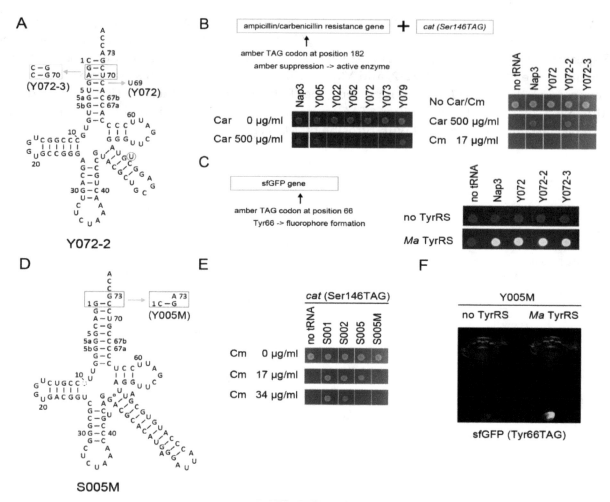

Figure 5. Development of synthetic tyrosine tRNAs using candidate tRNA chassis. (**A**) The cloverleaf structure of Y072-series tRNA variants. (**B**) The S072 and S073 variants conferred only marginal carbenicillin (Car) resistance (less than Car 500 μg/mL) to *E. coli* DH10B cells expressing the *bla(Ala182TAG)* reporter gene and the *cat(Ser146TAG)* reporter gene in the absence of any heterologous TyrRS enzyme. Amber suppressor variants of archaeal *M. jannaschii* tRNA^{Tyr} species (Nap3) and the Y005, Y079, and Y072-2 variants conferred up to 500 μg/mL Car resistance, indicating that these tRNAs were to a less extent charged by endogenous aaRSs. No Cm resistance was introduced by the Y072, Y072-2, and Y072-3 variants, indicating that they were orthogonal to SerRS. (**C**) In the presence of Ca. M. alvus TyrRS in *E. coli* DH10B, the Nap3 variant and the three Y072-series variants translated the amber codon in the *sfGFP(Tyr66TAG)* reporter gene with tyrosine, making the fluorophore of sfGFP. (**D**) The cloverleaf structure of the S005M and Y005M tRNA variants. (**E**) The S005M variant conferred a lower Cm resistance (less than Cm 17 μg/mL) to *E. coli* DH10B cells expressing the *cat(Ser146TAG)* reporter gene compared to the S002 and S005 variants. (**F**) In the presence of Ca. M. alvus TyrRS in *E. coli* DH10B, the Y005M variant translated the amber codon of the *sfGFP(Tyr66TAG)* reporter gene with tyrosine, making the fluorophore of sfGFP.

To improve the usefulness of the S005 chassis, I examined another chassis, S005M, which is a variant of A005-MS2-1 lacking the AlaRS identity elements (Figure 5D). It was found that the original S005 variant conferred 17 μg/mL Cm resistance to *E. coli* DH10B cells expressing the *cat(Ser146TAG)* reporter gene (Figure 5E). In contrast, the S005M variant having an MS2 hairpin at the tip of the V-arm did not confer 17 μg/mL Cm resistance (Figure 5E). Most likely, the S005M variant may have a stabilized bulge structure and a slightly reduced tRNA activity compared to S005. Although the marginal Ser-inserting activity of S005 can be eliminated by engineering the tip of the acceptor stem [22], the use of S005M facilitates engineering of the amino-acid identity owing to the ignorable Ser-inserting activity.

The S005M chassis was engineered to be coupled with *Ma* TyrRS, to develop Y005M (Figure 5D). In the presence of *Ma* TyrRS, Y005M translated the amber codon of the sfGFP(Tyr66TAG) reporter gene with tyrosine (Figure 5F). These results clearly demonstrated that SerRS-rejecting allo-tRNA chassis can be engineered and converted to synthetic Tyr tRNA species.

2.5. Enhancing Aminoacylation by Aptamer-Protein Interaction

Can an RNA aptamer moiety in tRNA work as an additional binding site for an aaRS fused with the aptamer-binding protein domain? To answer this, synthetic histidine (His) tRNA molecules [18] were developed by installing the major identity element for the *E. coli* histidyl-tRNA synthetase (HisRS) (G-1:C73) [2,26,50] into the S072/S073 chassis (Figure 6A). Two kinds of *cat* reporter genes, *cat(Ser146TAG)* and *cat(His193TAG)*, were used for examining the orthogonality and the His-inserting activity of designed tRNA variants. In the *cat(His193TAG)* gene, the essential catalytic His193 was changed to TAG [51]. It soon turned out that a S073 variant with a G-1:C73 pair (named H073) was not charged by the endogenous HisRS in *E. coli* expressing the *cat(His193TAG)* gene (Figure 6B). However, extending the V-stem length and modifying the V-arm loop sequence produced functional H072/H073 variants which conferred up to 34 µg/mL Cm resistance to cells (Figure 6B). Ser insertion was not mediated by these H072/H073 variants in *E. coli* cells expressing the *cat(Ser146TAG)* gene (Figure S4A). Point mutations in the V-arm loop sequence affected the His-inserting activity of H072 ψHis-1 (Figure 6A); UUUUGAU was better than the original UUGUGAU, while UUGAGAU and UUGCGAU were less suitable (Figure 6B and Figure S4B,C), implying that a good V-arm loop sequence on a moderately long V-arm may be rich in uracil residues. It was also found that a moderate overexpression of HisRS from a plasmid-encoded *hisS* gene enhanced the His-inserting activities by the synthetic His tRNAs (Figure 6C and Figure S4F). Up to 150 µg/mL Cm resistance was conferred by the H072-λboxB-0 variant (Figure 6A,C). On the other hand, the S005 chassis was useless for developing a synthetic His tRNA in the same manner (Figure S4D,E). Thus, it was revealed that the H072/H073 chassis were intrinsically compatible to *E. coli* HisRS and *E. coli* RNase P, which cleaves His tRNA precursors at the -1 position [52].

By using the H072 chassis and *E. coli* HisRS, several cognate pairs of RNA aptamers and protein domains [2,28,50,53–55] were examined (Figure S5A,B). The first preliminary experiment revealed that the cognate pair of the 7SK RNA stem-loop 4 (SL4) and the hLarp7 xRRM domain [53] fused to the *N*-terminus of *E. coli* HisRS (Figure 6D,E) might have enhanced the His-inserting activity by the tRNA-aptamer fusion in *E. coli* expressing the *cat(His193TAG)* gene (Figure S5C). The other pairs examined seemed to be rather inhibitory (Figure S5C). Three explanations are possible: 1) addition of an *N*-terminal domain might have reduced the expression level, stability, accessibility, or activity of the tethered HisRS, 2) too strong affinity between the cognate RNA-protein pair might have reduced the turnover rate of aminoacylation, and 3) a too short linker sequence SSGSNSNSGS between the *N*-terminal domain and HisRS might have kept the catalytic pocket of the tethered HisRS away from the CCA tail of tRNA. In contrast, the xRRM domain binds a 7SK SL4 RNA with a moderate affinity (Kd = 129 nM [53]). Furthermore, the *C*-terminus of the xRRM domain and the *N*-terminus of the *E. coli* HisRS may approach each other on the tRNA and can be fused by using the short linker (Figure 6E), according to crystal structural studies [50,53]. In another experiment, the H072-7SK-1 variant conferred up to 150 µg/mL resistance to *E. coli* DH10B cells expressing the *cat(His193TAG)* reporter gene and xRRM-HisRS (Figure 6F), while the U311A change in the 7SK SL4 loop (Figure 6D), which reportedly enhances the interaction (Kd = 51 nM [53]), slightly reduced the His-inserting activity (Figure 6F). The U311A change may have slightly impaired the turnover of the enzyme or slightly destabilized the tRNA. Lastly, the effect of the xRRM domain on aminoacylation was confirmed by using the H072-7SK-1 and H072-λboxB-0 (Figure 6A) variants. The expression of xRRM-HisRS greatly enhanced the His-inserting activity of H072-7SK-1 (Figure 6G), while H072-λboxB-0 was efficiently activated by the wildtype HisRS but not by xRRM-HisRS (Figure 6G). This meant that the appended SUMO-xRRM domains were rather inhibitory to histidylation in the absence of the cognate interaction between the xRRM domain and the aptamer moiety, probably due to the decreased expression level, stability, or accessibility of HisRS. Thus, it was concluded that the cognate interaction contributed to the enhancement of aminoacylation.

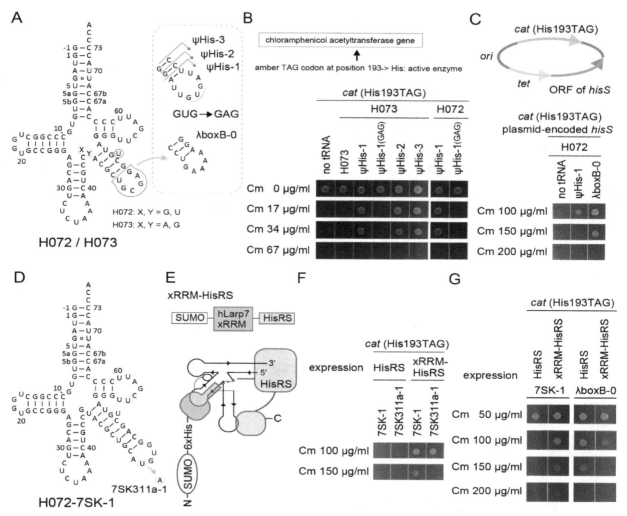

Figure 6. Development of synthetic histidine tRNAs using candidate tRNA chassis and hairpin aptamers. (**A**) The cloverleaf structures of H072 and H073 tRNA variants having an extended V-stem and a modified V-arm loop sequence. (**B**) The H073 ψHis-1/2/3 variants and the H072 ψHis-1 variant conferred up to Cm 34 µg/mL resistance to *E. coli* DH10B cells expressing the *cat(His193TAG)* reporter gene. The catalytic His193 residue is essential for the Cm acetyltransferase activity. A point mutation in the V-arm loop sequence (from GUG to GAG) eliminated the His-inserting activity of the ψHis-1 variants. (**C**) The H072 ψHis-1 and λboxB-0 variants conferred up to Cm 100 µg/mL resistance and 150 µg/mL resistance, respectively, to *E. coli* DH10B cells expressing the *cat(His193TAG)* reporter gene and the plasmid-encoded *hisS* gene (*E. coli* HisRS) shown. Note that the agarose plate containing Cm 100 µg/mL was incubated for another four hours to obtain the clear cell spot for H072 ψHis-1. (**D**) The cloverleaf structure of the H072-7SK-1 variant having a 7SK SL4 hairpin. (**E**) A chimeric *E. coli* HisRS variant fused with the xRRM domain of human Larp7. A binding model of H072-7SK-1 and this xRRM-HisRS is shown. The dashed line indicates linker residues SSGSNSNSGS. (**F**) Expression of xRRM-HisRS enhanced the His-inserting activities by the H072-7SK-1 and H072-7SK311a-1 variants. The H072-7SK-1 variant conferred up to 150 µg/mL resistance to *E. coli* DH10B cells expressing the *cat(His193TAG)* reporter gene and xRRM-HisRS. It was reported that the 7SK RNA SL4 binds the hLarp7 xRRM domain, while the 311a variant binds the xRRM domain more tightly. (**G**) The His-inserting activity of H072-7SK-1 was enhanced by xRRM-HisRS, while that of H072-λboxB-0 was not enhanced.

2.6. Recruiting aaRS via a Pseudo-Anticodon in the V-Arm

Can an aaRS variant having an additional heterologous anticodon binding domain (ABD) be recruited to tRNA via the V-arm which mimics the anticodon arm that can be recognized by the ABD? In other words, can a pseudo-anticodon arm moiety [1,56] on tRNAs work as an aptamer for the ABD and contribute to aminoacylation enhancement? Here, the full-length HisRSs, HisRS catalytic domains (CDs), and HisRS ABDs of two α-Proteobacteria species and their tRNA^{His} anticodon arm were used. It is well known that a large subgroup of α-Proteobacteria has a non-canonical tRNA^{His} lacking the G-1:C73 pair and has a non-canonical HisRS species that recognizes the non-canonical G1:U72 pair and A73 in addition to the GUG anticodon [17,57–60]. The two α-proteobacterial species are *Afifella pfennigii* DSM 17,143 (*Ap*) [17] and *Consotaella salsifontis* USBA 369 (*Cs*) [61], because the well-studied α-proteobacterial *Caulobacter crescentus* HisRS was not active enough in *E. coli* [60]. Two kinds of allo-tRNA chassis, S072 and S005, were engineered to have a typical α-proteobacterial tRNA^{His} anticodon arm as their V-arms, to develop H072-αψHis-3 and αH005-αψHis-3/4/5 variants (Figure 7A). The H072-αψHis-3 variant has the G-1:C73 element to be paired with *E. coli* HisRS variants N-terminally fused with an α-proteobacterial ABD using a 20 amino-acid linker (Figure 7B). On the other hand, the αH005-αψHis-3/4/5 variants have the G1:U72 and A73 elements to be paired with *Ap/Cs* HisRS N-terminally fused with a heterologous ABD using the same 20 amino-acid linker (Figure 7B). The C-terminal ABDs of chimeric *Ap/Cs* HisRS variants were kept intact or trimmed (Figure 7B). *ApEc*, *CsEc*, *ApCs*, and *CsAp* variants denote *Ap*ABD-*Ec*HisRS, *Cs*ABD-*Ec*HisRS, *Ap*ABD-*Cs*HisRS, and *Cs*ABD-*Ap*HisRS, respectively (Figure 7B).

By using the *cat(His193TAG)* reporter system, the His-inserting activities by combinations of the tRNA variants and HisRS variants were examined (Figure 7C). Addition of the N-terminal *Ap*ABD, but not *Cs*ABD, to the *E. coli* HisRS led to the enhancement of the His-inserting activity by H072-αψHis-3 (Figure 7C). Thus, the *Ap*ABD may be especially useful in *E. coli*, as explicitly predicted in a previous study [17]. Next, the αH005-αψHis-3/4/5 variants were combined with the chimeric *Ap/Cs* HisRS variants. The *ApCs* variant efficiently activated the αH005-αψHis-3/4/5 variants, while the *CsAp* variant was less active (Figure 7C and Figure S6). Among the αH005-αψHis-3/4/5 variants, αH005-αψHis-4 seems the most active, because αH005-αψHis-3 and αH005-αψHis-5 conferred slightly lower Cm resistance (Figure S6). In contrast, the chimeric *Ap/Cs* HisRS variants lacking the original C-terminal ABD were totally inactive (Figure 7C), probably because HisRS ABD is responsible not only for anticodon recognition but also for HisRS dimerization and for precise acceptor stem discrimination [18,62]. Thus, it is evident that the tRNA variants having a GUG pseudo-anticodon arm were charged with histidine by HisRS variants N-terminally fused with an *Ap*ABD.

To confirm the involvement of the GUG pseudo-anticodon in the tRNA-HisRS interaction, two lines of experiments were performed. First, the H072-αψHis-3 variant (having the GUG pseudo-anticodon) and the H072-7SK-1 variant (having no pseudo-anticodon arm) were compared (Figure 7D). To partially impair the anticodon recognition activity of the *Ap* ABD, two mutations (E484A-K493Q) which may partially impair anticodon recognition [50] were introduced to develop *Ap*^{AQ}*Ec* and *Ap*^{AQ}*Cs* HisRS variants (Figure 7B). As expected, *ApEc* enhanced the histidylation of H072-αψHis-3, whereas *Ap*^{AQ}*Ec* failed (Figure 7D). H072-αψHis-3 was activated at similar efficiencies by wildtype HisRS and xRRM-HisRS (Figure 7D). In contrast, the histidylation of H072-7SK-1 was significantly enhanced by xRRM-HisRS (Figure 6G,7D). Both *ApEc* and *Ap*^{AQ}*Ec* variants failed to enhance the histidylation of H072-7SK-1 (Figure 7D). Thus, it was revealed that the interaction between the GUG pseudo-anticodon arm and the *Ap* ABD moieties contributed to the enhancement of aminoacylation. However, compared to the 7SK-xRRM interaction (Figure 6E), this GUG-*Ap*ABD interaction exerted a modest effect on aminoacylation enhancement (Figure 7D).

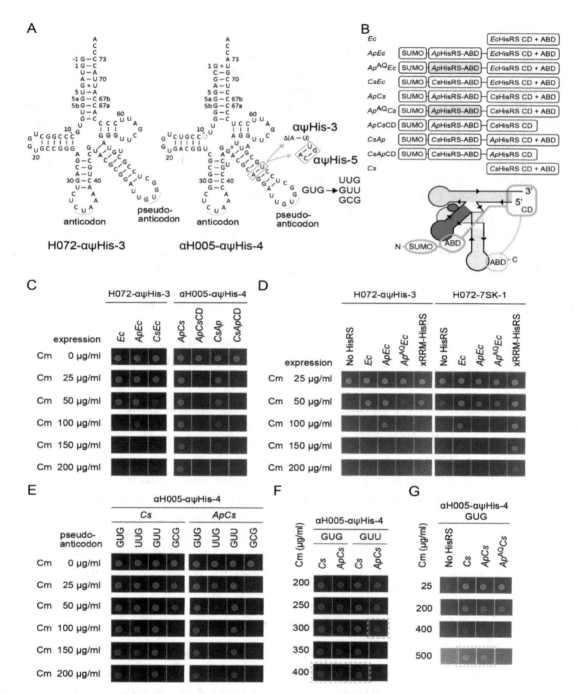

Figure 7. Recruiting HisRS via a pseudo-anticodon arm. (**A**) The cloverleaf structures of the H072-αψHis-3 and αH005-αψHis-3/4/5 variants with varying lengths of adaptors and with diverse pseudo-anticodon sequences. (**B**) The domain structures of HisRS and HisRS variants *N*-terminally fused with a heterologous anticodon binding domain (ABD). CD denotes catalytic domain. *Ec*, *Ap*, and *Cs* denote *E. coli*, *Afifella pfennigii*, and *Consotaella salsifontis*, respectively. *Ap*^{AQ} denotes an *Ap* ABD variant carrying E484A-K493Q mutations impairing anticodon recognition. (**C,D**) Cm resistance of *E. coli* DH10B (NEB 10-beta) cells expressing the *cat(His193TAG)* reporter gene and the indicated combinations of HisRS and tRNA variants. (**D**) The His-inserting activity of H072-αψHis-3 was enhanced by *ApEc* but not enhanced by *Ap*^{AQ}*Ec*, indicating that the recognition of the GUG pseudo-anticodon by the wildtype *Ap* ABD moiety was important. On the other hand, the His-inserting activity of H072-7SK-1 lacking the GUG pseudo-anticodon was not enhanced by both *ApEc* and *Ap*^{AQ}*Ec*. No HisRS indicates that only the genomic *hisS* gene was expressing. (**E–G**) Cm resistance of *E. coli* DH10B (NEB 10-beta) cells expressing the *cat(His193TAG)* reporter gene and the indicated combinations of HisRS variants and pseudo-anticodon variants of αH005-αψHis-4. The Cm concentrations are indicated. (**G**) Note that the agarose plate containing Cm 500 μg/mL was incubated for another four hours to obtain clear cell spots.

The second line of experiments were performed to examine the importance of the GUG pseudo-anticodon sequence. The GUG sequence in αH005-αψHis-4 was changed to UUG, GUU, and GCG (Figure 7A). According to a previous study, *C. crescentus* tRNA^{His} variants having these anticodon changes are poor substrates of *C. crescentus* HisRS [58]. Surprisingly, the wildtype *Cs* HisRS charged all pseudo-anticodon variants of αH005-αψHis-4 (Figure 7E). Thus, it is evident that the critical recognition elements of tRNA^{His} for *Cs* HisRS are the G1:U72 pair with the A73 discriminator base but not the GUG anticodon. In the presence of *Cs* HisRS, the original GUG variant and the GUU variant conferred at least Cm 400 μg/mL resistance (Figure 7F), while the UUG and GCG variants conferred only up to Cm 150 μg/mL and Cm 50 μg/mL resistance, respectively (Figure 7E). In the presence of *Cs* HisRS or *ApCs*, the original GUG variant conferred up to Cm 500 μg/mL resistance, while it still conferred at least Cm 400 μg/mL resistance in the presence of *Ap*^{AQ}*Cs* (Figure 7G). Thus, this GUG-*Ap*ABD interaction exerted only a modest effect on aminoacylation enhancement, probably because αH005-αψHis-4 is already a good substrate of wildtype *Cs* HisRS. In contrast, the UUG/GUU/GCG pseudo-anticodon variants were charged by *ApCs* much less efficiently than they were charged by *Cs* HisRS (Figure 7E,F). This indicated that the *N*-terminal appended domain was rather inhibitory to histidylation in the absence of the cognate interaction between the pseudo-anticodon and the appended ABD. In summary, it was shown that a pseudo-anticodon arm moiety can contribute to the enhancement of aminoacylation efficiency and tRNA-recognition specificity.

3. Discussion

In the present study, I showed that the SerRS-rejecting allo-tRNA chassis can be engineered to transfer Ala, Tyr, and His and fused with several kinds of hairpin RNAs. Although this study lacks any biochemical data, combinations of traditional in vivo reporter assay experiments clearly demonstrated that some of the RNA aptamers installed into the allo-tRNA chassis were bound by their cognate proteins or protein domains. Furthermore, the interactions between RNA aptamers and their cognate protein domains contributed to the enhancement of the tRNA-recognition specificity and the histidyl-tRNA formation by chimeric HisRS variants. Our next task should be simultaneous in vivo selection of the aptamer moieties and protein domains to optimize the V-stem length, the V-arm loop sequence, and the aptamer-recognizing residues of the protein domains. This optimization step may also be required for reducing the affinity of high-affinity aptamers and for preventing misrecognition of endogenous RNA molecules by the protein domains.

Some of the methods developed in this study may be applicable to genetic code expansion and genetic code reprogramming. For example, the pairs of Ca. M. alvus TyrRS and Y072/Y005M and the pairs of *Cs* HisRS variants and αH005-αψHis-4 variants may be orthogonal in *E. coli* and might be useful for incorporating non-canonical amino acids [17,58,60]. Alternatively, some of the endogenous tRNAs or tRNA-aaRS pairs might be replaceable with allo-tRNAs or orthogonal pairs composed of allo-tRNAs [60,63–65]. For reprogramming the genetic code, the allo-tRNA sequestration system may be useful for use in vivo and in vitro [66] and for the purification and elimination of allo-tRNA molecules from tRNA mixtures [66–69]. Application of the allo-tRNA chassis may not be limited to *E. coli* but may possibly need further optimization, especially for their use in archaea and eukaryotes due to the difference in the manner of tRNA^{Ser} recognition by SerRS [70,71].

In recent years, synthetic biologists have started to create synthetic tRNA-aminoacylation protein enzymes and ribozymes [18,72,73]. However, unlike most of the natural aaRS species, these artificial protein enzymes, including my HisRS chimeras, are composed of separate domains fused with a flexible linker [18,72]. Thus, such artificial enzymes were far less active and precise than natural enzymes [18,72]. On the other hand, the ribozymes required pre-activated amino acid substrates rather than free amino acid and ATP [73]. An apparent drawback of my allo-tRNA strategy is that the tip of the V-arm is distant from the CCA tail of tRNA. It may be helpful to use a long and rigid linker motif found in mitochondrial AlaRS species [74] or to use a long-D-arm binding protein domain to wrap around tRNA [37,38]. The use of T-box riboswitches as RNA scaffolds may be an alternative approach [73,75]. In summary, the novel allo-tRNA chassis, or orthogonal allo-tRNAs, would facilitate creation of synthetic tRNA-aminoacylation enzymes and ribozymes.

4. Materials and Methods

4.1. Cloning tRNA and Protein Genes

The nucleotide and amino-acid sequences are provided in the supplementary Excel file. All tRNA sequences (including the CCA tail) were cloned between the EcoRI and BglII sites in the pBAD-RSF5 plasmid [26] re-constructed in this study. The *cat* variant genes were developed from pACYC184 using Infusion (TAKARA Bio Inc., Shiga, Japan). The Ca. M. alvus TyrRS gene was cloned into pACYC184 by replacing the *cat* gene ORF. The HisRS ORFs were cloned immediately downstream of the *tet* gene of pACYC184-cat(193HisTAG). The *bla* variant gene was inserted downstream of the *tet* gene of pACYC184-cat(Ser146TAG) using Infusion. The other protein genes were cloned into the pGEX-6P-1 vector (a gift from Kazuhisa Nakayama, Addgene plasmid # 61838) and expressed under control of the *tac* promoter. Oligo DNAs and synthetic protein genes were purchased from Eurofin Genomics Co., Ltd. (Tokyo, Japan).

4.2. In Vivo Reporter Assays

E. coli DH10B and NEB 10-beta (NEB) cells were transformed with the indicated combinations of plasmids. Before spotting assays, transformed cells were pre-cultured in liquid LB media in the presence of 0.01% L-arabinose and 10 μM IPTG, if required, for a few hours [26]. In typical experiments, cell cultures were diluted to OD_{600} = 0.05 and then incubated for three hours. After spotting cell cultures (2 μL each) on LB agar plates, the plates were incubated at 37 °C overnight or for a longer time. For the induction of sfGFP genes, the final concentration of IPTG was 100 μM.

Supplementary Materials:
Figure S1: Cloverleaf structures of the eight starting amber suppressor (9/3) allo-tRNA variants other than the S001 variant (Figure 1B), Figure S2: Inefficient sequestration of allo-tRNA^{UTu1} (or S012), Figure S3: Sequestration of S012 derivatives, Figure S4: Examining synthetic tRNA^{His} variants, Figure S5: A preliminary screening experiment of several cognate pairs of RNA aptamers and binding protein domains towards enhancing the histidylation activity, Figure S6: Cm resistance of *E. coli* DH10B (NEB 10-beta) cells expressing the *cat(His193TAG)* reporter gene and the indicated combinations of HisRS and tRNA variants, Table S1: Screening of candidate chassis tRNAs by using *E. coli* DH10B cells expressing the *cat(Ser146TAG)* reporter gene on growth media containing chloramphenicol (Cm) at a fixed concentration of 34 μg/ml.

Acknowledgments: The author thanks Dieter Söll (Yale University) and Masayuki Su'etsugu (Rikkyo University) for supporting this study.

Abbreviations

aaRS	aminoacyl-tRNA synthetase
SerRS	seryl-tRNA synthetase
AlaRS	alanyl-tRNA synthetase
TyrRS	tyrosyl-tRNA synthetase
HisRS	histidyl-tRNA synthetase
SelA	selenocysteine synthase
PSTK	O-phosphoseryl-tRNA kinase
sfGFP	superfolder green fluorescent protein
IPTG	isopropyl β-D-1-thiogalactopyranoside
SECIS	selenocysteine insertion sequence
bp	base pair
cp	coat protein

SUMO	small ubiquitin-like modifier
WH	winged helix
SL	stem loop
ORF	open reading frame
NTD	*N*-terminal domain
CTD	*C*-terminal domain
ABD	anticodon binding domain
CD	catalytic domain

References

1. Levi, O.; Garin, S.; Arava, Y. RNA mimicry in post-transcriptional regulation by aminoacyl tRNA synthetases. *Wiley Interdiscip. Rev. RNA* **2020**, *11*, e1564. [CrossRef] [PubMed]
2. Levi, O.; Arava, Y. mRNA association by aminoacyl tRNA synthetase occurs at a putative anticodon mimic and autoregulates translation in response to tRNA levels. *PLoS Biol.* **2019**, *17*, e3000274. [CrossRef] [PubMed]
3. Soler, N.; Fourmy, D.; Yoshizawa, S. Structural insight into a molecular switch in tandem winged-helix motifs from elongation factor SelB. *J. Mol. Biol.* **2007**, *370*, 728–741. [CrossRef] [PubMed]
4. Gemmill, D.; D'Souza, S.; Meier-Stephenson, V.; Patel, T.R. Current approaches for RNA-labelling to identify RNA-binding proteins. *Biochem. Cell Biol.* **2020**, *98*, 31–41. [CrossRef]
5. Ono, H.; Kawasaki, S.; Saito, H. Orthogonal Protein-Responsive mRNA Switches for Mammalian Synthetic Biology. *ACS Synth. Biol.* **2020**, *9*, 169–174. [CrossRef]
6. van Gijtenbeek, L.A.; Kok, J. Illuminating Messengers: An Update and Outlook on RNA Visualization in Bacteria. *Front. Microbiol.* **2017**, *8*, 1161. [CrossRef]
7. Iioka, H.; Macara, I.G. Detection of RNA-Protein Interactions Using Tethered RNA Affinity Capture. *Methods Mol. Biol.* **2015**, *1316*, 67–73.
8. Masuda, I.; Igarashi, T.; Sakaguchi, R.; Nitharwal, R.G.; Takase, R.; Han, K.Y.; Leslie, B.J.; Liu, C.; Gamper, H.; Ha, T.; et al. A genetically encoded fluorescent tRNA is active in live-cell protein synthesis. *Nucleic Acids Res.* **2017**, *45*, 4081–4093. [CrossRef]
9. Ponchon, L.; Catala, M.; Seijo, B.; El Khouri, M.; Dardel, F.; Nonin-Lecomte, S.; Tisné, C. Co-expression of RNA-protein complexes in *Escherichia coli* and applications to RNA biology. *Nucleic Acids Res.* **2013**, *41*, e150. [CrossRef]
10. Lee, D.; McClain, W.H. Aptamer redesigned tRNA is nonfunctional and degraded in cells. *RNA* **2004**, *10*, 7–11. [CrossRef]
11. Preston, M.A.; Porter, D.F.; Chen, F.; Buter, N.; Lapointe, C.P.; Keles, S.; Kimble, J.; Wickens, M. Unbiased screen of RNA tailing activities reveals a poly(UG) polymerase. *Nat. Methods* **2019**, *16*, 437–445. [CrossRef] [PubMed]
12. Zheng, Y.G.; Wei, H.; Ling, C.; Martin, F.; Eriani, G.; Wang, E.D. Two distinct domains of the beta subunit of *Aquifex aeolicus* leucyl-tRNA synthetase are involved in tRNA binding as revealed by a three-hybrid selection. *Nucleic Acids Res.* **2004**, *32*, 3294–3303. [CrossRef] [PubMed]
13. Paul, A.; Warszawik, E.M.; Loznik, M.; Boersma, A.J.; Herrmann, A. Modular and Versatile Trans-Encoded Genetic Switches. *Angew. Chem. Int. Ed. Engl.* **2020**. [CrossRef]
14. Warner, K.D.; Chen, M.C.; Song, W.; Strack, R.L.; Thorn, A.; Jaffrey, S.R.; Ferré-D'Amaré, A.R. Structural basis for activity of highly efficient RNA mimics of green fluorescent protein. *Nat. Struct. Mol. Biol.* **2014**, *21*, 658–663. [CrossRef] [PubMed]
15. Volkov, I.L.; Johansson, M. Single-Molecule Tracking Approaches to Protein Synthesis Kinetics in Living Cells. *Biochemistry* **2019**, *58*, 7–14. [CrossRef] [PubMed]
16. Thyer, R.; Ellington, A.D. The Role of tRNA in Establishing New Genetic Codes. *Biochemistry* **2019**, *58*, 1460–1463. [CrossRef]
17. Cervettini, D.; Tang, S.; Fried, S.D.; Willis, J.C.W.; Funke, L.F.H.; Colwell, L.J.; Chin, J.W. Rapid discovery and evolution of orthogonal aminoacyl-tRNA synthetase-tRNA pairs. *Nat. Biotechnol.* **2020**, *38*, 989–999. [CrossRef]

18. Ding, W.; Zhao, H.; Chen, Y.; Zhang, B.; Yang, Y.; Zang, J.; Wu, J.; Lin, S. Chimeric design of pyrrolysyl-tRNA synthetase/tRNA pairs and canonical synthetase/tRNA pairs for genetic code expansion. *Nat. Commun.* **2020**, *11*, 3154. [CrossRef]

19. Wang, L.; Brock, A.; Herberich, B.; Schultz, P.G. Expanding the genetic code of *Escherichia coli*. *Science* **2001**, *292*, 498–500. [CrossRef]

20. Mukai, T.; Lajoie, M.J.; Englert, M.; Söll, D. Rewriting the Genetic Code. *Annu. Rev. Microbiol.* **2017**, *71*, 557–577. [CrossRef]

21. Arranz-Gibert, P.; Patel, J.R.; Isaacs, F.J. The Role of Orthogonality in Genetic Code Expansion. *Life* **2019**, *9*, 58. [CrossRef] [PubMed]

22. Zambaldo, C.; Koh, M.; Nasertorabi, F.; Han, G.W.; Chatterjee, A.; Stevens, R.C.; Schultz, P.G. An orthogonal seryl-tRNA synthetase/tRNA pair for noncanonical amino acid mutagenesis in *Escherichia coli*. *Bioorganic Med. Chem.* **2020**, *28*, 115662. [CrossRef] [PubMed]

23. Saint-Léger, A.; Bello, C.; Dans, P.D.; Torres, A.G.; Novoa, E.M.; Camacho, N.; Orozco, M.; Kondrashov, F.A.; Ribas de Pouplana, L. Saturation of recognition elements blocks evolution of new tRNA identities. *Sci. Adv.* **2016**, *2*, e1501860. [CrossRef] [PubMed]

24. Suzuki, T.; Miller, C.; Guo, L.T.; Ho, J.M.L.; Bryson, D.I.; Wang, Y.S.; Liu, D.R.; Söll, D. Crystal structures reveal an elusive functional domain of pyrrolysyl-tRNA synthetase. *Nat. Chem. Biol.* **2017**, *13*, 1261–1266. [CrossRef] [PubMed]

25. Willis, J.C.W.; Chin, J.W. Mutually orthogonal pyrrolysyl-tRNA synthetase/tRNA pairs. *Nat. Chem.* **2018**, *10*, 831–837. [CrossRef] [PubMed]

26. Mukai, T.; Vargas-Rodriguez, O.; Englert, M.; Tripp, H.J.; Ivanova, N.N.; Rubin, E.M.; Kyrpides, N.C.; Söll, D. Transfer RNAs with novel cloverleaf structures. *Nucleic Acids Res.* **2017**, *45*, 2776–2785. [CrossRef]

27. Ohtsuki, T.; Manabe, T.; Sisido, M. Multiple incorporation of non-natural amino acids into a single protein using tRNAs with non-standard structures. *FEBS Lett.* **2005**, *579*, 6769–6774. [CrossRef]

28. Mukai, T.; Englert, M.; Tripp, H.J.; Miller, C.; Ivanova, N.N.; Rubin, E.M.; Kyrpides, N.C.; Söll, D. Facile Recoding of Selenocysteine in Nature. *Angew. Chem. Int. Ed. Engl.* **2016**, *55*, 5337–5341. [CrossRef]

29. Mukai, T.; Sevostyanova, A.; Suzuki, T.; Fu, X.; Söll, D. A Facile Method for Producing Selenocysteine-Containing Proteins. *Angew. Chem. Int. Ed. Engl.* **2018**, *57*, 7215–7219. [CrossRef]

30. Varani, G.; McClain, W.H. The G·U wobble base pair. A fundamental building block of RNA structure crucial to RNA function in diverse biological systems. *EMBO Rep.* **2000**, *1*, 18–23. [CrossRef]

31. Rudinger, J.; Hillenbrandt, R.; Sprinzl, M.; Giegé, R. Antideterminants present in minihelix[Sec] hinder its recognition by prokaryotic elongation factor Tu. *EMBO J.* **1996**, *15*, 650–657. [CrossRef] [PubMed]

32. Ibba, M.; Losey, H.C.; Kawarabayasi, Y.; Kikuchi, H.; Bunjun, S.; Söll, D. Substrate recognition by class I lysyl-tRNA synthetases: A molecular basis for gene displacement. *Proc. Natl. Acad. Sci. USA* **1999**, *96*, 418–423. [CrossRef] [PubMed]

33. Fu, X.; Crnkovic, A.; Sevostyanova, A.; Söll, D. Designing seryl-tRNA synthetase for improved serylation of selenocysteine tRNAs. *FEBS Lett.* **2018**, *592*, 3759–3768. [CrossRef]

34. Naganuma, M.; Sekine, S.; Chong, Y.E.; Guo, M.; Yang, X.L.; Gamper, H.; Hou, Y.M.; Schimmel, P.; Yokoyama, S. The selective tRNA aminoacylation mechanism based on a single G•U pair. *Nature* **2014**, *510*, 507–511. [CrossRef] [PubMed]

35. Bessho, Y.; Shibata, R.; Sekine, S.; Murayama, K.; Higashijima, K.; Hori-Takemoto, C.; Shirouzu, M.; Kuramitsu, S.; Yokoyama, S. Structural basis for functional mimicry of long-variable-arm tRNA by transfer-messenger RNA. *Proc. Natl. Acad. Sci. USA* **2007**, *104*, 8293–8298. [CrossRef] [PubMed]

36. Lewendon, A.; Murray, I.A.; Shaw, W.V.; Gibbs, M.R.; Leslie, A.G. Evidence for transition-state stabilization by serine-148 in the catalytic mechanism of chloramphenicol acetyltransferase. *Biochemistry* **1990**, *29*, 2075–2080. [CrossRef]

37. Itoh, Y.; Bröcker, M.J.; Sekine, S.; Hammond, G.; Suetsugu, S.; Söll, D.; Yokoyama, S. Decameric SelA•tRNA[Sec] ring structure reveals mechanism of bacterial selenocysteine formation. *Science* **2013**, *340*, 75–78. [CrossRef]

38. Chiba, S.; Itoh, Y.; Sekine, S.; Yokoyama, S. Structural basis for the major role of O-phosphoseryl-tRNA kinase in the UGA-specific encoding of selenocysteine. *Mol. Cell* **2010**, *39*, 410–420. [CrossRef]

39. Lim, F.; Peabody, D.S. RNA recognition site of PP7 coat protein. *Nucleic Acids Res.* **2002**, *30*, 4138–4144. [CrossRef]

40. Nakanishi, H.; Saito, H. Caliciviral protein-based artificial translational activator for mammalian gene circuits with RNA-only delivery. *Nat. Commun.* **2020**, *11*, 1297. [CrossRef]

41. Lim, F.; Peabody, D.S. Mutations that increase the affinity of a translational repressor for RNA. *Nucleic Acids Res.* **1994**, *22*, 3748–3752. [CrossRef]

42. Gesnel, M.C.; Del Gatto-Konczak, F.; Breathnach, R. Combined use of MS2 and PP7 coat fusions shows that TIA-1 dominates hnRNP A1 for K-SAM exon splicing control. *J. Biomed. Biotechnol.* **2009**, *2009*, 1–6. [CrossRef] [PubMed]

43. Steer, B.A.; Schimmel, P. Major anticodon-binding region missing from an archaebacterial tRNA synthetase. *J. Biol. Chem.* **1999**, *274*, 35601–35606. [CrossRef]

44. Sakamoto, K.; Hayashi, A. Synthetic Tyrosine tRNA Molecules with Noncanonical Secondary Structures. *Int. J. Mol. Sci.* **2018**, *20*, 92. [CrossRef] [PubMed]

45. Mukai, T.; Reynolds, N.M.; Crnković, A.; Söll, D. Bioinformatic Analysis Reveals Archaeal tRNA^{Tyr} and tRNA^{Trp} Identities in Bacteria. *Life* **2017**, *7*, 8. [CrossRef] [PubMed]

46. Guo, J.; Melançon, C.E., 3rd; Lee, H.S.; Groff, D.; Schultz, P.G. Evolution of amber suppressor tRNAs for efficient bacterial production of proteins containing nonnatural amino acids. *Angew. Chem. Int. Ed. Engl.* **2009**, *48*, 9148–9151. [CrossRef]

47. Huang, W.; Petrosino, J.; Hirsch, M.; Shenkin, P.S.; Palzkill, T. Amino acid sequence determinants of beta-lactamase structure and activity. *J. Mol. Biol.* **1996**, *258*, 688–703. [CrossRef]

48. Wang, L.; Xie, J.; Deniz, A.A.; Schultz, P.G. Unnatural amino acid mutagenesis of green fluorescent protein. *J. Org. Chem.* **2003**, *68*, 174–176. [CrossRef]

49. Mukai, T.; Hoshi, H.; Ohtake, K.; Takahashi, M.; Yamaguchi, A.; Hayashi, A.; Yokoyama, S.; Sakamoto, K. Highly reproductive *Escherichia coli* cells with no specific assignment to the UAG codon. *Sci. Rep.* **2015**, *5*, 9699. [CrossRef]

50. Tian, Q.; Wang, C.; Liu, Y.; Xie, W. Structural basis for recognition of G-1-containing tRNA by histidyl-tRNA synthetase. *Nucleic Acids Res.* **2015**, *43*, 2980–2990. [CrossRef]

51. Kleanthous, C.; Cullis, P.M.; Shaw, W.V. 3-(Bromoacetyl)chloramphenicol, an active site directed inhibitor for chloramphenicol acetyltransferase. *Biochemistry* **1985**, *24*, 5307–5313. [CrossRef] [PubMed]

52. Burkard, U.; Willis, I.; Söll, D. Processing of histidine transfer RNA precursors. Abnormal cleavage site for RNase P. *J. Biol. Chem.* **1988**, *263*, 2447–2451. [PubMed]

53. Eichhorn, C.D.; Yang, Y.; Repeta, L.; Feigon, J. Structural basis for recognition of human 7SK long noncoding RNA by the La-related protein Larp7. *Proc. Natl. Acad. Sci. USA* **2018**, *115*, E6457–E6466. [CrossRef] [PubMed]

54. Baskerville, S.; Bartel, D.P. A ribozyme that ligates RNA to protein. *Proc. Natl. Acad. Sci. USA* **2002**, *99*, 9154–9159. [CrossRef] [PubMed]

55. Tawk, C.S.; Ghattas, I.R.; Smith, C.A. HK022 Nun Requires Arginine-Rich Motif Residues Distinct from λ N. *J. Bacteriol.* **2015**, *197*, 3573–3582. [CrossRef]

56. Torres-Larios, A.; Dock-Bregeon, A.C.; Romby, P.; Rees, B.; Sankaranarayanan, R.; Caillet, J.; Springer, M.; Ehresmann, C.; Ehresmann, B.; Moras, D. Structural basis of translational control by *Escherichia coli* threonyl tRNA synthetase. *Nat. Struct. Biol.* **2002**, *9*, 343–347. [CrossRef]

57. Ardell, D.H.; Andersson, S.G. TFAM detects co-evolution of tRNA identity rules with lateral transfer of histidyl-tRNA synthetase. *Nucleic Acids Res.* **2006**, *34*, 893–904. [CrossRef]

58. Yuan, J.; Gogakos, T.; Babina, A.M.; Söll, D.; Randau, L. Change of tRNA identity leads to a divergent orthogonal histidyl-tRNA synthetase/tRNA^{His} pair. *Nucleic Acids Res.* **2011**, *39*, 2286–2293. [CrossRef]

59. Ko, J.H.; Llopis, P.M.; Heinritz, J.; Jacobs-Wagner, C.; Söll, D. Suppression of amber codons in *Caulobacter crescentus* by the orthogonal *Escherichia coli* histidyl-tRNA synthetase/tRNA^{His} pair. *PLoS ONE* **2013**, *8*, e83630. [CrossRef]

60. Englert, M.; Vargas-Rodriguez, O.; Reynolds, N.M.; Wang, Y.S.; Söll, D.; Umehara, T. A genomically modified *Escherichia coli* strain carrying an orthogonal *E. coli* histidyl-tRNA synthetase•tRNA^{His} pair. *Biochim. Biophys. Acta Gen. Subj.* **2017**, *1861*, 3009–3015. [CrossRef] [PubMed]

61. Díaz-Cárdenas, C.; Bernal, L.F.; Caro-Quintero, A.; López, G.; David Alzate, J.; Gonzalez, L.N.; Restrepo, S.; Shapiro, N.; Woyke, T.; Kyrpides, N.C.; et al. Draft genome and description of *Consotaella salsifontis* gen. nov. sp. nov., a halophilic, free-living, nitrogen-fixing alphaproteobacterium isolated from an ancient terrestrial saline spring. *Int. J. Syst. Evol. Microbiol.* **2017**, *67*, 3744–3751.

62. Augustine, J.; Francklyn, C. Design of an Active Fragment of a Class II Aminoacyl-tRNA Synthetase and Its Significance for Synthetase Evolution. *Biochemistry* **1997**, *36*, 3473–3482. [CrossRef] [PubMed]

63. Iraha, F.; Oki, K.; Kobayashi, T.; Ohno, S.; Yokogawa, T.; Nishikawa, K.; Yokoyama, S.; Sakamoto, K. Functional replacement of the endogenous tyrosyl-tRNA synthetase-tRNA^{Tyr} pair by the archaeal tyrosine pair in *Escherichia coli* for genetic code expansion. *Nucleic Acids Res.* **2010**, *38*, 3682–3691. [CrossRef] [PubMed]

64. Italia, J.S.; Addy, P.S.; Wrobel, C.J.; Crawford, L.A.; Lajoie, M.J.; Zheng, Y.; Chatterjee, A. An orthogonalized platform for genetic code expansion in both bacteria and eukaryotes. *Nat. Chem. Biol.* **2017**, *13*, 446–450. [CrossRef]

65. Italia, J.S.; Latour, C.; Wrobel, C.J.J.; Chatterjee, A. Resurrecting the Bacterial Tyrosyl-tRNA Synthetase/tRNA Pair for Expanding the Genetic Code of Both *E. coli* and Eukaryotes. *Cell Chem. Biol.* **2018**, *25*, 1304–1312.e5. [CrossRef]

66. Attardi, D.G.; Tocchini-Valentini, G.P. Trans-acting RNA inhibits tRNA suppressor activity in vivo. *RNA* **2002**, *8*, 904–912. [CrossRef]

67. Cui, Z.; Wu, Y.; Mureev, S.; Alexandrov, K. Oligonucleotide-mediated tRNA sequestration enables one-pot sense codon reassignment *in vitro*. *Nucleic Acids Res.* **2018**, *46*, 6387–6400. [CrossRef]

68. Yokogawa, T.; Kitamura, Y.; Nakamura, D.; Ohno, S.; Nishikawa, K. Optimization of the hybridization-based method for purification of thermostable tRNAs in the presence of tetraalkylammonium salts. *Nucleic Acids Res.* **2010**, *38*, e89. [CrossRef]

69. Lee, K.B.; Hou, C.Y.; Kim, C.E.; Kim, D.M.; Suga, H.; Kang, T.J. Genetic Code Expansion by Degeneracy Reprogramming of Arginyl Codons. *Chembiochem* **2016**, *17*, 1198–1201. [CrossRef]

70. Bilokapic, S.; Maier, T.; Ahel, D.; Gruic-Sovulj, I.; Söll, D.; Weygand-Durasevic, I.; Ban, N. Structure of the unusual seryl-tRNA synthetase reveals a distinct zinc-dependent mode of substrate recognition. *EMBO J.* **2006**, *25*, 2498–2509. [CrossRef]

71. Wang, C.; Guo, Y.; Tian, Q.; Jia, Q.; Gao, Y.; Zhang, Q.; Zhou, C.; Xie, W. SerRS-tRNA^{Sec} complex structures reveal mechanism of the first step in selenocysteine biosynthesis. *Nucleic Acids Res.* **2015**, *43*, 10534–10545. [PubMed]

72. Giessen, T.W.; Altegoer, F.; Nebel, A.J.; Steinbach, R.M.; Bange, G.; Marahiel, M.A. A synthetic adenylation-domain-based tRNA-aminoacylation catalyst. *Angew. Chem. Int. Ed. Engl.* **2015**, *54*, 2492–2496. [CrossRef] [PubMed]

73. Ishida, S.; Terasaka, N.; Katoh, T.; Suga, H. An aminoacylation ribozyme evolved from a natural tRNA-sensing T-box riboswitch. *Nat. Chem. Biol.* **2020**, *16*, 702–709. [CrossRef]

74. Kuhle, B.; Chihade, J.; Schimmel, P. Relaxed sequence constraints favor mutational freedom in idiosyncratic metazoan mitochondrial tRNAs. *Nat. Commun.* **2020**, *11*, 969. [CrossRef]

75. Battaglia, R.A.; Grigg, J.C.; Ke, A. Structural basis for tRNA decoding and aminoacylation sensing by T-box riboregulators. *Nat. Struct. Mol. Biol.* **2019**, *26*, 1106–1113. [CrossRef]

2

Methods and Applications of in Silico Aptamer Design and Modeling

Andrey A. Buglak [1,2,*] ⦿, **Alexey V. Samokhvalov** [1] ⦿, **Anatoly V. Zherdev** [1] ⦿ and **Boris B. Dzantiev** [1] ⦿

[1]   A. N. Bach Institute of Biochemistry, Research Center of Biotechnology, Russian Academy of Sciences, Leninsky prospect 33, 119071 Moscow, Russia; 03alexeysamohvalov09@gmail.com (A.V.S.); zherdev@inbi.ras.ru (A.V.Z.); boris.dzantiev@mail.ru (B.B.D.)

[2]   Physical Faculty, St. Petersburg State University, 7/9 Universitetskaya naberezhnaya, 199034 St. Petersburg, Russia

*   Correspondence: andreybuglak@gmail.com

Abstract: Aptamers are nucleic acid analogues of antibodies with high affinity to different targets, such as cells, viruses, proteins, inorganic materials, and coenzymes. Empirical approaches allow the design of in vitro aptamers that bind particularly to a target molecule with high affinity and selectivity. Theoretical methods allow significant expansion of the possibilities of aptamer design. In this study, we review theoretical and joint theoretical-experimental studies dedicated to aptamer design and modeling. We consider aptamers with different targets, such as proteins, antibiotics, organophosphates, nucleobases, amino acids, and drugs. During nucleic acid modeling and in silico design, a full set of in silico methods can be applied, such as docking, molecular dynamics (MD), and statistical analysis. The typical modeling workflow starts with structure prediction. Then, docking of target and aptamer is performed. Next, MD simulations are performed, which allows for an evaluation of the stability of aptamer/ligand complexes and determination of the binding energies with higher accuracy. Then, aptamer/ligand interactions are analyzed, and mutations of studied aptamers made. Subsequently, the whole procedure of molecular modeling can be reiterated. Thus, the interactions between aptamers and their ligands are complex and difficult to understand using only experimental approaches. Docking and MD are irreplaceable when aptamers are studied in silico.

Keywords: aptamers; in silico design; molecular modeling; docking; molecular dynamics

1. Introduction

Aptamers are single-stranded nucleic acids (both DNA and RNA) with a high affinity toward target molecules. In the past few years, aptamers have been obtained for a wide range of targets, for example, cells, viruses, inorganic materials, metal ions, coenzymes, nucleobases, amino acids, antibiotics, pesticides, polypeptides, and hormones and other low-molecular-weight molecules [1–3]. Multiple studies have been performed that are dedicated to the development of biosensors functionalized with aptamers; the biosensors are based on various nanomaterials: quantum dots, metal nanoparticles (NPs), and single- and multiwalled carbon nanotubes [4–6].

The systematic evolution of ligands by exponential enrichment, or SELEX, is an experimental method for the determination of nucleic acid aptamers that specifically bind to a target molecule with high affinity and selectivity [7]. The construction of aptamers using SELEX technology includes several rounds of selection and enrichment processes (Figure 1), which are time and labor consuming and have a low cost-efficiency rate and output. SELEX may be combined with high-throughput sequencers, a process that is usually called HT-SELEX, or HTS [8,9].

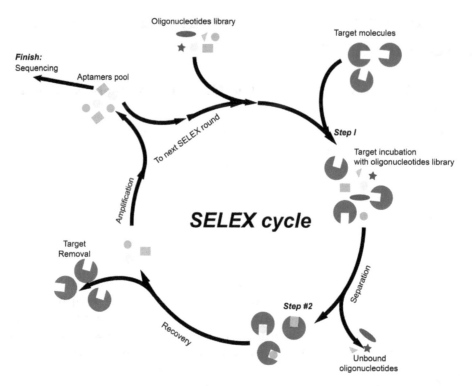

Figure 1. Schematic representation of SELEX aptamer selection and enrichment cycle.

Currently, there are several varieties of SELEX methodologies [10]. Two major steps exist in aptamer design selection and optimization. In the first step, several polynucleotides with probable binding affinity toward a target are screened by using the SELEX method and then selected. In the second step, aptamers with detected high affinity are shortened, modified, and stabilized. Both DNA and RNA aptamers can be designed. The usage of noncanonical nucleotides is possible [11].

An alternative strategy to SELEX that has been proposed in the past decade and a half is to apply computational methods of bioinformatics, namely, docking and molecular dynamics (MD), to design aptamers for various purposes. Of course, the in silico approach can be used along with SELEX and HTS to raise the efficacy of research. Some in silico tools are specialized for the analysis of HTS experimental data [12]. Rapid expansion and integration of next-generation sequencing (NGS) opens the possibility for conducting new high-throughput experiments and developing new screening strategies [13].

Hamada [14] highlighted several methods of oligonucleotide aptamer design: scalable clustering for RNA aptamers and motif-finding methods as well as aptamer optimization methods. An up-to-date review of aptamer design software is summarized in the paper by Emami et al. [15]. Additionally, new scoring functions are being developed, such as that in the study by Yan and Wang [16]. In silico aptamer design is usually performed with docking and MD. Sometimes, empirical research is carried out with only one technique, for example, MD. Sometimes, MD of oligonucleotide aptamers is accompanied by hybrid quantum mechanics/molecular mechanics (QM/MM) studies [17]. The quantitative structure–activity relationship (QSAR) method can be used along with other molecular modeling methods for aptamer design [18]. Thus, a full set of molecular modeling in silico tools can be exploited for aptamer design. Moreover, mathematical modeling of ligand/aptamer interaction kinetics is feasible [19,20].

A riboswitch is a noncoding RNA (ncRNA) that performs the function of genetic "switching" and regulates gene expression in response to a specific molecular target. Riboswitches are regulatory

RNA components, which are usually located in the 5'-untranslated region of certain mRNAs and control gene expression during transcription or translation. These components consist of a sensor domain and a neighboring actuator domain. The sensor domain is an aptamer that specifically binds to a ligand, and the actuator domain includes an intrinsic terminator or a ribosomal binding site for transcriptional or translational regulation, respectively. A large part of the research dedicated to the design of oligonucleotide aptamers belongs to the development of RNA riboswitches [21–25].

In silico tools provide a wide range of methods to a researcher. It is possible to design aptamers using simple compounds as targets as well as complex biopolymers, such as proteins. The single disadvantage is that the current level of molecular modeling makes it impossible to use cells as targets in aptamer design, which is a plus of the in vitro SELEX method. Using molecular modeling methods, it is possible to find new aptamers with improved affinity to the target and to identify structural patterns responsible for aptamer/target interaction, which is helpful because point mutations may improve aptamer affinity. A typical modeling workflow of an aptamer/ligand in silico study is presented in Figure 2. The workflow starts with secondary structure prediction and then proceeds to tertiary structure optimization. Subsequently, rigid or flexible docking of the target and aptamer is performed, and then the complexes with the lowest binding energies are selected. The next important, but not mandatory, step is to perform molecular dynamic simulations, which allow for an evaluation of the stability of the aptamer/ligand complex and determination of the binding energies with higher accuracy than would have been possible during the previous docking step. Then, aptamer/ligand interactions are analyzed, which allows researchers to make mutations or even chemical modifications to the studied aptamers. After the analysis step, the whole procedure of the in silico aptamer study can be repeated.

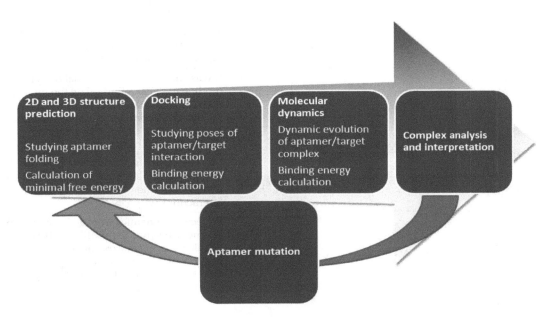

Figure 2. Typical workflow of aptamer in silico design and analysis.

In the next few sections, we summarize several studies dedicated to in silico aptamer design and modeling based on the type of aptamer targets used. Ligand molecules of aptamer design can be divided into several classes: peptides (especially surface glycoproteins, which allow the use of cells as targets in aptamer design), antibiotics, and organophosphates and other low-molecular-weight compounds. The aim and methodology of the study is different for various targets. For this reason, we grouped the studies based on the type of aptamer target used. The uniqueness of our review is due to the fact that we concentrated on concrete applications of in silico aptamer modeling and design, not on aptamer modeling methods nor modeling software.

2. Proteins as Targets of Aptamer Design

Proteins are probably the most popular targets of aptamer design and modeling. Short summaries of the computational methods and software used in these studies and descriptions of the main point of each study are presented in Table 1. Some of the proteins are autonomous analytes (e.g., thrombin), some of them are elements of polymolecular complexes, and some are cell surface proteins (e.g., hepatitis B surface antigen). The differences between protein structure and function influence the aptamer structure and design. In some studies, in silico design is dedicated to affinity modulation; in others, the aim is more therapeutic, and the main parameter is aptamer/ligand stability.

Table 1. Main features of aptamer design studies in which proteins were used as the target.

Source	Target	Computational Methods	Software	Description
[26]	Thrombin	Structure prediction, molecular dynamics (MD)	3D-DART, Amber 10	Molecular dynamics along with entropic fragment-based approach (EFBA) allowed designing a DNA aptamer, which was surpassed by an aptamer obtained using SELEX prior to it.
[27]	Thrombin	Structure prediction, MD	PyMOL 1.1, 3DNA, GROMACS 4.0	In silico calculations were accompanied by an in vitro thrombin inhibition assay. Two new thrombin aptamers, a 29-mer and a 31-mer with high inhibitory activity, were obtained.
[28,29]	Thrombin	MD	Amber 8	Novel triazole-modified and duplex-added aptamers showed potent thrombin-inhibiting activity.
[30]	Thrombin	MD	NAMD	DNA-coated nanopore for protein detection was investigated.
[31]	Thrombin	MD	Amber	The in silico-designed aptamer demonstrated seven times higher efficiency than previously known anti-thrombin aptamers.
[32]	Thrombin	MD	Amber 12	It was shown that the internal 8-aryl-guanine modification can manipulate the interactions between the DNA bases and the amino acid residues of thrombin. Nevertheless, guanine arylation at the G-tetrad reduced thrombin-binding affinity.
[33]	HIV1 integrase	Docking, MD	Hex, GROMACS	MD simulation was performed for the 93del/HIV1 integrase complex. HIV1 integrase interactions with the aptamer inhibited HIV1 integrase interactions with DNA.
[34]	HIV1 integrase	MD	Amber	Molecular dynamics were accompanied by nuclear magnetic resonance (NMR) spectroscopy and circular dichroism experiments. T30695 aptamer had a higher interaction energy (-116.4 kcal mol^{-1}) than the previously known 93del aptamer (-103.4 kcal mol^{-1}).
[35]	HIV1 reverse transcriptase (HIV1 RT)	MD	GROMACS 4.5	T1.1 RNA aptamer complex with HIV1 RT was more stable than that with a DNA substrate.
[36]	HIV1 RT	Structure prediction, docking, MD	Vfold2D, IsRNA, MDockPR, NAMD	The combination of in silico modeling and NMR allowed the identification of structural RNA elements critical for HIV1 RT inhibition and the determination of the role of UCAA motif in RT–aptamer interaction.
[18]	Influenza hemagglutinin	QSAR	CORAL	Experimental pIC50 values were used as a target parameter during QSAR modeling. The study resulted in the design of nine new aptamers with high inhibitory activity.

Table 1. *Cont.*

Source	Target	Computational Methods	Software	Description
[37]	SARS-CoV-2 spike glycoprotein	Structure prediction, docking, MD	SMART-Aptamer 2, MFold, RNAComposer	Two potent and selective DNA aptamers were designed with equilibrium dissociation constant (K_d) values of 5.8 and 19.9 nM.
[38]	Hepatitis B surface antigen (HBsAg)	Structure prediction, docking, MD	Mfold, RNAComposer, AutoDock Vina, GROMACS 5.1	It was determined that HBsAg/aptamer interactions were stabilized by the dynamic hydrogen bond formation between the active amino acid residues ("a" determinant region) and nucleotides.
[39]	*Streptococcus agalactiae* surface protein	Structure prediction, docking	Mfold, 3dRNA 2.0, AutoDock Vina	All seven RNA aptamers designed carried a hairpin. The best aptamer was a 40-mer with predicted ΔG equal to -14.7 kcal mol^{-1} and predicted affinity equal to -16.3 kcal mol^{-1}.
[40]	Prostate-specific membrane antigen (PSMA)	Structural prediction, docking	RNAstructure 4.6, Amber, MDockPP	Using the "rational truncation" technique, bases were removed from the aptamer to predict the secondary structure of the remaining oligonucleotide. Molecular docking allowed the identification of binding sites of the aptamers on PSMA.
[41]	Epithelial adhesion molecule (EpCAM)	Structure prediction, docking, MD	Vienna RNA, Rosetta, AutoDock Vina, GROMACS 5.0	Flow cytometry and fluorescence microscopy showed that in silico-designed RNA aptamer interacts specifically with the cells that express EpCAM but not with the EpCAM-negative cells.
[42]	EpCAM	Structure prediction, docking, MD	Mfold, Dot 2.0, NAMD 2	The binding modes of aptamers were first predicted and then optimized with MD and docking. Titration calorimetry experiments confirmed that the designed aptamers possessed high affinity to EpCAM.
[43]	Carcinoembryonic antigen (CEA)	Structure prediction, docking	Mfold, RNAComposer, ZDOCK	According to ZDOCK, parent sequence with ATG attached to the 3'-end and GAC sequence attached to the 5'-end had the highest score among the designed aptamers. The high affinity of the developed aptamers was confirmed experimentally by bilayer interferometry.
[44]	Transmembrane glycoprotein mucin 1 (MUC1)	Docking, MD	AutoDock Vina, Amber 16	MD, molecular mechanics Generalized Born surface area (MM-GBSA), and conformational analysis revealed novel anti-MUC1 aptamer that might be used in anti-cancer therapy.
[45]	Allophycocyanin	Structure prediction, statistical analysis	UNAFold 3.4, R	A joint theoretical-experimental approach, called closed loop aptameric directed evolution (CLADE), was used when 44,131 aptamers were analyzed using the DNA microarray technique. Statistical analysis was done using random forest, regression tree, and genetic programming.
[46]	Angiopoietin-2 (Ang2)	Structure prediction, docking	Centroid-Fold, RNAComposer, Discovery Studio 3.5	Surface plasmon resonance along with Zrank algorithm realized in DS 3.5 allowed finding an RNA aptamer with high target-binding affinity.
[47]	Ang2	Structure prediction, docking	SimRNA, AutoDock Vina	The calculated effective affinities of the Ang2/aptamer complexes were in agreement with the experiment.
[48]	Cytochrome p450	Docking, molecular dynamics	DOCK 6.5, SYBYL 8.1, Amber 9	A series of aptamers was designed and showed selective affinity toward cytochrome p450.

Table 1. *Cont.*

Source	Target	Computational Methods	Software	Description
[49]	Estrogen receptor alpha (ERα)	Docking	AutoDock Vina, Haddock, PatchDock	The aptamer was designed based on independent docking analysis in three different programs and was validated by measuring the thermodynamic parameters of ERα/aptamer interactions using isothermal titration calorimetry.
[50]	Angiotensin II	Structure prediction, docking	Mfold 3.1, RNAComposer, ZDOCK 3.0	The interactions of the aptamers with the protein were analyzed by means of surface plasmon resonance spectroscopy and were consistent with in silico data.
[51]	T-cell immunoglobulin mucin-3 (TIM-3)	Structure prediction, docking	RNAstructure 5.3, Rosetta, 3dRPC	Docking scoring parameters were analyzed along with experimental data. Binding sites and binding modes in protein/aptamer complexes were identified.

Abbreviations: angiopoietin-2 (Ang2); carcinoembryonic antigen (CEA); closed loop aptameric directed evolution (CLADE); dissociation constant (Kd); entropic fragment–based approach (EFBA); epithelial adhesion molecule (EpCAM); estrogen receptor alpha (ERα); molecular dynamics (MD); molecular mechanics Generalized Born surface area analysis (MM-GBSA); Hepatitis B surface antigen (HBsAg); HIV1 reverse transcriptase (HIV1 RT); nuclear magnetic resonance (NMR); prostate-specific membrane antigen (PSMA); T-cell immunoglobulin mucin-3 (TIM-3); transmembrane glycoprotein mucin 1 (MUC1).

The aim of numerous in silico investigations was to improve aptamer-binding affinity using different methodologies in theoretical and joint theoretical-experimental studies. A second goal was to find structural patterns responsible for aptamer/protein binding. Some studies were dedicated to the approbation of new computational techniques for aptamer design. The most popular protein target of in silico aptamer design is thrombin. Inhibition of thrombin is important because it allows modulation of coagulation. The HIV1 proteins integrase and reverse transcriptase are also popular and prospective ligands because the development of such aptamers allows antiretroviral therapy to be performed and inhibition of virus DNA integration into the host genome. However, thrombin and HIV1 proteins were popular targets in the past. In the past few years, epithelial cell adhesion molecule (EpCAM)—a marker for carcinoma—became a prospective target, and many more in silico studies dedicated to this glycoprotein should be expected in the future.

We divided proteins as targets of aptamer design into four classes: (1) coagulation-related proteins, (2) infection-related proteins, (3) cancer-related proteins, and (4) other proteins.

2.1. Coagulation-Related Proteins

Thrombin (EC 3.4.21.5) is a 37-kDa serine protease that converts soluble fibrinogen into insoluble fibrin. The regulation of thrombin enzymatic activity offers the possibility of controlling blood coagulation. Thrombin-binding oligonucleotide aptamers have been used to inhibit thrombin activity. A ssDNA aptamer was designed toward thrombin protein using the entropic fragment-based approach (EFBA) [26]. EFBA was used to determine the probability distribution of the nucleobase sequences that most likely interact with the target protein. At the same time, sequences and corresponding tertiary structures were defined. EFBA included three steps: (1) determination of the probability distribution of a favored first nucleobase, (2) determination of the probability distribution of preferred neighboring nucleobases given the probability distribution of the favored nucleobase that was established at step 1, and (3) application of the entropic criterion to define the preferred sequence length. Molecular dynamic simulations with a 5-ns trajectory were performed using Amber 10 software and allowed the binding energies to be obtained. The binding modes of EFBA 8-mer and its SELEX analogue were determined. Binding energies were determined using the molecular mechanics Poisson–Boltzmann surface area/generalized Born surface area (MM-PBSA/GBSA) algorithms [52]. The binding energy between the thrombin and SELEX aptamers was higher than the energy between the thrombin and EFBA aptamers. Because of this issue, the use of the EFBA method is not encouraging.

Thrombin is one of the most popular objects in aptamer chemistry because of the peculiarities of its geometry and its production of high-affinity aptamers to two epitopes [53]. Thus, a series of DNA aptamers towards human thrombin was developed [27]. The tertiary structure of aptamers was predicted using PyMOL 1.1 and 3DNA [54]. The GROMACS 4.0 program package [55] and two force fields, parm99 [56] and parmbsc0 [57], were used to perform molecular dynamic simulations. MD was done using the isothermal-isobaric NPT ensemble and midrange trajectories from 60 to 200 ns. Using a previously known 15-mer G-quadruplex, a 29-mer and a 31-mer were designed.

Thrombin aptamer TBA15 and its modifications were studied in yet another study [28,29]. Thrombin-binding aptamer TBA15 has the 5'-GGTTGGTGTGGTTGG-3' sequence. Chemical modification and the addition of a duplex module to the aptamer core structure were done to find whether chemical modification of the aptamer could improve its affinity to thrombin. The sequence of duplex-added TBA31 aptamer was 5'-CACTGGTAGGTTGGTGTGGTTGGGGCCAGTG-3'. The MD simulations were done using the Amber 8 program package along with ff99SB [58] and parmbsc0 [57] force fields. MD calculations were performed using the NPT ensemble while the trajectory length was up to 35 ns [28]. The MM-GBSA method was used to calculate the thrombin/aptamer-binding free energy. Novel triazole-modified and duplex-added aptamers showed pronounced thrombin-inhibiting activity. In another study by the same group of researchers, 5-nitoidole-modified TBA15 aptamer was studied, and it showed improved binding affinity to thrombin [59].

The Nanoscale Molecular Dynamics (NAMD) program [60] was used to perform the simulation with a CHARMM force field [61] for an 11-mer single-stranded DNA aptamer (PDB ID 2AVJ) and α-thrombin [30]. The sequence of this DNA was 5'-GGGGTTTGGGG-3'; it was used in a quadruplex form in MD analysis. This DNA aptamer had a TTT linker, which was responsible for forming the folds. It also provided the DNA its signature preferred loop conformation. As a whole, DNA-coated nanopore was investigated and showed advantages in protein detection. In yet another study, DNA aptamers were designed against thrombin using MD simulations [31]. The ff12SB force field [62], the NPT ensemble, and a 120-ns trajectory were used for the analysis. The in silico-designed aptamer demonstrated several times higher efficiency than did the aptamer that had undergone clinical trials earlier.

8-Aryl-guanine bases were intercalated into the G-tetrad and central TGT loop of the thrombin-binding aptamer to define their influence on the antiparallel G-quadruplex-folding and thrombin-binding affinity [32]. The initial aptamer/thrombin structure was obtained from X-ray crystallography (PDB ID 4DII). Aryl-modified guanines were optimized at the B3LYP/6-31G(d,p) level of theory and were used for parametrization of the aptamer topology. Detailed MD analysis, with the parm14SB force field [62], on the modified DNA/thrombin complexes occurred with no constraints and a 20-ns trajectory under the NPT ensemble. With modification at G8 of the central TGT loop, the aptamers produced the most stable G-quadruplex topology and exhibited the highest protein-binding affinity.

Thus, almost a decade of thrombin aptamer investigation and molecular design allowed the application of in vitro mutation, truncation, and chemical modification along with molecular modeling. This allowed improvement of the aptamer affinity to thrombin and brought multiple benefits in blood coagulation control.

2.2. Infection-Related Proteins

HIV proteins are popular targets of aptamer design. Thus, for example, interactions of HIV1 integrase with a DNA aptamer were investigated using rigid docking and MD [33]. The quadruplex aptamer 93del (Protein Data Bank ID 1Y8D) with the 5'-GGGGTGGGAGGAGGGT-3' sequence was docked into a positively charged cavity of HIV1 integrase with the *Hex* program [63]. The electrostatic potential calculations of HIV1 integrase and aptamer structures were done. A 35-ns MD simulation under the NPT ensemble was performed for the aptamer/HIV1 integrase complex. The hydrogen bonds formed between the aptamer and HIV1 integrase by interacting with key amino acid residues dioturbod HIV1 integrase interactions with DNA. In the next study, yet another anti-HIV aptamer,

T30695 (PDB ID 2LE6) with a sequence of $(GGGT)_4$ [34], was docked into the HIV1 integrase in a fashion similar to 93del. However, 93del had lower binding energy than T30695, which is due to the additional interactions made by single nucleotide loops. There were four single nucleotide thymine loops in T30695 and only two in 93del.

Another HIV1 enzyme, reverse transcriptase (HIV1 RT), interaction with an aptamer was investigated using molecular dynamics [35]. The purpose of the study was to gain insight into the conformational dynamics of HIV1 RT and its aptamer. A 100-ns molecular dynamics trajectory using an ff99SB force field [58] was reported. The binding free energies of the HIV1 RT with the RNA aptamer T1.1 (5'-GGGAGAUUCGGUUUUCAGUCGGGAAAAACUGAA-3' sequence) and the HIV1 RT with the DNA substrate were calculated using the g_mmpbsa tool [64]. The binding energy with the aptamer was higher than the binding energy with the DNA substrate. In yet another study, binding between the HIV1 RT and UCAA aptamer family was studied using a joint experimental-theoretical approach [36]. RNA aptamers were modeled in silico with the help of the NMR experimental spectra. The 2-D structure of the aptamers was derived from NMR spectroscopy and predicted from sequence by using the free energy-based approach in the Vfold2D program [65]. IsRNA [66] was used for coarse-grained MD; the top 10 conformations with the lowest energy were further used in protein/aptamer docking with the help of the MDockPP program [67]. The combination of in silico modeling and NMR spectroscopy allowed conclusions to be made about protein/RNA binding.

P. Kumar and A. Kumar designed aptamers against the influenza hemagglutinin using the Monte Carlo method [18]. They used a dataset of 98 oligonucleotides. During the QSAR Monte Carlo analysis, CORAL software was used [68]. Experimental pIC50 values (negative logarithm value calculated from the inhibitory activity of IC50 nM) were obtained from a previous publication [69]. The interesting point was that CORAL needed neither structure prediction nor docking to perform the analysis. The ssDNA nucleotide sequence was used in SMILES (simplified molecular input line entry specification) notation to evaluate the correlation between the structural parameters and pIC50 values. This research resulted in the design of nine new aptamers with pIC50 up to 9.86. Therefore, it was demonstrated that the QSAR Monte Carlo method is legitimate for the design of novel DNA aptamers.

In 2019, the outbreak of a novel coronavirus (SARS-CoV-2) reached a global pandemic level. Nevertheless, the mechanism of the coronavirus infection is not yet fully understood. Moreover, there are no common therapeutic agents and vaccines against this disease. High-binding affinity aptamers targeting the receptor-binding domain (RBD) of the SARS-CoV-2 spike glycoprotein were established [37] using SELEX in vitro experiments combined with in silico studies. SMART-Aptamer 2 [70] was used for the analysis of in vitro SELEX DNA pools. Moreover, molecular docking and MD were used for the study of aptamer/RBD binding. The K_d values of the two optimized RBD aptamers demonstrated their high affinity toward SARS-CoV-2. This makes the designed aptamers potent tools in SARS-CoV-2 diagnostics and antiviral therapeutics.

Bacterial surface proteins are also a popular target of aptamer design. Thus, a series of anti-hepatitis B surface antigen (HBsAg) aptamers was designed using docking and MD [38]. Three original SELEX aptamers with affinity to HBsAg were truncated into five short aptamers. The 2-D structures of the aptamers were obtained using Mfold, and the 3-D structures were determined with RNAComposer. Flexible docking was conducted. MD calculations with a 20-ns trajectory were performed using the CHARMM27 force field topology [71]. The affinities toward HBsAg were thoroughly investigated with an MM-PBSA algorithm. The results showed that truncated aptamers bind to HBsAg "a" determinant region (amino acid residues 99–169). Residues with the most effective interactions with all five aptamers were determined to be the active binding residues.

Aptamers against *Streptoccocus agalactiae* surface protein (PDB ID 2XTL) were studied [39]. The authors tested a computational approach for the aptamer design to find an RNA with the highest affinity to the protein. A previously known 56-mer RNA sequence was truncated, and the secondary structure was predicted using Mfold. The 2-D geometries were transformed into 3-D structures using the 3dRNA 2.0 web server [72]. RNA aptamers were docked into the protein using AutoDock Vina, and then the binding affinities were determined. The best aptamer was a 40-mer with high predicted affinity to the target.

2.3. Cancer-Related Proteins

A 41-mer RNA aptamer was designed using prostate-specific membrane antigen (PSMA) as a target [40]. The rational in silico design of aptamers was done using tertiary structure prediction and protein/RNA docking. The goal of the study was to develop a new aptamer with improved affinity to PSMA and to find nucleobases responsible for aptamer–PSMA interaction. RNAstructure 4.6 [73] was used to predict the secondary structure of RNA aptamers. Using a "rational truncation" technique, nucleobases were removed from the 5′ and 3′ ends of previously known PSMA RNA aptamer to predict the secondary structure of the remaining sequence to be as similar as possible to the initial full-length aptamer. RNA 3-D structure minimization was done using the Amber program [74]. Molecular docking allowed the identification of key nucleobases critical for binding to PSMA and the suppression of its enzymatic activity. The binding sites of the aptamer on PSMA were globally searched by using the fast Fourier transform-based molecular docking program MDockPP [75].

EpCAM is a 40-kDa transmembrane glycoprotein that participates in cell adhesion, migration, proliferation, differentiation, and cellular signaling [76]; it is the predominant surface antigen in human colon cancer. EpCAM 15-mer RNA aptamers were designed and studied by Bavi et al. [41]. Secondary RNA structures were predicted using the RNA Vienna program [77]. The 2-D geometries were converted into the 3-D form using Rosetta software. The resulting geometries were docked into rigid EpCAM. The molecular dynamic simulation of the complexes obtained was done using the Amber 99SBildn force field [78]. A 40-ns MD trajectory was used in further analysis. The protocol for calculating the binding free energy according to MM-PBSA methodology was described previously [79] and was exploited to calculate the binding free energies for EpCAM/aptamer complexes.

Further, several EpCAM aptamers have been designed in silico [42]. The binding modes of aptamers were optimized with MD and docking. The RNA aptamer structure was predicted using Mfold. Then, MD simulations with a 450-ns trajectory were performed using NAMD 2 [60] and the ff14SB force field [80]. The CHARMM36 force field was used for EpCAM MD simulations. These calculations were followed by docking in Dot 2.0 [81]. MD simulations were performed on the 10 best aptamer conformations predicted by the docking. Experiments confirmed that the two best aptamers possessed affinities higher than the previously patented nanomolar aptamer EP23.

DNA aptamers were designed against carcinoembryonic antigen (CEA; PDB ID 2QSQ) [43]. A DNA mutant library was developed via nucleobase substitution or addition to the parent (P) sequence (5′-ATACCAGCTTATTCAATT-3′). The Mfold and RNAComposer web servers were used for secondary and tertiary structure prediction, respectively. Molecular docking was performed using ZDOCK [82], which searched all possible binding modes in the translational and rotational space between a DNA aptamer and a CEA protein and calculated each conformation using an energy-based shape complementary scoring function. The study showed that an in silico post-SELEX screening approach was feasible for improving DNA aptamers. The high affinity of the developed aptamers to CEA was confirmed experimentally.

Aptamers towards transmembrane glycoprotein mucin 1 (MUC1) were investigated [44]. One of the goals of the study was to identify mutants in the T11–T13 region that are capable of a tight interaction with the APDTRPAPG epitope. MD simulations were carried out to examine the intermolecular association between the MUC1 epitope APDTRPAPG and a series of aptamers derived from the structure of the parent S2.2 MUC1 aptamer (5′-CAGTTGATCCTTTGGATACCCTG-3′ sequence). MD simulations were performed with the Amber 16 program package. MM-GBSA binding free energy calculations were done to evaluate the strength of MUC1–aptamer interactions. MD, MM-GBSA, and conformational analysis revealed the novel MUC1 aptamer with mutations located at T11 and T12 residues.

2.4. Other Proteins

Thus, Knight et al. [45] designed aptamers against a relatively large 110-kDa fluorescent protein allophycocyanin. The aim of the study was to apply machine learning to develop a novel method for the rapid design of aptamers with desired binding properties. A combined in vitro and in silico approach, called closed loop aptameric directed evolution (CLADE) (Figure 3), was applied. The ssDNA structure was predicted with UNAFold 3.4 software [83]. Statistical analysis was done using several statistical methods in R programming language. The random forest gave a high correlation coefficient score of 0.87 between in silico and in vitro binding.

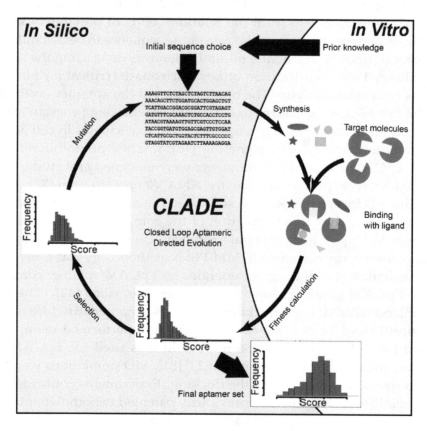

Figure 3. A scheme of a combined in silico and in vitro closed loop aptameric directed evolution (CLADE) approach.

Computational simulations of angiopoietin-2 (Ang2)–aptamer interactions were performed [46] by using ZDOCK and ZRANK docking functions in Discovery Studio 3.5. The purpose of the study was to introduce a computational approach to screen aptamers with high binding affinity to Ang2. Sixteen RNA aptamers were collected from previous studies and then truncated. The 2-D structures of the aptamers were generated with the CentroidFold web server (http://www.ncrna.org/centroidfold) [84]. The 3-D RNA structures were generated using the RNAComposer web server (http://rnacomposer.cs.put.poznan.pl) [85]. It was found that the aptamer with the highest target-binding affinity (5'-ACUAGCCUCAUCAGCUCAUGUGCCCCUCCGCCUGGAUCAC-3' sequence) can bind Ang2 with a low K_d equal to 2.2 nM.

A series of five anti-Ang2 aptamers was designed using RNA structure prediction and molecular docking methods [47]. The goal of the study was to test a novel strategy to validate the reliability of the 3-D structures of aptamers produced in silico. The work consisted of three phases: (1) production of a large set of conformations for each aptamer, (2) rigid docking into the Ang2 protein, and (3) characterization of the Ang2/aptamer complexes. Thus, the SimRNA software [86] was used for tertiary structure prediction. The docking was performed using AutoDock Vina [87]. The calculated binding scores of the Ang2/aptamer complexes were based on the calculation of the "effective affinity", which is the sum of the conformational energy (from SimRNA) and the docking energy (from AutoDock Vina). Effective affinities were in agreement with previous experimental studies.

Shcherbinin et al. [48] investigated and designed aptamers toward cytochrome p450. The aim of the study was to test a novel computational strategy and to find new cytochrome p450 aptamers. The study involved three phases: (1) finding a potential binding site, (2) designing novel aptamers, and (3) evaluating the experimental affinity. Thus, a series of 15-mer aptamers were designed using molecular docking and MD. Docking was performed using DOCK 6.5 [88]. During docking, amino acids and the structural part of the ligand were rigid, whereas nucleobases of the recognition part of the nucleic acid were flexible. The poses and conformations of oligonucleotides were analyzed using the SYBYL 8.1 package. The ff99SB force field was used for cytochrome, and the parmbsc0 was used for oligonucleotides. The binding free energies for the aptamer/protein complexes were calculated using MM-PBSA and normal mode analysis (NMA) methods implemented in Amber 9. MM-PBSA used ensembles obtained from MD simulations. Binding energy calculations consisted of the calculation of (1) interaction energy between cytochrome p450 and aptamer in the gas phase and (2) the solvation free energy. The gas phase interaction energy was the sum of van der Waals and electrostatic energies, whereas the solvation energy was the sum of nonpolar and polar energies. The stability of the complexes with cytochrome P450 was analyzed using MD simulation with a 10-ns trajectory under the NPT ensemble.

RNA aptamers towards 66-kDa protein estrogen receptor alpha (ERα) were investigated [49]. The purpose of the study was to find an aptamer with the highest possible affinity to ERα. A series of 18 aptamers were designed and docked into ERα using AutoDock Vina, Haddock, and PatchDock [87,89,90]. The docking was flexible in AutoDock Vina and Haddock, whereas in PatchDock, the docking was rigid. The strength of binding in the ERα/RNA complexes, the intermolecular H-bonds, and hydrophobic interactions were measured. H-bonding dominated in the case of the best aptamer; its binding energy was equal to $\Delta G = -11.1$ kcal mol^{-1}. The resulting 17-mer had a sequence of 5'-GGGGUCAAGGUGACCCC-3'. Target specificity of the best aptamer was confirmed experimentally with cytochemistry and solid-phase immunoassays.

A series of four ssDNA aptamers was designed to inhibit the activity of angiotensin II [50] after preliminary SELEX experiments had been done. The aim of the study was to present a combined method that could be used to predict the 3-D structure of an ssDNA aptamer and its interaction with the protein. In silico analysis was done using online web servers. The Mfold 3.1 program (http://unafold.rna.albany.edu) [91] was used to predict the secondary structure of the aptamers, and RNAComposer (http://unafold.rna.albany.edu) was used to model the tertiary structure. Molecular docking was performed in ZDOCK 3.0 (http://zdock.umassmed.edu) [82]. The best aptamer had ΔG equal to -19.1 kcal mol^{-1}. Protein–aptamer interactions were also analyzed experimentally.

Rabal and co-workers designed murine T-cell immunoglobulin mucin-3 (TIM-3) aptamers with both SELEX and in silico methods [51]. The tertiary structure was predicted using Rosetta [92]. Rigid docking of RNA aptamers into TIM-3 was done in the 3dRPC program [93]. Docking scoring parameters were analyzed along with experimental K_d values. Binding sites and binding modes in TIM-3/aptamer complexes were defined.

One can see that a full set of molecular modeling methods, including secondary and tertiary structure prediction, docking, MD, and QSAR, can be employed in designing aptamers against protein targets. Several proteins were especially popular in these studies, for example, thrombin and HIV1 reverse transcriptase, because of their significant biomedical importance. This provides multiple benefits to researchers who begin to study these aptamers. The next section is dedicated to antibiotic aptamers, which are also a popular subject of in silico studies.

3. Antibiotics as Targets of Aptamer Design

A large pool of in silico aptamer design studies is dedicated to low-molecular-weight compounds as targets. This interest is largely associated with the possibilities of using aptamers in analytical systems. High stability, ease of regeneration, and relatively low cost make aptamers a promising alternative to antibodies, which are mainly used for these purposes at present [94].

These practically important low-molecular-weight targets can be divided into three classes, namely, antibiotics, organophosphates, and other compounds. This section will be dedicated to antibiotics as targets of aptamer design (Table 2).

Table 2. Main features of aptamer design studies in which antibiotics were used as the target.

Source	Target	Computational Methods	Software	Description
[1]	Gentamicin, neomycin, tobramycin	Structure prediction, docking	Vienna RNA, Rosetta, Amber 10, AutoDock 4.0	The procedure for the selection of aptamers was rather complicated and included the free energy of secondary structure formation calculation, RNA geometry optimization, and rigid docking. The predicted binding energies were in good agreement with experimental values.
[22]	Tetracycline, streptomycin	Structure prediction	RNAFold	Riboswitches were designed using randomly generated spacers with a length from 6 to 20 bases. The in silico design was based on a minimal free energy calculation, which consisted of an antibiotic aptamer, a spacer, a complementary part for the aptamer, and a poly-U sequence at the 3'-end. In the presence of tetracycline, the expression of β-galactosidase was induced in *E. coli*, resulting in the increase of the enzyme's activity.
[95]	Neomycin-B	Structure prediction, MD	Mfold, GROMACS	Experimental NMR and titration colorimetry studies combined with MD simulations revealed that, despite the difference in nucleotide sequence, the structural and dynamical features of the studied aptamers were similar. The affinity of the aptamers toward other aminoglycosides was shown to be lower compared to the target.
[96]	Sulfadimethoxine	Structure prediction, MD	PSI-Blast, GROMACS 5.1	The aptamer's affinity to the target was determined through the calculation of binding Gibbs free energy using the MM-PBSA method. The designing procedure was done repeatedly and resulted in a creation of mutant aptamers with the improved affinity to sulfadimethoxine.

The common problem of in silico antibiotic aptamer design is in testing new methodology and finding new sequences with improved affinity to the target. Aminoglycosides are obviously the most popular targets of in silico aptamer modeling among antibiotics [1,22,95].

Chushak and Stone [1] designed RNA aptamers toward a set of antibiotics, namely, gentamicin, neomycin, and tobramycin. The primary aim of the study was to test a novel aptamer design strategy. RNA secondary structure generation was done in Vienna RNA web service [97]. The RNA geometries were optimized using the Amber 10 package and Amber99 force field. They studied 27-base RNA molecules with a 5'–GGC–N21–GUC-3' sequence and a central random region of 21 nucleobases. The constant trinucleotides at the 5'-end and 3'-end relate to sequences of well-known theophylline aptamers. The developed aptamer selection technique was rather complicated and included two levels. At the first level, the following constraints were applied to the aptamers: (1) the free energy of the secondary structure formation was set to -5.7 kcal mol^{-1} or lower, (2) more than 10 nucleotides must not form Watson–Crick pairs, and (3) the number of the structures was limited to 150. At the second level, molecular docking was applied to identify aptamers that bind to the target. The selected RNA molecules were placed into a pool of sequences for experimental screening and selection of high-affinity aptamers. The docking was done in AutoDock 4.0 (autodock.scripps.edu). RNA aptamers were used as rigid molecules, and no metal ions were added during the docking. The predicted binding energies were in good agreement with experimental values. Thus, for example, the predicted gentamicin binding to the NMR-determined aptamer geometry (PDB ID 1BYJ) possessed -11.4 kcal mol^{-1}, whereas the experimental value was equal to -10.9 kcal mol^{-1}.

Domin et al. [22] designed RNA riboswitches against tetracycline and streptomycin. RNAFold was used for the prediction of the secondary structure [77]. The aptamers were designed using randomly generated spacers, with a length from 6 to 20 bases, located between the antibiotic sensor sequence and the 3′-end part terminator. The resulting riboswitches contained a tetracycline aptamer "cb32sh" (in case of streptomycin "motif 1" aptamer). The design procedure consisted of a candidate generation and in silico minimal free energy calculation using RNAFold, which resulted in 25 tetracycline and 11 streptomycin riboswitch candidates. The activity of in silico-designed constructs was approved by in vitro experiments.

The performance of two neomycin aptamers was compared [95], also considered the binding to other aminoglycosides. These NEO1A and NEO2A aptamers possessed a 43% sequence similarity. MD simulations were performed using the Amber99SB force field. Minimized aptamer structures were equilibrated under NVT and NPT ensembles at 298 K for 100 ps. A 20-ns MD simulation was performed with constant pressure and temperature. It was shown that the two neomycin-B aptamers show similar binding affinity, activity, and selectivity, despite structural differences.

A computational study of sulfadimethoxine aptamers was performed [96]. The purpose of their study was to test new computational methodology and to find a new aptamer with improved affinity to sulfadimethoxine with the application in aptasensor design. The native aptamer had a 5′-GAGGGCAACGAGTGTTTATAGA-3′ sequence. The tertiary structure of the aptamer was predicted by applying the Blast/PSI-Blast sequence-finding method [98]. The authors performed a step-by-step mutation of the native aptamer, based on MD calculations. MD was performed using the CHARMM27 force field. First, the aptamer MD simulation was done in the presence of sulfadimethoxine. The aptamer's affinity to the target was determined through the calculation of binding Gibbs free energy using the MM-PBSA method. Then, a so-called conformational factor P_i [99] was evaluated as a measure of the target binding to each nucleotide of the aptamer. A nucleobase mutation was done for the residue with the smallest P_i value to the nucleobase with the highest P_i value, which led to the creation of a mutant sequence. MD simulation was done for the antibiotic/mutant complex, and the binding Gibbs energy and P_i were calculated again. The designing procedure was done repeatedly and resulted in the creation of five mutant aptamers with an improved affinity to sulfadimethoxine. The aptamer/target complexes were equilibrated in the NVT and NPT ensembles for 100 ps prior to 100-ns MD simulation. The described techniques improved the binding Gibbs energy from -24.9 kcal mol^{-1} for the native aptamer to -163.5 kcal mol^{-1} for the resulting 5′-AAGGGCAAGGAGGGTTCCTAGA-3′ sequence.

The calculation of conformational factor P_i and the subsequent mutation of nucleotides in the study by Khoshbin and Housaindokht [96] looks like a promising methodology of aptamer mutation and design.

4. Organophosphates as Targets of In Silico Aptamer Modeling

This section is dedicated to organophosphates as targets of aptamer design (Table 3). Some of the compounds regarded in this section are toxic substances, such as paraoxon. Others, such as free nucleotides (phosphorous ethers of nucleosides), are important biological compounds; they are structural elements of nucleic acids and coenzymes. For this reason, the popularity of mononucleotides as targets of aptamer design is explicable.

Table 3. Main features of aptamer design studies in which organophosphates were used as the target.

Source	Target	Computational Methods	Software	Description
[100]	Guanosine triphosphate (GTP)	Graph theory, matrix analysis	RAGPOOLS	An approach for engineering RNA pools used an exact set of starting sequences and certain mutation ratios in specific locations within a random region. To produce these key parameters, graph theory and matrix analysis were used. The initial aptamer pools acquired by the described methodology provided the selection of RNAs with higher affinity when compared to the in vitro pools.
[1]	Adenosine triphosphate (ATP), flavin mononucleotide (FMN)	Structure prediction, docking	Vienna RNA, Rosetta, Amber 10, AutoDock 4.0	Both 35- and 40-base RNA aptamers were designed toward FMN and ATP, respectively. The in silico-predicted binding energy of, for example, FMN was in agreement with the experimental binding energy, -7.7 kcal mol^{-1} and -8.6 kcal mol^{-1}, respectively.
[101]	ATP	Structure prediction	Vienna RNA, Mfold	Two methods of improvement of RNA/DNA aptamer complexity were created: random filtering and genetic filtering. One of the obtained 5-way junction aptamers demonstrated improved K_d values compared to those of native ATP aptamers.
[28]	Phosphatidylserine (PS)	Molecular dynamics (MD)	Amber 10	Molecular dynamics along with entropic fragment–based approach (EFBA) allowed designing a DNA 6-mer, which, however, possessed low binding energy.
[102]	PS	MD	Amber 11	The EFBA algorithm was applied to design a DNA aptamer that binds specifically to PS. This study identified the 5'-AAAGAC-3' sequence as a prospective ssDNA aptamer for PS detection.
[103]	Diazinon	Structure prediction, docking, MD	Mfold, RNAComposer, AutoDock 4.2, GROMACS 4.5	Flexible ligand/receptor docking along with MD calculations under the NVT ensemble (10-ns trajectory) showed that G-quadruplex–forming aptamer is reliable for diazinon sensing.
[104]	FMN	MD	Amber 12	The binding energy of FMN/RNA complex was evaluated using MM-GBSA. FMN/aptamer binding increased significantly when the system was immobilized on the surface of gold, which is in accordance with the experimental data.
[105]	Paraoxon	Structure prediction, docking, MD	HyperChem, Discovery Studio, AutoDock 4.2, GROMACS 5.0	The T17C mutation allowed the improvement of the affinity between aptamer and ligand (from -31.0 kcal mol^{-1} to -32.3 kcal mol^{-1}), and the T17C-C18T double mutation increased the effectiveness of ligand binding (-32.8 kcal mol^{-1}).

In one of the pioneer reports relating to in silico aptamer design [100], guanosine triphosphate (GTP) was used as a target. Enrichment of in vitro SELEX aptamer pools with complex aptamers can extend the probability of discovering new aptamers. An approach for designing RNA pools was also reported. The web server RAGPOOLS (RNA-As-Graph-Pools) was developed for the design and analysis of structured aptamer pools for SELEX (http://rubin2.biomath.nyu.edu/home.html).

Chushak and Stone [1] designed RNA aptamers toward a large set of compounds, which was discussed in the previous section. The aim of this study was to test the aptamer design methodology on multiple target compounds. They designed aptamers against mononucleotides FMN and ATP. They used docking to make binding energy predictions. Both 35- and 40-base RNA aptamers were designed toward FMN and ATP, respectively. In silico-predicted binding energies for FMN/aptamer and ATP/aptamer complexes were in agreement with the experiment.

Bioinformatic tools for ATP aptamer many-way junction creation were developed [101]. It is well known that RNA/DNA aptamers obtained using SELEX are not structurally diverse and mostly consist of simple topological geometries, such as stem loops; for this reason, many-way junctions are not frequent in ATP aptamers. The structural variety of the starting RNA/DNA pool can increase the probability of finding new aptamers with improved affinity [106]. Two methods of structural complexity and diversity improvement were developed: random filtering (RF) and genetic filtering (GF). RF starts from a random aptamer pool and calculates the number of junctions for each nucleotide sequence in the pool. Each 5-way junction sequence is then mutated at every single-stranded site one million times to evaluate the structural distribution of the respective RNA/DNA pool design. The pool with the highest percentage of five-way junctions was selected. Using the Vienna RNA program to fold a million 100-base random sequences, 76 5-way junction sequences were defined. These sequences were subjected to RF to create a 5-way junction-enhanced pool. The idea of GF is similar; its purpose was to create a pool of 1-way, 2-way, 3-way, 4-way, and 5-way junctions with an equal distribution of 20% each. A pool designed with GF was synthesized and subjected to a SELEX experiment for ATP aptamers. After eight rounds of selection, complex five-way junction topologies still accounted for a sizable percentage of the pool, confirming that RF and GF greatly improved generation of high-complexity structures and that these structures were maintained during the selection process. As a result, these techniques seem to be promising because one of the obtained five-way junction aptamers demonstrated improved K_d values compared to previously published ATP aptamers: 3.7 and 6.0 µM, respectively.

Tseng et al. designed a ssDNA aptamer towards cell membrane phospholipid phosphatidylserine (PS) using the entropic fragment-based approach (EFBA) [28]. The approbation of EFBA methodology was a primary aim of the study. Both GAMESS US and AM1 semi-empirical methods were used to optimize PS geometry prior to molecular dynamic simulation. EFBA was used to determine the nucleobase sequences that most likely interact with the target molecule. Molecular dynamic simulations allowed binding energies to be obtained. The experimentally determined binding energy between the PS and EFBA six-nucleotide DNA aptamer was low. This obstacle makes the EFBA technique less than promising.

In yet another investigation, the EFBA algorithm was applied to design a DNA aptamer that binds specifically to PS [102]. PS geometry was taken from the previous study [28]. MD simulations were performed using an ff03 force field [107] with a constant temperature of 300 K and a 10-ns trajectory. The binding energies for aptamer/PS complexes were determined both computationally and experimentally. This study identified a short 6-mer oligonucleotide sequence as a prospective ssDNA aptamer for PS detection.

Jokar and co-authors designed ssDNA aptamers against the widely known organophosphorus insecticide diazinon [103]. The aim of the study was to develop a biosensor for diazinon detection. The docking of diazinon into aptamers was performed using the Lamarckian genetic algorithm (LGA) for flexible ligand/receptor docking. Docking revealed that one of the most influential factors in the stability of the aptamer/diazinon complex was the number of H-bonds. V-rescale coupling was used to maintain constant temperature and pressure during MD simulations. Prior to MD simulation, equilibration under the NPT ensemble was used for 200 ps. Major MD calculations occurred under the NVT ensemble with a 10-ns trajectory; CHARMM22 and CHARMM27 force fields were used for the aptamers and diazinon, respectively. The QGRS Mapper server [108] indicated that only one ssDNA sequence was able to form G-quadruplexes. This G-quadruplex DNA demonstrated that it is a reliable candidate for diazinon sensing both theoretically and experimentally, once again indicating that aptamers with complicated topology are favorable for high affinity with targets.

Immobilization of the FMN aptamer on a gold surface was studied [104]. The aim of the research was not to describe the aptasensor but rather to evaluate the interaction energy between RNA and FMN using the MM-GBSA algorithm. MD calculations were performed with a force field composed of the ff99SBildn [109] and gaff [110] set of parameters to describe the RNA system interaction with FMN. MD simulation of the RNA/FMN complex was conducted under the NPT ensemble with a 100-ns trajectory. The ligand/aptamer binding increased significantly (from -24.6 to -11.6 kcal mol^{-1}) when the system was immobilized on a gold surface, which was in agreement with the experimental data.

In the next study, paraoxon aptamers were designed [105]. The goal of the study was to develop an approach to rational in silico design of aptamers for organophosphates using paraoxon as an example. The 3-D structure of a previously known 35-nucleobase aptamer (5'-AGCTTGCTGCAGCGATTCTTGATCGCCACAGAGCT-3' sequence) was predicted using HyperChem software and then was optimized with a 10-ns MD and energy minimization. Using the Discovery Studio program, 3-D aptamer structure mutation was performed. Binding energy ΔG and dissociation constant K_d were evaluated using the Lamarckian genetic algorithm [111]. For each aptamer/ligand complex, 50 conformations were achieved. The conformation with the lowest K_d was selected as the terminal one and was used as the starting geometry for further MD simulations. A constant temperature (300 K) and pressure (1 bar) were maintained with a V-rescale thermostat and a Berendsen barostat. Before running the main MD simulations, 100-ps equilibration was performed. MD with a 10-ns trajectory was performed using the MM-PBSA method to calculate the aptamer/paraoxon complex binding energy. Only the T17C mutation allowed for the improvement of the affinity between the aptamer and ligand (from -31.0 to -32.3 kcal mol^{-1}), whereas mutations at positions 18, 19, and 20 did not allow for any increase of the affinity. The double mutation T17C-C18T also increased the effectiveness of ligand binding (-32.8 kcal mol^{-1}). Thus, the aptamers with improved affinity to paraoxon were designed using only in silico approaches.

5. Different Low-Molecular-Weight Compounds as Targets of Aptamer Design

This section is dedicated to other targets—low-molecular-weight compounds neither antibiotics, nor organophosphates (Table 4). The typical aim of the studies in this section is the approbation of in silico aptamer design methodology and to verify that the improvement of aptamer affinity toward target molecules is not rare. The popular low-molecular-weight targets include nucleobases, amino acids, amino acid derivatives, therapeutical agents, and hormones.

RNA aptamers were designed toward multiple compounds by Chushak and Stone [1], whose work we reviewed in previous sections. Among other targets, they designed aptamers against arginine, codeine, guanine, isoleucine, and theophylline. They used rigid docking to make binding energy predictions. In most cases, the experimental and predicted binding energies were similar. For example, the experimental binding energy between guanine and its aptamer was equal to -7.8 kcal mol^{-1}, and the in silico value was equal to -7.7 kcal mol^{-1}; for isoleucine, the experimental binding energy was equal to -4.0 kcal mol^{-1}, and the in silico value was equal to -4.2 kcal mol^{-1}.

Lin with co-authors studied DNA aptamer (sequence 5'-GATCGAAACGTAGCGCCTTCGATC-3') binding with argininamide (Arm) [112]. The aim of the research was to examine critical nucleotides involved in aptamer/Arm binding. MD simulations were performed using the NPT ensemble with a 20-ns trajectory and the Amber94 force field. The initial complex geometry was taken from PDB (PDB ID 1OLD). The nucleotides C9, A12, C17, and Watson–Crick pair G10-C16 were important for the aptamer/ligand binding. The critical nucleotides in DNA/Arm binding provided valuable information for further DNA aptamer design.

Table 4. Main features of aptamer design studies in which low-molecular-weight compounds were used as the target.

Source	Target	Computational Methods	Software	Description
[1]	Arginine, codeine, guanine, isoleucine, theophylline	Structure prediction, docking	Vienna RNA, Rosetta, Amber 10, AutoDock 4.0	Rigid docking binding energy predictions were in good agreement with experimental values, which confirms good performance of the applied aptamer design methodology.
[112]	L-Argininamide (L-Arm)	MD	NAMD 2.6	G10, C16, C9, A12, and C17 bases were significant for aptamer/L-Arm binding, which is important for further aptamer design.
[113]	L-Arm, D-Arm, L-Arg, D-Arg, agmatine, ethyl-guanidine, L-Lys, N-methyl L-Arg	Docking	AutoDock 4.0	The interaction of eight arginine (Arg) like ligands with a DNA aptamer was analyzed. D-Arm possessed the highest affinity toward the aptamer. Theoretically defined binding energies and the K_d of ligands were in good agreement with experimentally determined values.
[114]	L-Arm	Structure prediction, MD	Discovery Studio 4.0, Amber 12	MD simulations of 50 ns were accompanied with UV spectroscopy and NMR. Thermal stabilizing effects occurred upon addition of the imidazole-tethered thymidines. Multiple imidazole moieties also maintained L-Arm binding capacity, which enhanced aptamer efficacy.
[115]	Theophylline	Structure prediction	RNAFold 2.0	The energy difference between the free energy of a riboswitch and a ligand-free aptamer was calculated. Several riboswitches were experimentally tested, and some of them showed ligand-dependent control of gene expression in *E. coli*, demonstrating that it is possible to design riboswitches for transcription regulation.
[116]	Theophylline	Structure prediction, MD	X3DNA, GROMACS 4.5	Six potent aptamers designed in silico were experimentally determined to bind theophylline with high affinity: K_d was equal to 0.16–0.52 μM, whereas K_d of the original theophylline/RNA complex was equal to 0.32 μM.
[117]	Acetamiprid	Structure prediction, docking, MD	Mfold, RNA Composer, AutoDock, NAMD 2.9	A DNA-based aptasensor was designed for the detection of acetamiprid. Docking revealed two loops as active sites in the aptamer. Circular dichroism spectroscopy and colorimetry confirmed aptamer folding due to stem-loop formation upon acetamiprid binding.
[118]	Patulin	Structure prediction	UNAFold 3.8	Microarray aptamer analysis was combined with in silico secondary structure prediction. In silico studies applied three conditions to the aptamers: (1) presence of a predicted secondary DNA structure producing one hairpin loop without a ligand, (2) hairpin loop with a length from 3 to 7 bases, and (3) stem length from 6 to 9 bases. As a result, a novel patulin aptamer was optimized.
[119]	17β-estradiol (E2)	Structure prediction, docking, MD	Mfold, RNAstructure, ZDOCK, RNAComposer, NAMD 2.10	Rigid docking of aptamers to E2 was used along with a 30-ns MD. It was demonstrated that E2 binds to a thymine loop region common to all E2-specific aptamers.
[120]	N-butanoyl-L-homoserine lactone (C4-HSL)	Structure prediction	RNAstructure 5.6, 3dRNA	The 2-D and 3-D RNA structure predictions showed that SELEX-designed aptamers possessed a conservative Y-shaped structural unit, which is probably responsible for C4-HSL binding.

In the next study, the binding characteristics of a DNA aptamer to Arm and other arginine-like targets were investigated [113]. The aim of the study was to provide valuable guidelines for the application of docking methodology and the prediction of aptamer/ligand binding energies. Docking was performed on seven NMR-obtained aptamer geometries (PDB ID 1OLD). The Lamarckian genetic algorithm (LGA) and Amber03 force field [57] were used. Global docking was done. Docking simulation was performed on the entire DNA geometry, and the geometry of the ligands was flexible, whereas the aptamer was rigid. The best docking poses were determined after an energy minimization of Arm/DNA complexes. Experiments reflected that D-Arm binds slightly stronger to the aptamer than does L-Arm; K_d was equal to 135 μM and 98 μM for L-Arm and D-Arm, respectively. This fact was partially confirmed by docking; the calculated K_d was equal to 343 and 643 pM for D-Arm and L-Arm, respectively, whereas the binding energies were −12.1 and −11.6 kcal mol^{-1}, respectively. L-Arg and D-Arg possessed significantly lower affinities toward the aptamer, which was evidenced both experimentally and computationally. The inability of L-lysine to bind to the DNA was computationally confirmed by low binding affinity. Therefore, theoretically defined binding energies of Arg-like ligands showed a good correlation with the experimentally evaluated binding energies.

L-Arm dsDNA aptamers with imidazole-tethered thymines were investigated using molecular dynamics [114]. The goal of the study was to investigate the influence of chemical modification on aptamer/ligand binding. The initial dsDNA geometry was predicted in Discovery Studio 4.0. Aptamer/dsDNA complexes were heated to 300 K, using the Langevin temperature equilibration protocol, during a 20-ps NVT run. Additionally, the system was equilibrated for 100 ps under the NPT ensemble prior to using a 50-ns MD simulation under the NPT ensemble. These MD simulations were accompanied by UV spectroscopy and NMR. It was demonstrated that thermal stabilizing effects occur upon addition of a single imidazole-tethered thymidine; the hydrogen bond forms between imidazolium residue and the Hoogsteen side of a guanosine residue on the opposite DNA strand. Moreover, multiple imidazolium moieties also increase the thermal stability of the DNA aptamer.

Using the well-known theophylline aptamer as a sensor, the actuator part of the riboswitch was designed [115]. The aim was to design a riboswitch capable of performing ligand-dependent control of *E. coli* gene expression. The designed riboswitches consisted of four parts: the theophylline aptamer, a 6- to 20-base spacer, a sequence complementary to the 3'-end of the aptamer (3'-part terminator), and a poly-U stretch at the 5'-end. The RNA secondary structure prediction and free energy evaluation were done. The energy difference between the free energy of the folded full-length riboswitch and an aptamer constrained to form the ligand-binding complex was calculated. A total of six riboswitches were experimentally tested. The riboswitch with the lowest free energy value produced a stable structure that could not be disturbed on theophylline binding; for this reason, this construct was always in an OFF state. For other structures, equilibrium was shifted toward the aptamer/ligand binding (always in an ON state). The functional riboswitches are the golden mean; they demonstrated intermediate stability and allowed the desired ligand-dependent rearrangement of the constructs, switching between ON and OFF states. As a result, several riboswitches showed ligand-dependent control of gene expression in *E. coli*, demonstrating that it is possible to design riboswitches for transcription regulation.

The aim of the study by Zhou et al. was to test a so-called in silico SELEX approach for the theophylline aptamer design [116]. The approach consisted of two phases. First, secondary structure-based sequence screening occurred, whose purpose was to select the sequences that can form an acceptable RNA motif. Second, sequence enrichment regarding theophylline binding by MD was virtually screened. The original theophylline/RNA complex was obtained from PDB (PDB ID 1O15). The x3DNA program was used to make in silico mutations. The final round of MD with a 100-ns trajectory included 24 RNA sequences. Binding energies of the target/aptamer complexes were calculated using the MM-PBSA algorithm. Six potent aptamers, which were derived from a space containing 413 sequences, were experimentally determined to bind the theophylline with high affinity (see Table 4 for details). These results demonstrated the high potential of in silico SELEX as a method for aptamer design and optimization.

The aim of Jokar et al. was to design an ssDNA-based aptasensor for the detection of the insecticide acetamiprid [117]. Docking was performed in AutoDock, which revealed two loops as active sites in the aptamer. Circular dichroism spectroscopy and colorimetry confirmed aptamer folding due to stem-loop formation upon acetamiprid binding. Stability of the DNA/ligand complex was demonstrated by MD simulations with a 100-ns trajectory.

Tomita and co-authors designed aptamer sensors based on a microarray assay that was combined with the computational secondary structure prediction [118]. The aim was to optimize the patulin aptamer. This was performed using in silico structure prediction. Three rules were applied to the in silico aptamer selection (see Table 4). The microarray aptamer library included 10^4 sequences, whereas the in silico library consisted of more than 800,000 structures. As a result of this joint theoretical-experimental research, a new patulin aptamer was designed.

Hilder and Hodgkiss designed DNA aptamers towards 17β-estradiol (E2) [119]. The 2-D aptamer structures were determined using Mfold or RNAstructure, and, subsequently, the 3-D structures were defined using RNAComposer. Rigid docking of aptamers to E2 was performed. The stability of the top-ranked complexes was checked by MD with a 30-ns trajectory. A CHARMM36 force field was used along with the CHARMM27 for DNA aptamer structures. The free energy of binding was calculated using the free energy perturbation (FEP) method, which resulted in excellent agreement between the computations and the experiment for the best E2 aptamer (5'-GCCGTTTGGGCCCAAGTTCGGC-3' sequence); the computational K_d value was equal to 10.9 nM, and the experimental value was 11 nM.

Zhao et al. investigated N-butanoyl-L-homoserine lactone (C4-HSL) aptamers using SELEX along with in silico tertiary structure prediction [120]. The purpose was to screen DNA aptamers against C4-HSL for the inhibition of biofilm formation of *Pseudomonas aeruginosa*. The 3-D RNA structure prediction showed that the designed aptamers possessed a highly conserved Y-shaped structural unit, which is probably responsible for C4-HSL binding. In vitro biofilm inhibition experiments showed that the activity of *P. aeruginosa* was efficiently reduced to about one-third by the aptamers.

6. Conclusions

In this study, we reviewed theoretical and joint theoretical-experimental studies on DNA and RNA aptamers. We regarded several classes of aptamer targets, namely, proteins; antibiotics; organophosphates; and other low-molecular-weight compounds, which included nucleobases, amino acids, and drugs. During aptamer modeling and in silico design, a full set of molecular modeling methods might be used, namely, docking, molecular dynamics, quantum-chemical calculations, and even the quantitative structure–activity relationship (QSAR). Only a few of the developments carried out to date have used the QM/MM method, which can give fruitful results in the analysis of the structural patterns of the aptamer/ligand complex. Additionally, the application of the QSAR technique is very rare; only one found paper described its use in aptamer designing [18]. QSAR and machine learning can provide serious benefits for aptamer design and modeling.

To date, the methods of computational analysis of the structure and properties of aptamers have shown their effectiveness. The in silico predictions are confirmed in vitro and make it possible to increase the affinity of aptamers in relation to their target analytes, as well as to ensure the stabilization of required conformations. In the framework of further work, the integral application of considered here methods of in silico design is in demand for efficient control of the binding properties of aptamers in different reaction media, changes in their cross-reactivity with respect to structurally related compounds, construction of oligonucleotide chains uniting several spatially close binding sites for multispecific aptamers, and also for regulators of the binding properties of aptamers.

Author Contributions: Writing—original draft preparation, A.A.B.; visualization, A.A.B. and A.V.S.; writing—review and editing, A.V.Z. and B.B.D.; supervision, B.B.D. All authors have read and agreed to the published version of the manuscript.

References

1. Chushak, Y.; Stone, M.O. In silico selection of RNA aptamers. *Nucleic Acids Res.* **2009**, *37*, e87. [CrossRef] [PubMed]

2. Kruspe, S.; Mittelberger, F.; Szameit, K.; Hahn, U. Aptamers as drug delivery vehicles. *ChemMedChem* **2014**, *9*, 1998–2011. [CrossRef] [PubMed]

3. Cai, S.; Yan, J.; Xiong, H.; Liu, Y.; Peng, D.; Liu, Z. Investigations on the interface of nucleic acid aptamers and binding targets. *Analyst* **2018**, *143*, 5317–5338. [CrossRef] [PubMed]

4. Yuce, M.; Kurt, H. How to make nanobiosensors: Surface modification and characterisation of nanomaterials for biosensing applications. *RSC Adv.* **2017**, *7*, 49386–49403. [CrossRef]

5. Ren, Q.; Ga, L.; Lu, Z.; Ai, J.; Wang, T. Aptamer-functionalized nanomaterials for biological applications. *Mater. Chem. Front.* **2020**, *4*, 1569–1585. [CrossRef]

6. Villalonga, A.; Pérez-Calabuig, A.M.; Villalonga, R. Electrochemical biosensors based on nucleic acid aptamers. *Anal. Bioanal. Chem.* **2020**, *412*, 55–72. [CrossRef]

7. Tuerk, C.; Gold, L. Systematic evolution of ligands by exponential enrichment: RNA ligands to bacteriophage T4 DNA polymerase. *Science* **1990**, *249*, 505–510. [CrossRef]

8. Zhuo, Z.; Yu, Y.; Wang, M.; Li, J.; Zhang, Z.; Liu, J.; Wu, X.; Lu, A.; Zhang, G.; Zhang, B. Recent advances in SELEX technology and aptamer applications in biomedicine. *Int. J. Mol. Sci.* **2017**, *18*, 2142. [CrossRef]

9. Komarova, N.; Kuznetsov, A. Inside the black box: What makes SELEX better? *Molecules* **2019**, *24*, 3598. [CrossRef]

10. Bayat, P.; Nosrati, R.; Alibolandi, M.; Rafatpanah, H.; Abnous, K.; Khedri, M.; Ramezani, M. SELEX methods on the road to protein targeting with nucleic acid aptamers. *Biochimie* **2018**, *154*, 132–155. [CrossRef]

11. Antipova, O.M.; Zavyalova, E.G.; Golovin, A.V.; Pavlova, G.V.; Kopylov, A.M.; Reshetnikov, R.V. Advances in the application of modified nucleotides in SELEX technology. *Biochemistry* **2018**, *83*, 1161–1172. [CrossRef] [PubMed]

12. Hoinka, J.; Przytycka, T. AptaPLEX—A dedicated, multithreaded demultiplexer for HT-SELEX data. *Methods* **2016**, *106*, 82–85. [CrossRef] [PubMed]

13. McKeague, M.; Wong, R.S.; Smolke, C.D. Opportunities in the design and application of RNA for gene expression control. *Nucleic Acids Res.* **2016**, *44*, 2987–2999. [CrossRef] [PubMed]

14. Hamada, M. In silico approaches to RNA aptamer design. *Biochimie* **2018**, *145*, 8–14. [CrossRef]

15. Emami, N.; Pakchin, P.S.; Ferdousi, R. Computational predictive approaches for interaction and structure of aptamers. *J. Theor. Biol.* **2020**, *497*, 110268. [CrossRef]

16. Yan, Z.; Wang, J. SPA-LN: A scoring function of ligand-nucleic acid interactions via optimizing both specificity and affinity. *Nucleic Acids Res.* **2017**, *45*, e110. [CrossRef]

17. Li, X.; Chung, L.W.; Li, G. Multiscale simulations on spectral tuning and the photoisomerization mechanism in fluorescent RNA spinach. *J. Chem. Theory Comput.* **2016**, *12*, 5453–5464. [CrossRef]

18. Kumar, P.; Kumar, A. Nucleobase sequence based building up of reliable QSAR models with the index of ideality correlation using Monte Carlo method. *J. Biomol. Struct. Dyn.* **2020**, *38*, 3296–3306. [CrossRef]

19. Boushaba, K.; Levine, H.; Hamilton, M.N. A mathematical feasibility argument for the use of aptamers in chemotherapy and imaging. *Math. Biosci.* **2009**, *220*, 131–142. [CrossRef]

20. Chen, X.; Ellington, A.D. Design principles for ligand-sensing, conformation-switching ribozymes. *PLoS Comput. Biol.* **2009**, *5*, e1000620. [CrossRef]

21. Avihoo, A.; Gabdank, I.; Shapira, M.; Barash, D. In silico design of small RNA switches. *IEEE Trans. Nanobiosci.* **2007**, *6*, 4–11. [CrossRef] [PubMed]

22. Domin, G.; Findeiß, S.; Wachsmuth, M.; Will, S.; Stadler, P.F.; Mörl, M. Applicability of a computational design approach for synthetic riboswitches. *Nucleic Acids Res.* **2017**, *45*, 4108–4119. [CrossRef]

23. Findeiß, S.; Etzel, M.; Will, S.; Mörl, M.; Stadler, P.F. Design of artificial riboswitches as biosensors. *Sensors* **2017**, *17*, 1990. [CrossRef] [PubMed]

24. Gong, S.; Wang, Y.; Wang, Z.; Zhang, W. Computational methods for modeling aptamers and designing riboswitches. *Int. J. Mol. Sci.* **2017**, *18*, 2442. [CrossRef] [PubMed]

25. Boussebayle, A.; Torka, D.; Ollivaud, S.; Braun, J.; Bofill-Bosch, C.; Dombrowski, M.; Groher, F.; Hamacher, K.; Suess, B. Next-level riboswitch development-implementation of Capture-SELEX facilitates identification of a new synthetic riboswitch. *Nucleic Acids Res.* **2019**, *47*, 4883–4895. [CrossRef] [PubMed]

26. Tseng, C.Y.; Ashrafuzzaman, M.; Mane, J.Y.; Kapty, J.; Mercer, J.R.; Tuszynski, J.A. Entropic fragment-based approach to aptamer design. *Chem. Biol. Drug Des.* **2011**, *78*, 1–13. [CrossRef]

27. Zavyalova, E.; Golovin, A.; Reshetnikov, R.; Mudrik, N.; Panteleyev, D.; Pavlova, G.; Kopylov, A. Novel modular DNA aptamer for human thrombin with high anticoagulant activity. *Curr. Med. Chem.* **2011**, *18*, 3343–3350. [CrossRef]

28. Varizhuk, A.M.; Tsvetkov, V.B.; Tatarinova, O.N.; Kaluzhny, D.N.; Florentiev, V.L.; Timofeev, E.N.; Shchyolkina, A.K.; Borisova, O.F.; Smirnov, I.P.; Grokhovsky, S.L.; et al. Synthesis, characterization and in vitro activity of thrombin-binding DNA aptamers with triazole internucleotide linkages. *Eur. J. Med. Chem.* **2013**, *67*, 90–97. [CrossRef]

29. Tatarinova, O.; Tsvetkov, V.; Basmanov, D.; Barinov, N.; Smirnov, I.; Timofeev, E.; Kaluzhny, D.; Chuvilin, A.; Klinov, D.; Varizhuk, A.; et al. Comparison of the 'chemical' and 'structural' approaches to the optimization of the thrombin-binding aptamer. *PLoS ONE* **2014**, *9*, e89383. [CrossRef]

30. Mahmood, M.A.; Ali, W.; Adnan, A.; Iqbal, S.M. 3D structural integrity and interactions of single-stranded protein-binding DNA in a functionalized nanopore. *J. Phys. Chem. B* **2014**, *118*, 5799–5806. [CrossRef]

31. Rangnekar, A.; Nash, J.A.; Goodfred, B.; Yingling, Y.G.; LaBean, T.H. Design of potent and controllable anticoagulants using DNA aptamers and nanostructures. *Molecules* **2016**, *21*, 202. [CrossRef] [PubMed]

32. Van Riesen, A.J.; Fadock, K.L.; Deore, P.S.; Desoky, A.; Manderville, R.A.; Sowlati-Hashjin, S.; Wetmore, S.D. Manipulation of a DNA aptamer-protein binding site through arylation of internal guanine residues. *Org. Biomol. Chem.* **2018**, *16*, 3831–3840. [CrossRef] [PubMed]

33. Sgobba, M.; Olubiyi, O.; Ke, S.; Haider, S. Molecular dynamics of HIV1-integrase in complex with 93del—A structural perspective on the mechanism of inhibition. *J. Biomol. Struct. Dyn.* **2012**, *29*, 863–877. [CrossRef] [PubMed]

34. Do, N.Q.; Lim, K.W.; Teo, M.H.; Heddi, B.; Phan, A.T. Stacking of G-quadruplexes: NMR structure of a G-rich oligonucleotide with potential anti-HIV and anticancer activity. *Nucleic Acids Res.* **2011**, *39*, 9448–9457. [CrossRef]

35. Aeksiri, N.; Songtawee, N.; Gleeson, M.P.; Hannongbua, S.; Choowongkomon, K. Insight into HIV-1 reverse transcriptase-aptamer interaction from molecular dynamics simulations. *J. Mol. Model.* **2014**, *20*, 2380. [CrossRef]

36. Nguyen, P.D.M.; Zheng, J.; Gremminger, T.J.; Qiu, L.; Zhang, D.; Tuske, S.; Lange, M.J.; Griffin, P.R.; Arnold, E.; Chen, S.J.; et al. Binding interface and impact on protease cleavage for an RNA aptamer to HIV-1 reverse transcriptase. *Nucleic Acids Res.* **2020**, *48*, 2709–2722. [CrossRef]

37. Song, Y.; Song, J.; Wei, X.; Huang, M.; Sun, M.; Zhu, L.; Lin, B.; Shen, H.; Zhu, Z.; Yang, C. Discovery of aptamers targeting the receptor-binding domain of the SARS-CoV-2 spike glycoprotein. *Anal. Chem.* **2020**, *92*, 9895–9900. [CrossRef]

38. Sabri, M.Z.; Abdul Hamid, A.A.; Sayed Hitam, S.M.; Abdul Rahim, M.Z. In silico screening of aptamers configuration against hepatitis B surface antigen. *Adv. Bioinform.* **2019**, *2019*, 6912914. [CrossRef]

39. Soon, S.; Nordin, N.A. In silico predictions and optimization of aptamers against Streptococcus agalactiae surface protein using computational docking. *Mater. Today Proc.* **2019**, *16*, 2096–2100. [CrossRef]

40. Rockey, W.M.; Hernandez, F.J.; Huang, S.Y.; Cao, S.; Howell, C.A.; Thomas, G.S.; Liu, X.Y.; Lapteva, N.; Spencer, D.M.; McNamara, J.O.; et al. Rational truncation of an RNA aptamer to prostate-specific membrane antigen using computational structural modeling. *Nucleic Acid Ther.* **2011**, *21*, 299–314. [CrossRef]

41. Bavi, R.; Liu, Z.; Han, Z.; Zhang, H.; Gu, Y. In silico designed RNA aptamer against epithelial cell adhesion molecule for cancer cell imaging. *Biochem. Biophys. Res. Commun.* **2019**, *509*, 937–942. [CrossRef] [PubMed]

42. Bell, D.R.; Weber, J.K.; Yin, W.; Huynh, T.; Duan, W.; Zhou, R. In silico design and validation of high-affinity RNA aptamers targeting epithelial cellular adhesion molecule dimers. *Proc. Natl. Acad. Sci. USA* **2020**, *117*, 8486–8493. [CrossRef] [PubMed]

43. Wang, Q.L.; Cui, H.F.; Du, J.F.; Lv, Q.Y.; Songa, X. In silico post-SELEX screening and experimental characterizations for acquisition of high affinity DNA aptamers against carcinoembryonic antigen. *RSC Adv.* **2019**, *9*, 6328–6334. [CrossRef]

44. Santini, B.L.; Zúñiga-Bustos, M.; Vidal-Limon, A.; Alderete, J.B.; Águila, S.A.; Jiménez, V.A. In silico design of novel mutant anti-MUC1 aptamers for targeted cancer therapy. *J. Chem. Inf. Model.* **2020**, *60*, 786–793. [CrossRef] [PubMed]

45. Knight, C.G.; Platt, M.; Rowe, W.; Wedge, D.C.; Khan, F.; Day, P.J.; McShea, A.; Knowles, J.; Kell, D.B. Array-based evolution of DNA aptamers allows modelling of an explicit sequence-fitness landscape. *Nucleic Acids Res.* **2009**, *37*, e6. [CrossRef] [PubMed]

46. Hu, W.P.; Kumar, J.V.; Huang, C.J.; Chen, W.Y. Computational selection of RNA aptamer against angiopoietin-2 and experimental evaluation. *Biomed. Res. Int.* **2015**, *2015*, 658712. [CrossRef] [PubMed]

47. Cataldo, R.; Ciriaco, F.; Alfinito, E. A validation strategy for in silico generated aptamers. *Comput. Biol. Chem.* **2018**, *77*, 123–130. [CrossRef]

48. Shcherbinin, D.S.; Gnedenko, O.V.; Khmeleva, S.A.; Usanov, S.A.; Gilep, A.A.; Yantsevich, A.V.; Shkel, T.V.; Yushkevich, I.V.; Radko, S.P.; Ivanov, A.S.; et al. Computer-aided design of aptamers for cytochrome p450. *J. Struct. Biol.* **2015**, *191*, 112–119. [CrossRef]

49. Ahirwar, R.; Nahar, S.; Aggarwal, S.; Ramachandran, S.; Maiti, S.; Nahar, P. In silico selection of an aptamer to estrogen receptor alpha using computational docking employing estrogen response elements as aptamer-alike molecules. *Sci. Rep.* **2016**, *6*, 21285. [CrossRef]

50. Heiat, M.; Najafi, A.; Ranjbar, R.; Latifi, A.M.; Rasaee, M.J. Computational approach to analyze isolated ssDNA aptamers against angiotensin II. *J. Biotechnol.* **2016**, *230*, 34–39. [CrossRef]

51. Rabal, O.; Pastor, F.; Villanueva, H.; Soldevilla, M.M.; Hervas-Stubbs, S.; Oyarzabal, J. In silico aptamer docking studies: From a retrospective validation to a prospective case study-TIM3 aptamers binding. *Mol. Ther. Nucleic Acids.* **2016**, *5*, e376. [CrossRef] [PubMed]

52. Genheden, S.; Ryde, U. The MM/PBSA and MM/GBSA methods to estimate ligand-binding affinities. *Expert Opin. Drug Discov.* **2015**, *10*, 449–461. [CrossRef] [PubMed]

53. Lietard, J.; Abou Assi, H.; Gómez-Pinto, I.; González, C.; Somoza, M.M.; Damha, M.J. Mapping the affinity landscape of Thrombin-binding aptamers on 2′F-ANA/DNA chimeric G-Quadruplex microarrays. *Nucleic Acids Res.* **2017**, *45*, 1619–1632. [CrossRef] [PubMed]

54. Lu, X.; Olson, W.K. 3DNA: A software package for the analysis, rebuilding, and visualization of three-dimensional nucleic acid structures. *Nucleic Acids Res.* **2003**, *31*, 5108–5121. [CrossRef] [PubMed]

55. Pronk, S.; Pall, S.; Schulz, R.; Larsson, P.; Bjelkmar, P.; Apostolov, R.; Shirts, M.R.; Smith, J.C.; Kasson, P.M.; van der Spoel, D.; et al. GROMACS 4.5: A high-throughput and highly parallel open source molecular simulation toolkit. *Bioinformatics* **2013**, *29*, 845–854. [CrossRef]

56. Sorin, E.J.; Pande, V.S. Exploring the helix-coil transition via all-atom equilibrium ensemble simulations. *Biophys. J.* **2005**, *88*, 2472–2493. [CrossRef]

57. Pérez, A.; Marchán, I.; Svozil, D.; Sponer, J.; Cheatham, T.E.; Laughton, C.A.; Orozco, M. Refinement of the AMBER force field for nucleic acids: Improving the description of alpha/gamma conformers. *Biophys. J.* **2007**, *92*, 3817–3829. [CrossRef]

58. Hornak, V.; Abel, R.; Okur, A.; Strockbine, B.; Roitberg, A.; Simmerling, C. Comparison of multiple amber force fields and development of improved protein backbone parameters. *Proteins* **2006**, *65*, 712–725. [CrossRef]

59. Tsvetkov, V.; Varizhuk, A.; Pozmogova, G.; Smirnov, I.; Kolganova, N.; Timofeev, E. A universal base in a specific role: Tuning up a thrombin aptamer with 5-nitroindole. *Sci. Rep.* **2015**, *5*, 16337. [CrossRef]

60. Phillips, J.C.; Braun, R.; Wang, W.; Gumbart, J.; Tajkhorshid, E.; Villa, E.; Chipot, C.; Skeel, R.D.; Kale, L.; Schulten, K. Scalable molecular dynamics with NAMD. *J. Comput. Chem.* **2005**, *26*, 1781–1802. [CrossRef]

61. Brooks, B.R.; Brooks, C.L., 3rd; Mackerell, A.D., Jr.; Nilsson, L.; Petrella, R.J.; Roux, B.; Won, Y.; Archontis, G.; Bartels, C.; Boresch, S.; et al. CHARMM: The biomolecular simulation program. *J. Comput. Chem.* **2009**, *30*, 1545–1614. [CrossRef] [PubMed]

62. Maier, J.A.; Martinez, C.; Kasavajhala, K.; Wickstrom, L.; Hauser, K.E.; Simmerling, C. ff14SB: Improving the accuracy of protein side chain and backbone parameters from ff99SB. *J. Chem. Theory Comput.* **2015**, *11*, 3696–3713. [CrossRef] [PubMed]

63. Ritchie, D.W.; Venkatraman, V. Ultra-fast FFT protein docking on graphics processors. *Bioinformatics* **2010**, *26*, 2398–2405. [CrossRef] [PubMed]

64. Kumari, R.; Kumar, R.; Open Source Drug Discovery Consortium; Lynn, A. g_mmpbsa—A GROMACS tool for high-throughput MM-PBSA calculations. *J. Chem. Inf. Model.* **2014**, *54*, 1951–1962. [CrossRef]

65. Xu, X.; Zhao, P.; Chen, S.J. Vfold: A web server for RNA structure and folding thermodynamics prediction. *PLoS ONE* **2014**, *9*, e107504. [CrossRef]

66. Zhang, D.; Chen, S.J. IsRNA: An iterative simulated reference state approach to modeling correlated interactions in RNA folding. *J. Chem. Theor. Comput.* **2018**, *14*, 2230–2239. [CrossRef]

67. Xu, X.; Qiu, L.; Yan, C.; Ma, Z.; Grinter, S.Z.; Zou, X. Performance of MDockPP in CAPRI rounds 28–29 and 31–35 including the prediction of water-mediated interactions. *Proteins* **2017**, *85*, 424–434. [CrossRef]
68. Benfenati, E.; Toropov, A.A.; Toropova, A.P.; Manganaro, A.; Gonella Diaza, R. coral software: QSAR for anticancer agents. *Chem. Biol. Drug Des.* **2011**, *77*, 471–476. [CrossRef]
69. Musafia, B.; Oren-Banaroya, R.; Noiman, S. Designing anti-influenza aptamers: Novel quantitative structure activity relationship approach gives insights into aptamer-virus interaction. *PLoS ONE* **2014**, *9*, e97696. [CrossRef]
70. Song, J.; Zheng, Y.; Huang, M.; Wu, L.; Wang, W.; Zhu, Z.; Song, Y.; Yang, C. A Sequential Multidimensional Analysis Algorithm for Aptamer Identification based on Structure Analysis and Machine Learning. *Anal. Chem.* **2020**, *92*, 3307–3314. [CrossRef]
71. Vanommeslaeghe, K.; Hatcher, E.; Acharya, C.; Kundu, S.; Zhong, S.; Shim, J.; Darian, E.; Guvench, O.; Lopes, P.; Vorobyov, I.; et al. CHARMM general force field: A force field for drug-like molecules compatible with the CHARMM all-atom additive biological force fields. *J. Comput. Chem.* **2010**, *31*, 671–690. [CrossRef] [PubMed]
72. Wang, J.; Wang, J.; Huang, Y.; Xiao, Y. 3dRNA v2.0: An updated Web server for RNA 3D structure prediction. *Int. J. Mol. Sci.* **2019**, *20*, 4116. [CrossRef] [PubMed]
73. Bellaousov, S.; Reuter, J.S.; Seetin, M.G.; Mathews, D.H. RNAstructure: Web servers for RNA secondary structure prediction and analysis. *Nucleic Acids Res.* **2013**, *41*, 471–474. [CrossRef] [PubMed]
74. Case, D.A.; Cheatham, T.E., 3rd; Darden, T.; Gohlke, H.; Luo, R.; Merz, K.M., Jr.; Onufriev, A.; Simmerling, C.; Wang, B.; Woods, R.J. The Amber biomolecular simulation programs. *J. Comput. Chem.* **2005**, *26*, 1668–1688. [CrossRef] [PubMed]
75. Huang, S.Y.; Zou, X. MDockPP: A hierarchical approach for protein-protein docking and its application to CAPRI rounds 15-19. *Proteins* **2010**, *78*, 3096–3103. [CrossRef]
76. Patriarca, C.; Macchi, R.M.; Marschner, A.K.; Mellstedt, H. Epithelial cell adhesion molecule expression (CD326) in cancer: A short review. *Cancer Treat. Rev.* **2012**, *38*, 68e75. [CrossRef]
77. Lorenz, R.; Bernhart, S.H.; Höner Zu Siederdissen, C.; Tafer, H.; Flamm, C.; Stadler, P.F.; Hofacker, I.L. ViennaRNA package 2.0. *Algorithms Mol. Biol.* **2011**, *6*, 26. [CrossRef]
78. Aliev, A.E.; Kulke, M.; Khaneja, H.S.; Chudasama, V.; Sheppard, T.D.; Lanigan, R.M. Motional timescale predictions by molecular dynamics simulations: Case study using proline and hydroxyproline sidechain dynamics. *Proteins* **2014**, *82*, 195e215. [CrossRef]
79. Bavi, R.; Kumar, R.; Choi, L.; Woo Lee, K. Exploration of novel inhibitors for bruton's tyrosine kinase by 3D QSAR modeling and molecular dynamics simulation. *PLoS ONE* **2016**, *11*, e0147190. [CrossRef]
80. Cheatham, T.E., 3rd; Case, D.A. Twenty-five years of nucleic acid simulations. *Biopolymers* **2013**, *99*, 969–977. [CrossRef]
81. Roberts, V.A.; Thompson, E.E.; Pique, M.E.; Perez, M.S.; Ten Eyck, L.F. DOT2: Macromolecular docking with improved biophysical models. *J. Comput. Chem.* **2013**, *34*, 1743–1758. [CrossRef] [PubMed]
82. Pierce, B.G.; Wiehe, K.; Hwang, H.; Kim, B.H.; Vreven, T.; Weng, Z. ZDOCK server: Interactive docking prediction of protein-protein complexes and symmetric multimers. *Bioinformatics* **2014**, *30*, 1771–1773. [CrossRef] [PubMed]
83. Markham, N.R.; Zuker, M. UNAFold: Software for nucleic acid folding and hybridization. *Methods Mol. Biol.* **2008**, *453*, 3–31. [PubMed]
84. Sato, K.; Hamada, M.; Asai, K.; Mituyama, T. CentroidFold: A web server for RNA secondary structure prediction. *Nucleic Acids Res.* **2009**, *37*, 277–280. [CrossRef]
85. Popenda, M.; Szachniuk, M.; Antczak, M.; Purzycka, K.J.; Lukasiak, P.; Bartol, N.; Blazewicz, J.; Adamiak, R.W. Automated 3D structure composition for large RNAs. *Nucleic Acids Res.* **2012**, *40*, e112. [CrossRef]
86. Boniecki, M.J.; Lach, G.; Dawson, W.K.; Tomala, K.; Lukasz, P.; Soltysinski, T.; Rother, K.M.; Bujnicki, J.M. SimRNA: A coarse-grained method for RNA folding simulations and 3D structure prediction. *Nucleic Acids Res.* **2016**, *44*, e63. [CrossRef]
87. Trott, O.; Olson, A.J. AutoDock Vina: Improving the speed and accuracy of docking with a new scoring function: Efficient optimization and multithreading. *J. Comput. Chem.* **2010**, *31*, 455–461. [CrossRef]
88. Allen, W.J.; Balius, T.E.; Mukherjee, S.; Brozell, S.R.; Moustakas, D.T.; Lang, P.T.; Case, D.A.; Kuntz, I.D.; Rizzo, R.C. DOCK 6: Impact of new features and current docking performance. *J. Comput. Chem.* **2015**, *36*, 1132–1156. [CrossRef]

89. Vries, S.J.; de Dijk, M.; van Bonvin, A.M. The HADDOCK web server for data-driven biomolecular docking. *Nat. Protoc.* **2010**, *5*, 883–897. [CrossRef]

90. Schneidman-Duhovny, D.; Inbar, Y.; Nussinov, R.; Wolfson, H.J. PatchDock and SymmDock: Servers for rigid and symmetric docking. *Nucleic Acids Res.* **2005**, *33*, 363–367. [CrossRef]

91. Zuker, M. Mfold web server for nucleic acid folding and hybridization prediction. *Nucleic Acids Res.* **2003**, *31*, 3406–3415. [CrossRef] [PubMed]

92. Cheng, C.Y.; Chou, F.C.; Das, R. Modeling complex RNA tertiary folds with Rosetta. *Methods Enzymol.* **2015**, *553*, 35–64. [PubMed]

93. Huang, Y.; Liu, S.; Guo, D.; Li, L.; Xiao, Y. A novel protocol for three-dimensional structure prediction of RNA-protein complexes. *Sci. Rep.* **2013**, *3*, 1887. [CrossRef] [PubMed]

94. Bauer, M.; Strom, M.; Hammond, D.S.; Shigdar, S. Anything you can do, I can do better: Can aptamers replace antibodies in clinical diagnostic applications? *Molecules* **2019**, *24*, 4377. [CrossRef] [PubMed]

95. Ilgu, M.; Yan, S.; Khounlo, R.M.; Lamm, M.H.; Nilsen-Hamilton, M. Common secondary and tertiary structural features of aptamer-ligand interaction shared by RNA aptamers with different primary sequences. *Molecules* **2019**, *24*, 4535. [CrossRef] [PubMed]

96. Khoshbin, Z.; Housaindokht, M.R. Computer-aided aptamer design for sulfadimethoxine antibiotic: Step by step mutation based on MD simulation approach. *J. Biomol. Struct. Dyn.* **2020**, 1–14. [CrossRef] [PubMed]

97. Gruber, A.R.; Lorenz, R.; Bernhart, S.H.; Neubock, R.; Hofacker, I.L. The Vienna RNA Websuite. *Nucleic Acids Res.* **2008**, *36*, W70–W74. [CrossRef] [PubMed]

98. Bhagwat, M.; Aravind, L. PSI-BLAST tutorial. *Methods Mol. Biol.* **2007**, *395*, 177–186.

99. Housaindokht, M.R.; Bozorgmehr, M.R.; Bahrololoom, M. Analysis of ligand binding to proteins using molecular dynamics simulations. *J. Theor. Biol.* **2008**, *254*, 294–300. [CrossRef]

100. Kim, N.; Gan, H.H.; Schlick, T. A computational proposal for designing structured RNA pools for in vitro selection of RNAs. *RNA* **2007**, *13*, 478–492. [CrossRef]

101. Luo, X.; McKeague, M.; Pitre, S.; Dumontier, M.; Green, J.; Golshani, A.; Derosa, M.C.; Dehne, F. Computational approaches toward the design of pools for the in vitro selection of complex aptamers. *RNA* **2010**, *16*, 2252–2262. [CrossRef] [PubMed]

102. Ashrafuzzaman, M.; Tseng, C.Y.; Kapty, J.; Mercer, J.R.; Tuszynski, J.A. A computationally designed DNA aptamer template with specific binding to phosphatidylserine. *Nucleic Acid Ther.* **2013**, *23*, 418–426. [CrossRef] [PubMed]

103. Jokar, M.; Safaralizadeh, M.H.; Hadizadeh, F.; Rahmani, F.; Kalani, M.R. Apta-nanosensor preparation and in vitro assay for rapid diazinon detection using a computational molecular approach. *J. Biomol. Struct. Dyn.* **2017**, *35*, 343–353. [CrossRef] [PubMed]

104. Ruan, M.; Seydou, M.; Noel, V.; Piro, B.; Maurel, F.; Barbault, F. Molecular dynamics simulation of a RNA aptasensor. *J. Phys. Chem. B* **2017**, *121*, 4071–4080. [CrossRef]

105. Belinskaia, D.A.; Avdonin, P.V.; Avdonin, P.P.; Jenkins, R.O.; Goncharov, N.V. Rational in silico design of aptamers for organophosphates based on the example of paraoxon. *Comput. Biol. Chem.* **2019**, *80*, 452–462. [CrossRef]

106. Carothers, J.M.; Oestreich, S.C.; Davis, J.H.; Szostak, J.W. Informational complexity and functional activity of RNA structures. *J. Am. Chem. Soc.* **2004**, *126*, 5130–5137. [CrossRef]

107. Duan, Y.; Wu, C.; Chowdhury, S.; Lee, M.C.; Xiong, G.; Zhang, W.E.I.; Yang, R.; Cieplak, P.; Luo, R.A.Y.; Lee, T.; et al. A point-charge force field for molecular mechanics simulations of proteins based on condensed-phase quantum mechanical calculations. *J. Comput. Chem.* **2003**, *24*, 1999–2012. [CrossRef]

108. Kikin, O.; D'Antonio, L.; Bagga, P.S. QGRS Mapper: A web-based server for predicting G-quadruplexes in nucleotide sequences. *Nucleic Acids Res.* **2006**, *34*, W676–W682. [CrossRef]

109. Lindorff-Larsen, K.; Piana, S.; Palmo, K.; Maragakis, P.; Klepeis, J.; Dror, R.O.; Shaw, D.E. Improved side-chain torsion potentials for the amber ff99SB protein force field. *Proteins* **2010**, *78*, 1950–1958. [CrossRef]

110. Wang, J.; Wolf, R.M.; Caldwell, J.W.; Kollman, P.A.; Case, D.A. Development and testing of a general amber force field. *J. Comput. Chem.* **2004**, *25*, 1157–1174. [CrossRef]

111. Morris, G.M.; Goodsell, D.S.; Halliday, R.S.; Huey, R.; Hart, W.E.; Belew, R.K.; Olson, A.J. Automated docking using a Lamarckian genetic algorithm and an empirical binding free energy function. *J. Comput. Chem.* **1998**, *19*, 1639–1662. [CrossRef]

112. Lin, P.H.; Tsai, C.W.; Wu, J.W.; Ruaan, R.C.; Chen, W.Y. Molecular dynamics simulation of the induced-fit binding process of DNA aptamer and L-argininamide. *Biotechnol. J.* **2012**, *7*, 1367–1375. [CrossRef]

113. Albada, H.B.; Golub, E.; Willner, I. Computational docking simulations of a DNA-aptamer for argininamide and related ligands. *J. Comput. Aided Mol. Des.* **2015**, *29*, 643–654. [CrossRef]

114. Verdonck, L.; Buyst, D.; de Vries, A.M.; Gheerardijn, V.; Madder, A.; Martins, J.C. Tethered imidazole mediated duplex stabilization and its potential for aptamer stabilization. *Nucleic Acids Res.* **2018**, *46*, 11671–11686. [CrossRef]

115. Wachsmuth, M.; Findeiß, S.; Weissheimer, N.; Stadler, P.F.; Mörl, M. De novo design of a synthetic riboswitch that regulates transcription termination. *Nucleic Acids Res.* **2013**, *41*, 2541–2551. [CrossRef] [PubMed]

116. Zhou, Q.; Xia, X.; Luo, Z.; Liang, H.; Shakhnovich, E. Searching the sequence space for potent aptamers using SELEX in silico. *J. Chem. Theory Comput.* **2015**, *11*, 5939–5946. [CrossRef] [PubMed]

117. Jokar, M.; Safaralizadeh, M.H.; Hadizadeh, F.; Rahmani, F.; Kalani, M.R. Design and evaluation of an apta-nano-sensor to detect acetamiprid in vitro and in silico. *J. Biomol. Struct. Dyn.* **2016**, *34*, 2505–2517. [CrossRef]

118. Tomita, Y.; Morita, Y.; Suga, H.; Fujiwara, D. DNA module platform for developing colorimetric aptamer sensors. *Biotechniques* **2016**, *60*, 285–292. [CrossRef]

119. Hilder, T.A.; Hodgkiss, J.M. The bound structures of 17β-estradiol-binding aptamers. *Chemphyschem* **2017**, *18*, 1881–1887. [CrossRef]

120. Zhao, M.; Li, W.; Liu, K.; Li, H.; Lan, X. C4-HSL aptamers for blocking qurom sensing and inhibiting biofilm formation in *Pseudomonas aeruginosa* and its structure prediction and analysis. *PLoS ONE* **2019**, *14*, e0212041. [CrossRef]

3

Selection, Characterization and Interaction Studies of a DNA Aptamer for the Detection of *Bifidobacterium bifidum*

Lujun Hu [1], Linlin Wang [1], Wenwei Lu [1,2], Jianxin Zhao [1,2], Hao Zhang [1,2] and Wei Chen [1,2,3,*]

[1] State Key Laboratory of Food Science and Technology, School of Food Science and Technology, Jiangnan University, Wuxi 214122, China; 7130112038@vip.jiangnan.edu.cn (L.H.); wanglllynn09@163.com (L.W.); luwenwei@jiangnan.edu.cn (W.L.); jxzhao@jiangnan.edu.cn (J.Z.); zhanghao@jiangnan.edu.cn (H.Z.)
[2] International Joint Research Center for Probiotics & Gut Health, Jiangnan University, Wuxi 214122, China
[3] Beijing Innovation Centre of Food Nutrition and Human Health, Beijing Technology and Business University, Beijing 100048, China
* Correspondence: chenwei66@jiangnan.edu.cn

Academic Editor: Julian Alexander Tanner

Abstract: A whole-bacterium-based SELEX (Systematic Evolution of Ligands by Exponential Enrichment) procedure was adopted in this study for the selection of an ssDNA aptamer that binds to *Bifidobacterium bifidum*. After 12 rounds of selection targeted against *B. bifidum*, 30 sequences were obtained and divided into seven families according to primary sequence homology and similarity of secondary structure. Four FAM (fluorescein amidite) labeled aptamer sequences from different families were selected for further characterization by flow cytometric analysis. The results reveal that the aptamer sequence CCFM641-5 demonstrated high-affinity and specificity for *B. bifidum* compared with the other sequences tested, and the estimated K_d value was 10.69 ± 0.89 nM. Additionally, sequence truncation experiments of the aptamer CCFM641-5 led to the conclusion that the 5'-primer and 3'-primer binding sites were essential for aptamer-target binding. In addition, the possible component of the target *B. bifidum*, bound by the aptamer CCFM641-5, was identified as a membrane protein by treatment with proteinase. Furthermore, to prove the potential application of the aptamer CCFM641-5, a colorimetric bioassay of the sandwich-type structure was used to detect *B. bifidum*. The assay had a linear range of 10^4 to 10^7 cfu/mL ($R^2 = 0.9834$). Therefore, the colorimetric bioassay appears to be a promising method for the detection of *B. bifidum* based on the aptamer CCFM641-5.

Keywords: *Bifidobacterium bifidum*; aptamer; SELEX; sequence truncation; colorimetric bioassay

1. Introduction

Aptamers are highly structured single-stranded oligonucleotides obtained from an in vitro evolution process called Systematic Evolution of Ligands by Exponential Enrichment (SELEX) according to their binding abilities to target molecules [1,2]. Compared with traditional antibodies, aptamers have many advantages such as low molecular weight, ease of synthesis and modifications, and comparable stability during long-term storage [3–5]. Whole-bacterium SELEX was specifically developed to separate aptamers against live bacteria, and it is a particularly promising selection strategy for the identification of bacteria. Bacterium-based aptamer selection methods have been implemented to select ssDNA aptamers against many bacteria, including *Campylobacter jejuni*, *Escherichia coli*, *Lactobacillus acidophilus*, *Mycobacterium tuberculosis*, *Vibrio parahemolyticus*, *Streptococcus pyogenes*, and *Staphylococcus aureus* without previous knowledge of a specific target molecule [6–12].

Many studies have found that *Bifidobacterium bifidum* has the potential to prevent inflammatory bowel disease and necrotizing enterocolitis, reduce cholesterol activity, treat infantile eczema, modulate the host innate immune response, show preventive potential for diarrheal disease in infants, and exert a key role in the evolution and maturation of the immune system of the host [13–19]. *B. bifidum* has also been granted QPS (quality and presumption of safety) status by the European Food Safety Authority. Thus, *B. bifidum* is often used in probiotic products along with other lactic acid bacteria [20–22], and identification of *B. bifidum* is vital for its industrial use. The conventional approaches for the identification of *B. bifidum* are laborious and time-consuming. There is thus a need of developing alternative methods for the identification of *B. bifidum*. Many molecular methods have been developed for the identification of *B. bifidum* [23–26], but they increase the analysis cost in that they require specialized instruments and highly trained personnel [24,27,28]. Therefore, aptamers may be an alternative method for the detection of *B. bifidum*.

Enzyme linked aptamer assay (ELAA), a variant of the classical ELISA (enzyme linked immunosorbent assay) uses aptamers instead of antibodies [29,30], and uses an enzyme as the signal readout element and an aptamer as the recognition element. ELAA can realize high-throughput with a 96-well microplate and is convenient because the signal readout requires only simple instruments (or even no instruments, when read with the naked eye). Therefore, ELAA has been used in many bioanalytical applications for target-specific detection of some substances such as *M. tuberculosis*, ochratoxin A, cocaine and thrombin [31–34]. However, ELAA has not been reported in the detection of *B. bifidum*.

In this study, we used an improved whole-bacterium SELEX strategy to select an ssDNA aptamer specific for *B. bifidum*. In addition, truncation experiments were carried out to narrow down the sequence region of the potential aptamers essentially for their binding abilities to the target *B. bifidum*. In addition, *B. bifidum* was treated with proteinases to determine whether the targets of the aptamer were the membrane proteins on the cell surface of *B. bifidum*. Furthermore, to confirm the potential application of the candidate aptamer, we developed a colorimetric assay that was a high-throughput, sensitive and specific method for the detection of *B. bifidum*.

2. Results and Discussion

2.1. SELEX Optimization

To separate aptamers that specifically recognize *B. bifidum*, we gradually increased the selective pressure by increasing bovine serum albumin (BSA) and tRNA from a 10-fold molar excess of each in the starting round of selection to a maximum 120-fold molar excess in the 12th round, and by increasing the number of washes (from twice for the first six cycles to three times for the last six cycles). In addition, the suspended cell solutions were transferred to fresh microcentrifuge tubes to remove sequences that bound to the tube walls between each incubation and elution step. Counter-selection against a mixture of unrelated *Bifidobacterium* species, including *B. longum*, *B. animalis*, *B. breve*, and *B. adolescentis*, was employed in the 9th and 11th rounds. In addition, 2.5 μL DMSO (dimethyl sulfoxide) was chosen for the total volume of 50 μL during the PCR amplification of SELEX.

2.2. Determination of Affinity and Specificity

Fluorescently labeled aptamer sequences were incubated with *B. bifidum* and tested via flow cytometric analysis. After the 12th round of selection, 30 sequences were obtained after the aptamer pools were cloned and sequenced. These sequences were then grouped into seven families according to the homology of the DNA sequences and the similarity of the secondary structure (data shown in Table S1 and Figure S1 in the Supplemental Materials). Four sequences were selected for further screening on the basis of their repetitiveness, predicted secondary structure and free energy of formation (Table 1).

Table 1. Tested aptamer sequences [a].

Name	Sequence (5′ → 3′)
CCFM641-2	GCCTGGCCAGGTGCCCCGATATAGCGACGCCTTGCCCGGC
CCFM641-4	GCCCCGGACGGCGGGAAGCCTCGTACCCCCCGTGAGCGGC
CCFM641-5	TGCGTGAGCGGTAGCCCCGTACGACCCACTGTGGTTGGGC
CCFM641-12	GTCACACCGGCCGTCTCCGGTGTGGGACGCCCGCTGTGGC

[a] The primer sequences are AGCAGCACAGAGGTCAGATG at the 5′ end and CCTATGCGTGCTACCGTGAA at the 3′ end.

The results displayed in Figure 1 demonstrated that CCFM641-5 showed a stronger binding affinity for *B. bifidum* than the other three aptamers. To further evaluate the binding ability of the aptamer CCFM641-5 to the target *B. bifidum*, we performed binding assays by varying the concentrations of the aptamer (from 0 to 100 nM) and using a constant number of cells (10^8 cells) for each assay. Saturation curves were fit from these data and the dissociation constant K_d values were determined via nonlinear regression analysis. The dissociation constant K_d between CCFM641-5 and *B. bifidum* was calculated to be 10.69 ± 0.89 nM. Therefore, aptamer CCFM641-5 was chosen for the specificity detection. The predicted secondary structure and binding saturation curve of aptamer CCFM641-5 for *B. bifidum* are shown in Figure 2.

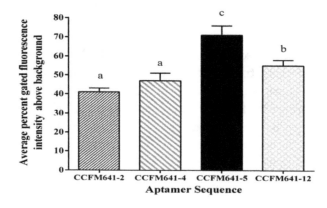

Figure 1. Binding affinity of aptamers for *B. bifidum*. The 5′-FAM-labeled individual aptamers were incubated with *B. bifidum* at 37 °C for 45 min (as described in the text). The values of aptamer binding represent the mean $\pm$ SD of three independent experiments. Bars with different letters are significantly different ($p < 0.05$).

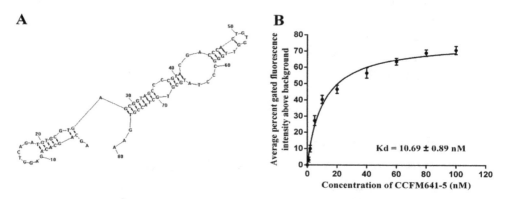

Figure 2. The secondary structure and binding ability of aptamer CCFM641-5 against *B. bifidum*. (**A**) The secondary structure of aptamer CCFM641-5. The secondary structure was predicted using RNAstructure 3.0. (**B**) The binding saturation curve of aptamer CCFM641-5 with *B. bifidum*. A nonlinear regression curve was fit according to the data from flow cytometric analysis using GraphPad Prism 5.0. The values of aptamer binding and K_d represent the mean $\pm$ SD of three independent experiments.

To determine the specificity of the candidate aptamer CCFM641-5 for the target *B. bifidum*, the FAM (fluorescein amidite) labeled aptamer CCFM641-5 was also tested against a variety of other bacterial species, including *B. longum*, *B. animalis*, *B. breve*, *B. adolescentis* and *L. plantarum*. As shown in Figure 3, the aptamer CCFM641-5 displayed preferential binding ability to *B. bifidum* over the other bacteria tested. In addition, the results from qPCR shown in Figure S2 demonstrated that CCFM641-5 displayed a stronger binding ability to *B. bifidum* than the other bacterial species. Taken together, this preferential binding demonstrated the excellent specificity of the aptamer CCFM641-5 for *B. bifidum*.

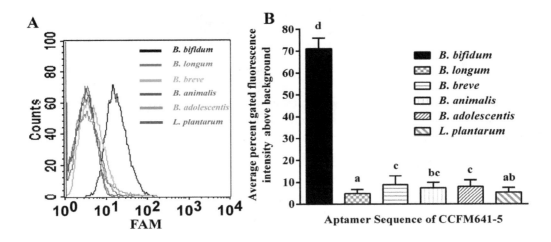

Figure 3. Characterization of the specificity of aptamer CCFM641-5 for *B. bifidum*. Selected aptamer sequence CCFM641-5 preferentially bound to *B. bifidum* over other species of bacteria. (**A**) Flow cytometric analysis of aptamer CCFM641-5 binding for different species of bacteria which are shown with differently colored curves; (**B**) Histogram of the percent gated fluorescence intensity above background for aptamer CCFM641-5. The values of aptamer binding represent the mean ± SD of three independent experiments. Bars with different letters are significantly different ($p < 0.05$).

2.3. Aptamer Truncations and Their Effects on the Binding Ability to B. bifidum

In general, not all nucleic acids of the aptamers are necessary for binding affinity between the aptamers and the targets [35]. To determine the minimal sequence necessary for binding affinity between *B. bifidum* and the aptamer CCFM641-5, the aptamer was truncated to narrow down the sequence region responsible for target binding affinity. Either specific primer binding site at the ends of aptamer CCFM641-5 (CCFM641-5F and CCFM641-5R) or both sites (CCFM641-5FR) were removed (Table 2). As displayed in Figure 4, all the aptamer variants bound to *B. bifidum* with a lower binding affinity compared to the full-length aptamer CCFM641-5. Taken together, the results indicate that the 3'-primer and 5'-primer binding sites of aptamer CCFM641-5 are important for its binding affinity for *B. bifidum*, even if the aptamer affinity is still largely preserved for the 5' truncation.

Table 2. Full-length aptamer CCFM641-5 and truncated aptamer variants [a].

Name	Sequence (5' → 3')
CCFM641-5	AGCAGCACAGAGGTCAGATGTGCGTGAGCGGTAGCCCCGTACGACCCACTGTGGTTGG GCCCTATGCGTGCTACCGTGAA
CCFM641-5F	TGCGTGAGCGGTAGCCCCGTACGACCCACTGTGGTTGGGC CCTATGCGTGCTACCGTGAA
CCFM641-5R	AGCAGCACAGAGGTCAGATGTGCGTGAGCGGTAGCCCCGTACGACCCACTGTGGTTGGGC
CCFM641-5FR	TGCGTGAGCGGTAGCCCCGTACGACCCACTGTGGTTGGGC

[a] The underlined sequences AGCAGCACAGAGGTCAGATG and CCTATGCGTGCTACCGTGAA are the primer binding sites.

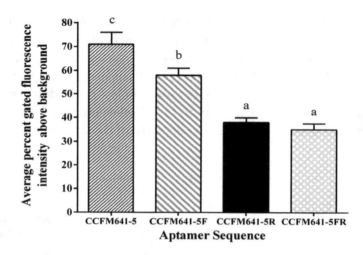

Figure 4. Binding abilities of the truncated aptamer variants to *B. bifidum* compared to the full-length aptamer CCFM641-5. The values of aptamer binding represent the mean ± SD of three independent experiments. Bars with different letters are significantly different ($p < 0.05$).

2.4. Proteinase Treatment for Bacteria

To evaluate that the targets of the aptamer are membrane proteins on the *B. bifidum* cell surface, we treated *B. bifidum* with proteinases including trypsin and proteinase K for a short time before adding the aptamer CCFM641-5 to these treated bacteria. As revealed in Figure 5, after the bacteria were treated with trypsin or proteinase K for 2 and 10 min, respectively, in phosphate-buffere saline (PBS, pH 7.2) at 37 °C, aptamer CCFM641-5 lowered its binding ability to *B. bifidum*. It can be deduced that the binding entities of aptamer CCFM641-5 had been broken by the proteinases, suggesting that the target molecules are in fact membrane proteins.

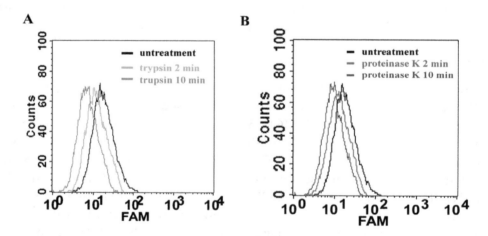

Figure 5. Binding assays of aptamer CCFM641-5 to trypsin-treated or proteinase K-treated *B. bifidum*. (**A**) Flow cytometric analysis of aptamer CCFM641-5 binding affinity for trypsin-treated *B. bifidum*; (**B**) Flow cytometric analysis of aptamer CCFM641-5 binding ability to proteinase K-treated *B. bifidum*.

2.5. Colorimetric Detection of B. bifidum

In this study, the candidate aptamer CCFM641-5 was used not only to capture but also to detect *B. bifidum* in the configuration of the colorimetric assay. In the assay, a sandwich-type structure of aptamer/target/aptamer was established. To evaluate the specificity of this method, one blank sample and five samples including *B. bifidum*, *B. longum*, *B. animalis*, *B. breve* and *B. adolescentis* were measured. The assays of all samples were carried out under the same conditions, and the concentrations of all bacteria were between 10^3 and 10^8 cfu/mL. As shown in Figure 6, the optical density (OD) values at 450 nm of the blank sample and the four bacteria other than *B. bifidum* did not change when the concentrations of bacteria were increased, whereas the OD values at 450 nm of *B. bifidum* increased as concentrations of the bacteria increased. Particularly, the results shown in Figure 7 displayed a good linear relationship between the amounts of *B. bifidum* ranging from 10^4 to 10^7 cfu/mL and the OD values at 450 nm, with a regression coefficient of 0.9834. The limit of detection of the proposed method was estimated to be 10^4 cfu/mL at a signal to noise ratio of 3.

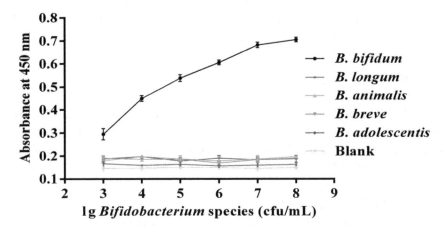

Figure 6. The absorbance at 450 nm measured for different species of bifidobacteria at concentrations ranging from 10^3 to 10^8 cfu/mL. The absorbance at 450 nm represents the mean ± SD of three independent experiments.

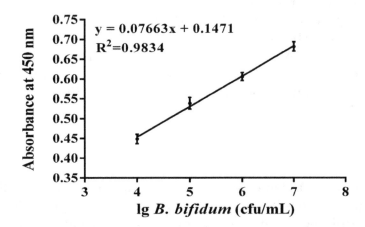

Figure 7. The calibration curve between the concentrations of *B. bifidum* and the intensity of the signals. The OD values were determined by the microplate reader at 450 nm wavelength. The absorbance at 450 nm represents the mean ± SD of three independent experiments.

3. Materials and Methods

3.1. Reagents and Apparatus

B. bifidum ATCC 29521 was adopted as the target for whole-bacterium SELEX. Other *Bifidobacterium* species used in the study included: *B. longum* ATCC 15697, *B. breve* ATCC 15700, *B. animalis* JCM 11658, and *B. adolescentis* ATCC 15705, which were supplied by the American Type Culture Collection (ATCC) or the Japan Collection of Microorganisms (JCM). All of the *Bifidobacterium* species were grown in de Man–Rogosa–Sharpe (MRS) broth with 0.05% of L-cysteine hydrochloride monohydrate at 37 °C. *L. plantarum* ST-III (CGMCC No. 0847) used in this study was cultured in MRS broth at 37 °C (Merck KGaA, Darmstadt, Germany). All bacterial strains were cultured to the logarithmic phase under anaerobic conditions.

The starting ssDNA library and the PCR primers used for PCR amplifications were supplied by Integrated DNA Technologies (IDT, Coralville, IA, USA); tRNA, DMSO and streptavidin-HRP (horseradish peroxidase) were purchased from Sigma (St. Louis, MO, USA); BSA and all PCR chemicals were ordered from Invitrogen China (Shanghai, China); trypsin and proteinase K were purchased from TaKaRa (TaKaRa, Dalian, China); and DNA-BIND 96-well plates were obtained from Corning (Corning, New York, NY, USA). Water was filtered with a Milli-Q water purification system (Millipore, Bedford, MA, USA). All reagents were of analytical grade. Purification treatments were carried out with an Eppendorf 5424R centrifuge. PCR amplification was carried out in a Bio-Rad T 100 Thermal Cycler (Bio-Rad Laboratories, Hercules, CA, USA).

3.2. DNA Library and PCR Amplification

The 80-nt ssDNA library consisted of a central random region of 40 nucleotides, where equimolar amounts of A, G, C and T are available at each position, flanked by two constant primer binding sequences, and the ssDNA library was synthesized with the following sequence: 5′-AGCAGCACAGAGGTCAGATG-N40-CCTATGCGTGCTACCGTGAA-3′. The ssDNA library and following aptamer pools were amplified with sense (5′-AGCAGCACAGAGGTCAGATG-3′) and antisense primer (5′-TTCACGGTAGCACGCATAGG-3′).

The PCR amplification mixture was as follows: 1× PCR amplification buffer, 10 µM forward and reverse primer, 25 mM dNTPs, 5 U/µL of Taq DNA polymerase, 2 µL of the template and 2.5 µL of DMSO in a total volume of 50 µL. PCR amplification was initiated with pre-denaturation at 95 °C for 6 min, followed by 25 cycles with 30 s denaturation at 95 °C, 30 s hybridization at 69 °C, 20 s extension at 72 °C, and finally elongation for 5 min at 72 °C.

PCR amplification products were detected by 8% nondenaturing polyacrylamide gel electrophoresis (PAGE) in 1× TBE buffer (90 mM Tris/89 mM boric acid/2.0 mM EDTA, pH 8.0) (Bio-Rad Protean III, Hercules, CA, USA) at 200 V for 25 min. The polyacrylamide gels were then stained with ethidium bromide, destained with gel running buffer solution, and photographed under UV light. A Qiagen MinElute PCR Purification Kit was used to purify all PCR amplification products (Qiagen Inc., Valencia, CA, USA).

3.3. Aptamer Selection

The SELEX procedure for aptamer selection was implemented using the method described previously with some modifications [8,36]. *B. bifidum* was grown in liquid culture media and collected when it reached the logarithmic phase. The cell mixtures were centrifuged at 5000× *g* and 4 °C for 5 min and washed twice with 500 µL PBS at room temperature. The original ssDNA library/pool was heat denatured at 95 °C for 10 min and rapidly chilled for 10 min in an ice bath before incubation. *B. bifidum* cells totaling 10^8 cfu/mL were incubated with 2 nmol of the ssDNA library for the initial round or 100 pmol of the aptamer pool for subsequent rounds (600 µL for the initial round, 350 µL for subsequent rounds). An excess of tRNA and BSA were put into the incubation buffer. All incubations were implemented in PBS at 37 °C for 45 min with slight agitation. The cells were centrifuged as described above before washing three times in PBS with 0.05% BSA. The cells were then resuspended with 100 µL of 1× PCR reaction buffer, heat denatured at 95 °C for 10 min, snap cooled for 10 min in an ice bath, and extracted by centrifugation as described above, and the supernatant was used as the template for PCR amplification to acquire the ssDNA pool for the following round of selection.

The ssDNA pool from the 12th round of selection was amplified and then cloned using the TOPO TA Cloning Kit (Invitrogen, Shanghai, China). Individual colonies were picked randomly and their inserts were sequenced. DNAMAN software was adopted to analyze the aptamer sequences and RNAstructure 3.0 was used for predicting a secondary structure for each sequence [10,12,37].

3.4. Flow-Cytometric Analysis

A FACSCalibur flow cytometer with a PowerMacG4 workstation and CellQuest Pro software (BD Biosciences, San Jose, CA, USA) was employed to assess the binding ability of the individual aptamer sequences to different species of bacteria (B. bifidum, B. breve , B. longum, B. animalis, B. adolescentis, and L. plantarum) in separate experiments. The aptamers were labeled with FAM fluorophore at the $5'$ end. In the binding assays, 10^8 cells were incubated with the fluorescently labeled aptamer pool (100 nM) at 37 °C for 45 min with slight agitation. The cells were then washed in PBS, collected by centrifugation, and resuspended in PBS for prompt flow cytometric assays. Forward scatter, side scatter, and fluorescence intensity were measured, and the gated fluorescence intensity above the background (cells with no aptamers) was quantified. BD CellQuest Pro software was adopted to analyze data from the FACSCalibur and to create histogram overlays. Binding dissociation constant K_d values were obtained from the binding curves created with GraphPad Prism 5.0 software by varying the aptamer concentration (0 to 100 nM) with a fixed number of cells (10^8 cells).

3.5. Aptamer CCFM641-5 Binding Assays by Quantitative PCR (qPCR)

To further determine binding affinities of aptamer CCFM641-5 for different bacterial species, qPCR was performed as previously described with some modifications [7]. In the binding assays, 10^7 bacterial cells were incubated with the aptamer (50 nM) at 37 °C for 45 min with slight agitation. The cells were then washed in PBS, collected by centrifugation, and resuspended with 50 μL of $1\times$ PCR reaction buffer. The ssDNA aptamers recovered in the supernatant were used as the template for quantification by SYBR Green-based qPCR using a CFX96 real-time PCR detection system (Bio-Rad Laboratories, Hercules, CA, USA). All qPCR amplifications were carried out in 20 μL volume using 96-well plates in triplicate.

3.6. Aptamer Truncation

Truncation experiments were performed to determine whether all nucleotides of the aptamer sequence CCFM641-5 are necessary [38,39]. The specific primer binding sites of the aptamers were first removed. If DNA aptamer variants possessed high binding affinity for B. bifidum, the aptamer variants were then truncated from the $5'$ end or $3'$ end. In the experiment, the truncated aptamer variants were FAM-labeled at the $5'$ end and tested for their binding abilities to B. bifidum with flow cytometric assays as described above.

3.7. Proteinase Treatment for Bacteria

The procedure used in this study was based on a previously published method [9,40] with some modifications, listed as follows. B. bifidum (10^8 cells) was collected by centrifugation, washed twice with PBS and incubated with 1 mL of 0.25% trypsin or 0.1 mg/mL proteinase K in PBS at 37 °C for 2 and 10 min. After incubation, the mixture was washed with PBS, and the treated bacteria were incubated with FAM-labeled aptamer for further binding assays as described above in flow-cytometric analysis.

3.8. Colorimetric Bioassay on the Basis of the Selected Aptamer

A colorimetric sandwich-type assay for the detection of B. bifidum was developed by ELAA. First, the amino-modified candidate aptamer was dissolved in binding buffer (0.01 mol/L PBS) and 100 μL of diluted aptamer (40 pmol/well) was added into each well of the DNA-BIND 96-well plate for incubation at 37 °C for 1 h. The wells were washed three times with washing buffer (0.01 mol/L PBS with 0.05% Tween-20) to remove unbound aptamer. The microplate wells were then blocked with blocking buffer (0.01 mol/L PBS with 3% BSA) for 1 h at 37 °C to prevent the appearance of nonspecific adsorption.

Then, a series of different concentrations of B. bifidum cells were added into each well for incubation at 37 °C for 45 min and the biotinylated aptamer and streptavidin-HRP were mixed at 37 °C for 30 min at the same time. After the microtiter plates were washed three times with the washing buffer, and 100 μL samples of the above biotinylated aptamer and streptavidin-HRP complexes were added to each well for reaction for 45 min at 37 °C. After washing, 200 μL TMB–H_2O_2 (tetramethyl benzidine-hydrogen peroxide) working solutions were added into each well for reaction without direct light exposure. After incubating for 15 min, the reaction was terminated with 50 μL 2 M H_2SO_4, and the OD at 450 nm was measured with the microplate reader.

4. Conclusions

This study is the first report of the use of whole-bacterium SELEX to identify ssDNA aptamers that are specific for *B. bifidum*. The results show that the aptamer CCFM641-5 bound tightly to *B. bifidum* with a K_d value in the nanomolar range, and could bind *B. bifidum* specifically over other bacterial species. According to the results of the present study, we demonstrate that the DNA aptamer CCFM641-5 can be used to capture and detect *B. bifidum*. Thus, the work described in this study testified the ability of this method to screen a good aptamer probe for the detection of *B. bifidum* and has the potential to contribute greatly to the development of the detection of *B. bifidum*.

In addition, the results from the experiments with aptamer truncations indicated that the 3′-primer and 5′-primer binding sites were important for an optimal binding affinity of the aptamer CCFM641-5 for *B. bifidum*. The experiments with proteinase treatment suggest that the component bound by the aptamer CCFM641-5 is likely protein on the *B. bifidum* cell surface.

Furthermore, we developed a colorimetric assay to detect *B. bifidum* which did not rely on expensive instrumentation, but on the basis of the aptamer CCFM641-5. The method is sensitive and specific and could be adopted to detect *B. bifidum* cells at concentrations as low as 10^4 cfu/mL. Therefore, the colorimetric bioassay based on the aptamer CCFM641-5 is a promising method for the detection of *B. bifidum*.

Acknowledgments: This work was supported by the Program of National Natural Science Foundation for the Youth of China (No. 31501454).

Author Contributions: Lujun Hu and Wei Chen conceived and designed the experiments; Lujun Hu and Linlin Wang performed the experiments; Lujun Hu and Wei Chen analyzed the data; Wenwei Lu, Jianxin Zhao and Hao Zhang contributed reagents/materials/analysis tools; and Lujun Hu and Wei Chen wrote the paper.

References

1.   Tuerk, C.; Gold, L. Systematic evolution of ligands by exponential enrichment: RNA ligands to bacteriophage T4 DNA polymerase. *Science* **1990**, *249*, 505–510. [CrossRef] [PubMed]

2.   Ellington, A.D.; Szostak, J.W. In vitro selection of RNA molecules that bind specific ligands. *Nature* **1990**, *346*, 818–822. [CrossRef] [PubMed]

3.   Chen, M.; Yu, Y.; Jiang, F.; Zhou, J.; Li, Y.; Liang, C.; Dang, L.; Lu, A.; Zhang, G. Development of Cell-SELEX technology and its application in cancer diagnosis and therapy. *Int. J. Mol. Sci.* **2016**, *17*, 2079. [CrossRef] [PubMed]

4.   Famulok, M.; Hartig, J.S.; Mayer, G. Functional aptamers and aptazymes in biotechnology, diagnostics, and therapy. *Chem. Rev.* **2007**, *107*, 3715–3743. [CrossRef] [PubMed]

5.   Navani, N.K.; Li, Y. Nucleic acid aptamers and enzymes as sensors. *Curr. Opin. Chem. Biol.* **2006**, *10*, 272–281. [CrossRef] [PubMed]

6.   Dwivedi, H.P.; Smiley, R.D.; Jaykus, L.A. Selection and characterization of DNA aptamers with binding selectivity to *Campylobacter jejuni* using whole-cell SELEX. *Appl. Microbiol. Biotechnol.* **2010**, *87*, 2323–2334. [CrossRef] [PubMed]

7.   Marton, S.; Cleto, F.; Krieger, M.A.; Cardoso, J. Isolation of an aptamer that binds specifically to *E. coli*. *PLoS ONE* **2016**, *11*, e0153637. [CrossRef] [PubMed]

8. Hamula, C.L.A.; Zhang, H.; Guan, L.L.; Li, X.F.; Le, X.C. Selection of aptamers against live bacterial cells. *Anal. Chem.* **2008**, *80*, 7812–7819. [CrossRef] [PubMed]

9. Chen, F.; Zhou, J.; Luo, F.L.; Mohammed, A.B.; Zhang, X.L. Aptamer from whole-bacterium SELEX as new therapeutic reagent against virulent *Mycobacterium tuberculosis*. *Biochem. Biophys. Res. Commun.* **2007**, *357*, 743–748. [CrossRef] [PubMed]

10. Duan, N.; Wu, S.; Chen, X.; Huang, Y.; Wang, Z. Selection and identification of a DNA aptamer targeted to *Vibrio parahemolyticus*. *J. Agric. Food Chem.* **2012**, *60*, 4034–4038. [CrossRef] [PubMed]

11. Hamula, C.L.A.; Le, X.C.; Li, X.F. DNA aptamers binding to multiple prevalent M-types of *Streptococcus pyogenes*. *Anal. Chem.* **2011**, *83*, 3640–3647. [CrossRef] [PubMed]

12. Cao, X.; Li, S.; Chen, L.; Ding, H.; Xu, H.; Huang, Y.; Li, J.; Liu, N.; Cao, W.; Zhu, Y.; Shen, B.; Shao, N. Combining use of a panel of ssDNA aptamers in the detection of *Staphylococcus aureus*. *Nucleic Acids Res.* **2009**, *37*, 4621–4628. [CrossRef] [PubMed]

13. Kim, N.; Kunisawa, J.; Kweon, M.N.; Ji, G.E.; Kiyono, H. Oral feeding of *Bifidobacterium bifidum* (BGN4) prevents CD4+ CD45RB^high T cell-mediated inflammatory bowel disease by inhibition of disordered T cell activation. *Clin. Immunol.* **2007**, *123*, 30–39. [CrossRef] [PubMed]

14. Repa, A.; Thanhaeuser, M.; Endress, D.; Weber, M.; Kreissl, A.; Binder, C.; Berger, A.; Haiden, N. Probiotics (*Lactobacillus acidophilus* and *Bifidobacterium bifidum*) prevent NEC in VLBW infants fed breast milk but not formula. *Pediatr. Res.* **2015**, *77*, 381–388. [CrossRef] [PubMed]

15. Zanotti, I.; Turroni, F.; Piemontese, A.; Mancabelli, L.; Milani, C.; Viappiani, A.; Prevedini, G.; Sanchez, B.; Margolles, A.; Elviri, L.; et al. Evidence for cholesterol-lowering activity by *Bifidobacterium bifidum* PRL2010 through gut microbiota modulation. *Appl. Microbiol. Biotechnol.* **2015**, *99*, 6813–6829. [CrossRef] [PubMed]

16. Lin, R.J.; Qiu, L.H.; Guan, R.Z.; Hu, S.J.; Liu, Y.Y.; Wang, G.J. Protective effect of probiotics in the treatment of infantile eczema. *Exp. Ther. Med.* **2015**, *9*, 1593–1596. [CrossRef] [PubMed]

17. Turroni, F.; Taverniti, V.; Ruas-Madiedo, P.; Duranti, S.; Guglielmetti, S.; Lugli, G.A.; Gioiosa, L.; Palanza, P.; Margolles, A.; van Sinderen, D.; et al. *Bifidobacterium bifidum* PRL2010 modulates the host innate immune response. *Appl. Environ. Microbiol.* **2014**, *80*, 730–740. [CrossRef] [PubMed]

18. Saavedra, J.M.; Bauman, N.A.; Oung, I.; Perman, J.A.; Yolken, R.H. Feeding of *Bifidobacterium bifidum* and *Streptococcus thermophilus* to infants in hospital for prevention of diarrhoea and shedding of rotavirus. *Lancet* **1994**, *344*, 1046–1049. [CrossRef]

19. López, P.; González-Rodríguez, I.; Gueimonde, M.; Margolles, A.; Suárez, A. Immune response to *Bifidobacterium bifidum* strains support Treg/Th17 plasticity. *PLoS ONE* **2011**, *6*, e24776. [CrossRef] [PubMed]

20. Theunissen, J.; Britz, T.J.; Torriani, S.; Witthuhn, R.C. Identification of probiotic microorganisms in South African products using PCR-based DGGE analysis. *Int. J. Food Microbiol.* **2005**, *98*, 11–21. [CrossRef] [PubMed]

21. Mazzola, G.; Aloisio, I.; Biavati, B.; Di Gioia, D. Development of a synbiotic product for newborns and infants. *LWT-Food Sci. Technol.* **2015**, *64*, 727–734. [CrossRef]

22. Stanton, C.; Gardiner, G.; Meehan, H.; Collins, K.; Fitzgerald, G.; Lynch, P.B.; Ross, R.P. Market potential for probiotics. *Am. J. Clin. Nutr.* **2001**, *73*, 476–483.

23. Dinoto, A.; Marques, T.M.; Sakamoto, K.; Fukiya, S.; Watanabe, J.; Ito, S.; Yokota, A. Population dynamics of *Bifidobacterium* species in human feces during raffinose administration monitored by fluorescence in situ hybridization-flow cytometry. *Appl. Environ. Microbiol.* **2006**, *72*, 7739–7747. [CrossRef] [PubMed]

24. Matsuki, T.; Watanabe, K.; Fujimoto, J.; Kado, Y.; Takada, T.; Matsumoto, K.; Tanaka, R. Quantitative PCR with 16S rRNA-gene-targeted species-specific primers for analysis of human intestinal bifidobacteria. *Appl. Environ. Microbiol.* **2004**, *70*, 167–173. [CrossRef] [PubMed]

25. Mullié, C.; Odou, M.F.; Singer, E.; Romond, M.B.; Izard, D. Multiplex PCR using 16S rRNA gene-targeted primers for the identifcation of bifidobacteria from human origin. *FEMS Microbiol. Lett.* **2003**, *222*, 129–136. [CrossRef]

26. Vincent, D.; Roy, D.; Mondou, F.; Déry, C. Characterization of bifidobacteria by random DNA amplification. *Int. J. Food Microbiol.* **1998**, *43*, 185–193. [CrossRef]

27. Torres-Chavolla, E.; Alocilja, E.C. Aptasensors for detection of microbial and viral pathogens. *Biosens. Bioelectron.* **2009**, *24*, 3175–3182. [CrossRef] [PubMed]

28. Langendijk, P.S.; Schut, F.; Jansen, G.J.; Raangs, G.C.; Kamphuis, G.R.; Wilkinson, M.H.; Welling, G.W. Quantitative fluorescence in situ hybridization of *Bifidobacterium* spp. with genus-specific 16S rRNA-targeted probes and its application in fecal samples. *Appl. Environ. Microbiol.* **1995**, *61*, 3069–3075. [PubMed]

29. Ikebukuro, K.; Kiyohara, C.; Sode, K. Novel electrochemical sensor system for protein using the aptamers in sandwich manner. *Biosens. Bioelectron.* **2005**, *20*, 2168–2172. [CrossRef] [PubMed]

30. Zhao, J.; Zhang, Y.; Li, H.; Wen, Y.; Fan, X.; Lin, F.; Tan, L.; Yao, S. Ultrasensitive electrochemical aptasensor for thrombin based on the amplification of aptamer–AuNPs–HRP conjugates. *Biosens. Bioelectron.* **2011**, *26*, 2297–2303. [CrossRef] [PubMed]

31. Aimaiti, R.; Qin, L.; Cao, T.; Yang, H.; Wang, J.; Lu, J.; Huang, X.; Hu, Z. Identification and application of ssDNA aptamers against $H_{37}Rv$ in the detection of *Mycobacterium tuberculosis*. *Appl. Microbiol. Biotechnol.* **2015**, *99*, 9073–9083. [CrossRef] [PubMed]

32. Barthelmebs, L.; Jonca, J.; Hayat, A.; Prieto-Simon, B.; Marty, J.L. Enzyme-Linked Aptamer Assays (ELAAs), based on a competition format for a rapid and sensitive detection of Ochratoxin A in wine. *Food Control* **2011**, *22*, 737–743. [CrossRef]

33. Nie, J.; Deng, Y.; Deng, Q.P.; Zhang, D.W.; Zhou, Y.L.; Zhang, X.X. A self-assemble aptamer fragment/target complex based high-throughput colorimetric aptasensor using enzyme linked aptamer assay. *Talanta* **2013**, *106*, 309–314. [CrossRef] [PubMed]

34. Park, J.H.; Cho, Y.S.; Kang, S.; Lee, E.J.; Lee, G.H.; Hah, S.S. A colorimetric sandwich-type assay for sensitive thrombin detection based on enzyme-linked aptamer assay. *Anal. Biochem.* **2014**, *462*, 10–12. [CrossRef] [PubMed]

35. Jayasena, S.D. Aptamers: An emerging class of molecules that rival antibodies in diagnostics. *Clin. Chem.* **1999**, *45*, 1628–1650. [PubMed]

36. Liu, G.; Yu, X.; Xue, F.; Chen, W.; Ye, Y.; Yang, X.; Lian, Y.; Yan, Y.; Zong, K. Screening and preliminary application of a DNA aptamer for rapid detection of Salmonella O8. *Microchim. Acta* **2012**, *178*, 237–244. [CrossRef]

37. Reuter, J.S.; Mathews, D.H. RNAstructure: Software for RNA secondary structure prediction and analysis. *BMC Bioinform.* **2010**, *11*, 129. [CrossRef] [PubMed]

38. Shangguan, D.; Tang, Z.; Mallikaratchy, P.; Xiao, Z.; Tan, W. Optimization and modifications of aptamers selected from live cancer cell lines. *ChemBioChem* **2007**, *8*, 603–606. [CrossRef] [PubMed]

39. Stoltenburg, R.; Schubert, T.; Strehlitz, B. In vitro selection and interaction studies of a DNA aptamer targeting protein A. *PLoS ONE* **2015**, *10*, e0134403. [CrossRef] [PubMed]

40. Shangguan, D.; Li, Y.; Tang, Z.; Cao, Z.C.; Chen, H.W.; Mallikaratchy, P.; Sefah, K.; Yang, C.J.; Tan, W. Aptamers evolved from live cells as effective molecular probes for cancer study. *Proc. Natl. Acad. Sci. USA* **2006**, *103*, 11838–11843. [CrossRef] [PubMed]

4

Charomers—Interleukin-6 Receptor Specific Aptamers for Cellular Internalization and Targeted Drug Delivery

Ulrich Hahn

Chemistry Department, Institute for Biochemistry and Molecular Biology, MIN-Faculty, Universität Hamburg, Martin-Luther-King-Platz 6, D-20146 Hamburg, Germany; uli.hahn@uni-hamburg.de

Abstract: Interleukin-6 (IL-6) is a key player in inflammation and the main factor for the induction of acute phase protein biosynthesis. Further to its central role in many aspects of the immune system, IL-6 regulates a variety of homeostatic processes. To interfere with IL-6 dependent diseases, such as various autoimmune diseases or certain cancers like multiple myeloma or hepatocellular carcinoma associated with chronic inflammation, it might be a sensible strategy to target human IL-6 receptor (hIL-6R) presenting cells with aptamers. We therefore have selected and characterized different DNA and RNA aptamers specifically binding IL-6R. These IL-6R aptamers, however, do not interfere with the IL-6 signaling pathway but are internalized with the receptor and thus can serve as vehicles for the delivery of different cargo molecules like therapeutics. We succeeded in the construction of a chlorin e6 derivatized aptamer to be delivered for targeted photodynamic therapy (PDT). Furthermore, we were able to synthesize an aptamer intrinsically comprising the cytostatic 5-Fluoro-2′-deoxy-uridine for targeted chemotherapy. The α6β4 integrin specific DNA aptamer IDA, also selected in our laboratory is internalized, too. All these aptamers can serve as vehicles for targeted drug delivery into cells. We call them charomers—in memory of Charon, the ferryman in Greek mythology, who ferried the deceased into the underworld.

Keywords: aptamers; charomers; targeted drug delivery; targeted chemotherapy; photodynamic therapy; interleukin-6 receptor

---

1. Introduction

The multifunctional cytokine interleukin-6 (IL-6) consists of 183 amino acids and is in case of e.g., a skin lesion secreted by violated cells to signal this violation to recipient cells, thus inducing an inflammation followed by the healing process. The IL-6 signal is recognized by a highly specific IL-6 receptor (IL-6R) which is presented at the surfaces of certain cells. At least two further molecules of the nearly ubiquitously occurring glycosylated transmembrane protein gp130 are needed to result in the active complex for initiating signal transduction from outside the cell, finally into the nucleus to regulate corresponding gene expression (for review see [1–3]). One prerequisite of many receptors is their ability to exhibit a mechanism for desensitizing. IL-6R achieves this by internalization.

IL-6 mediated signal transduction is involved in many disease processes and is thus of high medical relevance. In some cases, one might wish to have a tool at hand to interrupt this signaling pathway. Candidates therefore are antibodies or even better, aptamers. Highlighting advantages and disadvantages of aptamers can be omitted in a special issue on aptamers and thus we can step directly into the projects which should be described here.

Our original plan, initiated by Stefan Rose-John, was to select aptamers specific for IL-6R aiming at getting a tool at hand to block IL-6 mediated signal transduction. Attempts to select aptamers with high specificity for IL-6R were successful for canonical and modified RNA (dissociation constants from 20 nM to 55 nM [4–6]) as well as for DNA aptamers (dissociation constant 490 nM [7]).

All these aptamers, however, did not inhibit IL-6 signaling at all but most RNA aptamers were internalized and thus could function as vehicles for cargo delivery into target cells.

Another kind of cell surface proteins chosen as targets for the selection of aptamers in our laboratory was α6β4 integrin. This is presented by epithelial cells, Schwann cells, keratinocytes and endothelial cells [8,9]. The α6β4 integrin can bind to laminin, which leads to the assembly of hemidesmosomes followed by stable adhesion via connecting the intracellular keratin cytoskeleton to the basement membrane [10,11]. The selected α6β4 integrin specific aptamer IDA was also internalized.

In addition to the aptamers discussed so far, a number of others have been selected and characterized that can also be used to shuttle a variety of drugs, liposomes and (nano) particles into cells. Among those are aptamers targeting prostate-specific membrane antigen (PSMA) [12] which served for the directed delivery of an appropriate siRNA where it was connected to [13]. Aptamers specific for mucin-1 [14], nucleolin [15] transferrin receptor [16] or αvβ integrin [17]—just to list some as representatives—served as vehicles for different kinds of drug delivery approaches.

We have recently presented an overview on aptamers to be used as drug delivery vehicles [18,19]; readers are also referred to excellent reviews of the systemic administration of aptamer-based therapeutics by Burnett and Rossi [20] and Catuogno et al. [21], Sun et al. [22], Gilboa et al. [23], Jiang et al. [24] and not least, recently by Kruspe and Giangrande [25,26].

For all those internalized aptamers exhibiting the capability for cargo delivery I here would like to introduce the term "charomers".

In this brief review, however, solely aptamers selected in our laboratory and suitable as charomers will be dealt with in the following.

2. Interleukin-6-Recetor (IL-6R) Specific Aptamers

2.1. G-Quadruplex Forming Interleukin-6 Receptor (IL-6R) Specific Dimeric RNA Aptamers of 19 or 34 Nucleotides

2.1.1. AIR-3A—An Aptamer Specific for IL-6R and Consisting of RNA

The first IL-6R specific aptamers selected in our laboratory consisted of RNA. Sequencing of the enriched pool revealed six individual clones all comprising a very similar consensus sequence (Figure 1; [4]).

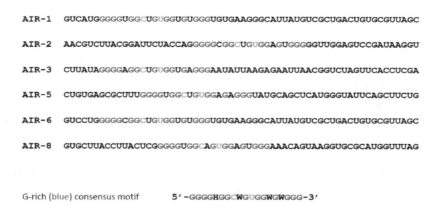

Figure 1. Alignment of interleukin-6 receptor (IL-6R) aptamer sequences from enriched pools. Consensus sequence is given below (conserved Gs in **blue** and conserved Cs and Us in **green**); H encodes A, C, or U and W encodes A or U, respectively. Flanking primer binding sites or constant regions of starting pool are omitted.

Minimal variants of each of these clones presenting each individual consensus motif were synthesized and analyzed for their capacity to bind IL-6R. AIR-3A (an aptamer specific for IL-6R and consisting of RNA; Figure 2) turned out to be the best candidate and was thus used for all further investigations [4]. Its high G-content was a strong hint of a G-quadruplex topology of this aptamer. Biophysical analyses like circular dichroism spectroscopy (CD) and UV-melting studies proved that AIR-3A adopted a parallel G-quadruplex structure (Figure 3).

AIR-3A	**5'-GGGGAGGCUGUGGUGAGGG-3'**
G17U	**5'-GGGGAGGCUGUGGUGAUGG-3'**
G18U	**5'-GGGGAGGCUGUGGUGAGUG-3'**
G17U/G18U	**5'-GGGGAGGCUGUGGUGAUUG-3'**

Figure 2. Nucleotide sequence of AIR-3A, the minimized active version of the IL-6R specific RNA aptamer AIR-3 and inactive AIR-3A variants; replaced nucleotides in red. A dissociation constant of about 20 nM was determined if AIR-3A was incubated with recombinant soluble human receptor (shIL-6R) in filter retention assays [4]. If the aptamer was incubated with IL-6R-presenting bone marrow-derived pro-B (BaF3) cells, the K_d turned out to be about 2 nM [27]. Variants with one (G17U or G18U) or two Gs replaced by Us (G17U/G18U), respectively, did not bind to any target at all.

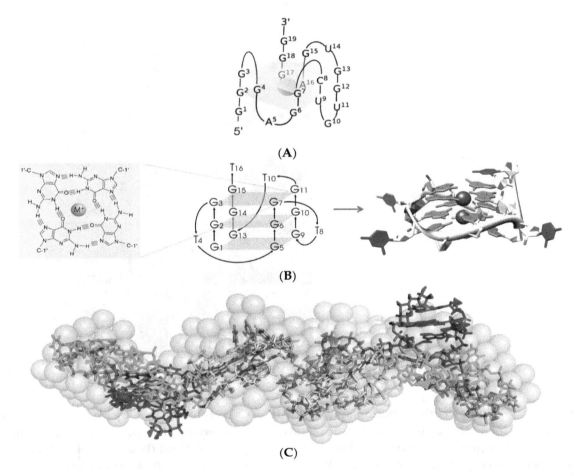

(A)

(B)

(C)

Figure 3. Aptamers AIR-3A, AID-1 as well as RAID3 all exhibit a G-quadruplex structure; the RNA aptamers AIR-3A and RAID3 were shown to dimerize. Circular dichroism (CD) spectroscopic investigations and UV-melting analyses revealed a G-quadruplex structure for both the RNA aptamers AIR-3A (**A**) [4] and RAID3 (**C**) [6], as well as for the DNA aptamer AID-1 (**B**) [7]. Balls in B represent structure stabilizing metal ions; gray semitransparent spheres in C symbolize a model deduced from synchrotron-based small-angle X-ray scattering (SAXS) analyses which could be superimposed with an ab initio model of an aptamer dimer.

2.1.2. RAID3—An RNA Aptamer for Interleukin-6 receptor Domain 3

Another IL-6R specific 34 nt long RNA aptamer selected in our laboratory was RAID3 (RNA Aptamer for Interleukin-6 receptor Domain 3) [6]. It also exhibited a G-quadruplex structure and, most remarkably, could post-selectively be modified by replacing all pyrimidines by their 2′-fluoro analogs, resulting in the aptamer RAID3 2′-F-Py. Both mentioned aptamers did not show significant differences in their target binding ability (K_d about 50 nM both). RAID3 2′-F-Py, however, exhibited an exceptional stability over a period of two days in Dulbecco's modified Eagle's medium supplemented with 10% fetal bovine serum (DMEM 10% FBS) at 37 °C. Not to forget that even the unmodified aptamer, RAID3, had a relatively long half-life of up to five minutes under the same conditions [6].

2.2. AIR-3A and RAID3 Are Internalized by IL-6R Presenting Cells and thus Charomers Allowing Their Usage as Vehicles for Targeted Drug Delivery

AIR-3A and also RAID3 both turned out not to interfere with IL-6 initiated signal transduction. IL-6R, however, was internalized [28] as are many other receptors or cell surface proteins. Therefore, it was obvious to assume that a considerably tight binding ligand might be internalized too, together with the receptor. This could be demonstrated for some of the IL-6R specific RNA aptamers selected in our laboratory (Figure 4 and [4,6]).

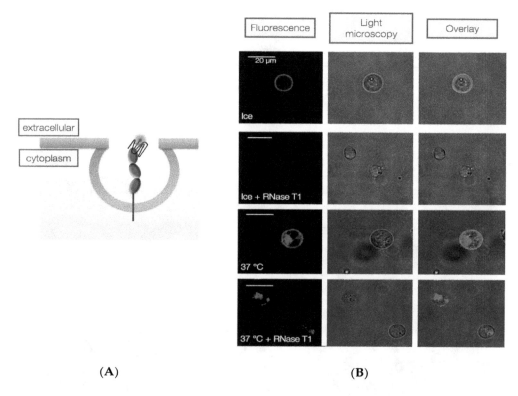

(A)   (B)

Figure 4. Fluorescently labeled aptamer AIR-3A is internalized by IL-6R presenting BAF/gp130/IL6R/TNF cells. (**A**) Schematic presentation of internalization process of G-quadruplex forming fluorescently labeled aptamer bound to the receptor IL-6R and (**B**) confocal laser scanning and light microscopy of IL-6R presenting cells after 30 min incubation with Atto645N-labeled AIR-3A at 37 °C and on ice (control, as internalization does not occur at 0 °C). Another control included an incubation with G specific ribonuclease (RNase) T1 which degraded surface bound RNA aptamers [4].

In Greek mythology, a ferryman named Charon ferried the dead from the world to the underworld. In memory of this ferryman and in honor of one of the first cloning vectors based on the bacteriophage lambda—which was invented by Blattner et al. in 1977 and named "Charon phages" [29]—and in search of an acronym, we named our internalized and drug delivering aptamers "charomers".

2.3. Charomer Mediated Targeted Photodynamic Therapy (PDT)

Chlorin e6 (c-e6) is a photoactivatable agent that generates singlet oxygen upon irradiation (Figure 5A). It is approved for ex vivo and in vivo application and thus very well suited for photodynamic therapy (PDT [30–33]). If pure c-e6 is applied to target cells it is non-specifically internalized and intracellularly accumulated. We have covalently linked c-e6 to the 3′-terminus of the IL-6R specific RNA aptamer AIR-3A which was then incubated with IL-6R presenting cells for appropriate times. After illumination of treated cells with light of 660 nm cell vitality dropped considerably under 50% and apoptosis increased significantly [34].

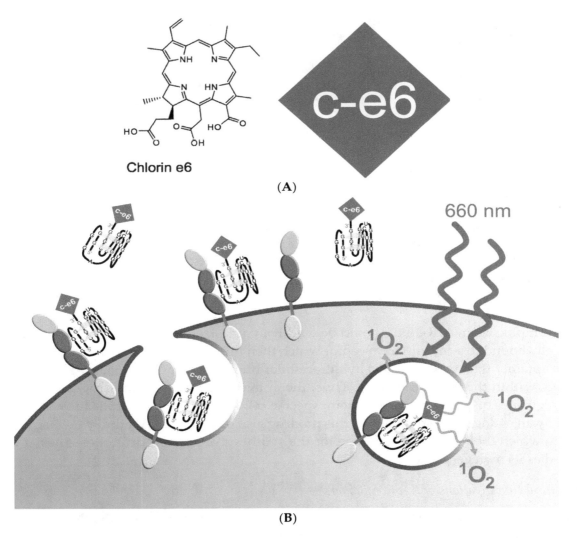

Figure 5. Chlorin e6 (c-e6) derivatized charomer AIR-3A-ce6 was internalized by IL-6R presenting cells leading to their destruction after illumination with appropriate light [34]. Schematic drawing of the aptamer mediated targeted photodynamic therapy (PDT). C-e6 (**A**) was covalently linked to the IL-6R specific aptamer (here schematically depicted as a G-quadruplex structured molecule); (**B**) This aptamer derivative was incubated with appropriate cells which did not survive after illumination with light of 660 nm wavelength (**red** waved lines) which is absorbed by ce-6 leading to the generation of singlet oxygen ($^1O_2$, **orange** waved lines) [34]; different colored ellipses of receptors symbolize different domains, extracellular domains in blue and green, intracellular part yellow.

2.4. Charomer Mediated Targeted Chemotherapy

The base analogue 5-fluorouracil (5-FU; Figure 6A) is a warts therapeutic [35] and known since 60 years as chemo therapeutic or cancer drug [36,37]. It is also used in different kinds of application forms [38]. We have enzymatically incorporated 5-fluoro-2′-deoxyuridine (5-FdU; Figure 6B) into aptamer AIR-3, the initially selected IL-6R specific "long version" of AIR-3A (Figures 1 and 2). AIR-3 was chosen as it exhibits significantly more Us than AIR-3A (Figures 1 and 6C). The resulting aptamer, AIR-3-FdU, still bound IL-6R with a dissociation constant of about 150 nM and IL-6R presenting BaF3 hIL-6R cells with a remarkable K_d of about 20 nM [27]. Furthermore, when incubated with target cells AIR-3-FdU also was internalized, finally resulting in a decrease of cell proliferation to about 75%.

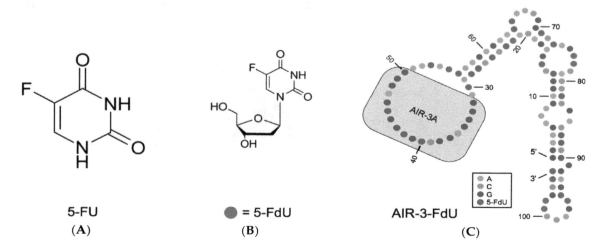

Figure 6. Charomer AIR-3-FdU, a tool for directed cancer drug delivery [27]. Shown are structures of 5-fluorouracil (5-FU; (**A**)), 5-fluoro-2′-deoxyuridine (5-FdU; (**B**)) and AIR-3-FdU (**C**) a derivative of the IL-6R specific RNA aptamer AIR-3 with each U replaced by 5-FdU (red dots); AIR-3-FdU structure is deduced from a model predicted for the originally selected aptamer AIR-3A with the software Mfold [39]; grey area emphasizes the minimized aptamer version AIR-3A.

To re-emphasize it, AIR-3-FdU could be readily synthesized in an enzymatic one step reaction. It specifically bound to a cell surface receptor which then most likely was transferred to the lysosome. When the aptamer then was degraded by intracellular nucleases, the active drug 5-FdU was released exclusively within the target cells [27]. Thus, the aptamer did act as a prodrug as it fulfilled two main prerequisites of a drug delivery system: specific cell targeting and controlled release of the drug triggered by an endogenous stimulus. As this prodrug also could be enzymatically reverse transcribed into DNA, which then served as template for the synthesis of new prodrug molecules, it thus also functioned as its own gene.

2.5. Structural Investigations of IL-6R Aptamers

Remarkably, all IL-6R specific aptamers selected in our laboratory—regardless whether consisting of DNA or RNA—at least partly comprise a G-quadruplex structure (Figure 3; [4–7,40]). One might get the impression that this structural motive is a prerequisite for a nucleic acid aptamer for binding to IL-6R [40]. This is especially striking in case of the SELEX-selected IL-6R specific only 16 nt long DNA aptamer d(GGGT)$_4$ whose RNA counterpart r(GGGU)$_4$ also behaves very similar with respect to IL-6R binding and inhibition of HIV-1 integrase and HIV-1 infection [7].

Further structural investigations of the different aptamers discussed here included small-angle X-ray scattering (SAXS) analyses, structure probing, electrophoretic mobility shift assays and microscale thermophoresis [6,40]. In all cases the investigated aptamers were shown to form dimers (Figure 3).

3. Integrin α6β4 Specific DNA Aptamer IDA—Another Charomer

In another project in our laboratory we selected IDA, a 77 nt long integrin α6β4 specific DNA aptamer [41]. The initial motivation for the selection of IDA was to get a tool at hand to inhibit α6β4 integrin mediated cell-cell-interactions. Especially as this particular interaction can constitute a pivotal step in transendothelial migration during metastasis formation [8,11,42,43]. This aptamer actually binds its target (K_d about 140 nM) and also blocks the integrin-laminin-interaction but it is also internalized very effectively. Under appropriate conditions 98% of fluorescently labelled aptamer was internalized within 10 min (Figure 7; [41]).

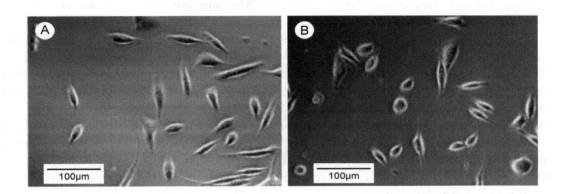

Figure 7. Fluorescently labeled aptamer IDA is internalized by integrin presenting PC-3 cells. Cells were incubated with fluorescently labelled non-specific DNA (**A**) and aptamer IDA (**B**) as described [41]. Scanning microscopic analysis after treating both samples with DNase showed clearly labelled molecules inside cells.

4. Conclusions

The nucleic acid charomers described here are targeting two different cell surface transmembrane proteins exhibiting different functions. The initial motivation for selecting these aptamers was to get tools at hand to inhibit the best-known functions of their targets—receiving signals from other cells or mediating unfavorable cellular interactions. As the targeted proteins are not solely presented but also internalized by the producing cells, it was not surprising that the mentioned and quite strongly binding nucleic acid aptamers were concurrently internalized, too. The possibility to fuse different kinds of cargo molecules [25,44] or even larger particles [21,45,46] with these internalized aptamers makes them charomers.

The inhibitory efficiency of the charomers reported here may not yet be very satisfactory but they can be precursors of a potentially very helpful new class of therapeutics and possibly their effect could be enhanced by the combination of several different charomers, covalently linked to each other or just in an appropriate mixture. Also attempts to improve stability and pharmacokinetic of the charomers might increase their utility. One can think about many different possibilities of new selection strategies or chemical modifications [6,47,48]. Lastly, the applicability in the living organism has to be demonstrated.

And now one final idea. If one imagines how many nucleic acid molecules can be found in the environment (early described by Karl and Bailiff [49]) one easily can imagine that not only a few of them will find their way from the environment into a cell just due to an accidentally sufficient affinity to a surface protein which is internalized. Maybe this is another noteworthy passway for gene exchange across species barriers.

Acknowledgments: I am grateful to Cindy Meyer, Florian Mittelberger, Sven Kruspe, Katharina Redder (née Berg) and Eileen Waldmann (née Magbanua) for providing and helping with Figures.

References

1.   Schaper, F.; Rose-John, S. Interleukin-6: Biology, signaling and strategies of blockade. *Cytokine Growth Factor Rev.* **2015**, *26*, 475–487. [CrossRef] [PubMed]

2. Garbers, C.; Aparicio-Siegmund, S.; Rose-John, S. The IL-6/gp130/STAT3 signaling axis: Recent advances towards specific inhibition. *Curr. Opin. Immunol.* **2015**, *34*, 75–82. [CrossRef] [PubMed]

3. Rothaug, M.; Becker-Pauly, C.; Rose-John, S. The role of interleukin-6 signaling in nervous tissue. *Biochim. Biophys. Acta* **2016**, *1863*, 1218–1227. [CrossRef] [PubMed]

4. Meyer, C.; Eydeler, K.; Magbanua, E.; Zivkovic, T.; Piganeau, N.; Lorenzen, I.; Grotzinger, J.; Mayer, G.; Rose-John, S.; Hahn, U. Interleukin-6 receptor specific RNA aptamers for cargo delivery into target cells. *RNA Biol.* **2012**, *9*, 57–65. [CrossRef] [PubMed]

5. Meyer, C.; Berg, K.; Eydeler-Haeder, K.; Lorenzen, I.; Grotzinger, J.; Rose-John, S.; Hahn, U. Stabilized Interleukin-6 receptor binding RNA aptamers. *RNA Biol.* **2014**, *11*, 57–65. [CrossRef] [PubMed]

6. Mittelberger, F.; Meyer, C.; Waetzig, G.H.; Zacharias, M.; Valentini, E.; Svergun, D.I.; Berg, K.; Lorenzen, I.; Grotzinger, J.; Rose-John, S.; et al. RAID3—An interleukin-6 receptor-binding aptamer with post-selective modification-resistant affinity. *RNA Biol.* **2015**, *12*, 1043–1053. [CrossRef] [PubMed]

7. Magbanua, E.; Zivkovic, T.; Hansen, B.; Beschorner, N.; Meyer, C.; Lorenzen, I.; Grotzinger, J.; Hauber, J.; Torda, A.E.; Mayer, G.; et al. d(GGGT) 4 and r(GGGU) 4 are both HIV-1 inhibitors and interleukin-6 receptor aptamers. *RNA Biol.* **2013**, *10*, 216–227. [CrossRef] [PubMed]

8. Stewart, R.L.; O'Connor, K.L. Clinical significance of the integrin α6β4 in human malignancies. *Lab. Investig.* **2015**, *95*, 976–986. [CrossRef] [PubMed]

9. Mercurio, A.M.; Rabinovitz, I.; Shaw, L.M. The α6β4 integrin and epithelial cell migration. *Curr. Opin. Cell Biol.* **2001**, *13*, 541–545. [CrossRef]

10. Litjens, S.H.; de Pereda, J.M.; Sonnenberg, A. Current insights into the formation and breakdown of hemidesmosomes. *Trends Cell Biol.* **2006**, *16*, 376–383. [CrossRef] [PubMed]

11. Giancotti, F.G. Targeting integrin β4 for cancer and anti-angiogenic therapy. *Trends Pharmacol. Sci.* **2007**, *28*, 506–511. [CrossRef] [PubMed]

12. Lupold, S.E.; Hicke, B.J.; Lin, Y.; Coffey, D.S. Identification and characterization of nuclease-stabilized RNA molecules that bind human prostate cancer cells via the prostate-specific membrane antigen. *Cancer Res.* **2002**, *62*, 4029–4033. [PubMed]

13. Bagalkot, V.; Gao, X. siRNA-aptamer chimeras on nanoparticles: Preserving targeting functionality for effective gene silencing. *ACS Nano* **2011**, *5*, 8131–8139. [CrossRef] [PubMed]

14. Ferreira, C.S.; Matthews, C.S.; Missailidis, S. DNA aptamers that bind to MUC1 tumour marker: Design and characterization of MUC1-binding single-stranded DNA aptamers. *Tumour Biol.* **2006**, *27*, 289–301. [CrossRef] [PubMed]

15. Bates, P.J.; Kahlon, J.B.; Thomas, S.D.; Trent, J.O.; Miller, D.M. Antiproliferative activity of G-rich oligonucleotides correlates with protein binding. *J. Biol. Chem.* **1999**, *274*, 26369–26377. [CrossRef] [PubMed]

16. Wilner, S.E.; Wengerter, B.; Maier, K.; de Lourdes Borba Magalhaes, M.; Del Amo, D.S.; Pai, S.; Opazo, F.; Rizzoli, S.O.; Yan, A.; Levy, M. An RNA alternative to human transferrin: A new tool for targeting human cells. *Mol. Ther. Nucleic Acids* **2012**, *1*, e21. [CrossRef] [PubMed]

17. Mi, J.; Zhang, X.; Giangrande, P.H.; McNamara, J.O., 2nd; Nimjee, S.M.; Sarraf-Yazdi, S.; Sullenger, B.A.; Clary, B.M. Targeted inhibition of αvβ3 integrin with an RNA aptamer impairs endothelial cell growth and survival. *Biochem. Biophys. Res. Commun.* **2005**, *338*, 956–963. [CrossRef] [PubMed]

18. Kruspe, S.; Mittelberger, F.; Szameit, K.; Hahn, U. Aptamers as drug delivery vehicles. *ChemMedChem* **2014**, *9*, 1998–2011. [CrossRef] [PubMed]

19. Meyer, C.; Hahn, U.; Rentmeister, A. Cell-specific aptamers as emerging therapeutics. *J. Nucleic Acids* **2011**, *2011*. [CrossRef] [PubMed]

20. Burnett, J.C.; Rossi, J.J. RNA-based therapeutics: Current progress and future prospects. *Chem. Biol.* **2012**, *19*, 60–71. [CrossRef] [PubMed]

21. Catuogno, S.; Esposito, C.L.; de Franciscis, V. Aptamer-mediated targeted delivery of therapeutics: An update. *Pharmaceuticals* **2016**, *9*, 69. [CrossRef] [PubMed]

22. Sun, H.; Zhu, X.; Lu, P.Y.; Rosato, R.R.; Tan, W.; Zu, Y. Oligonucleotide aptamers: New tools for targeted cancer therapy. *Mol. Ther. Nucleic Acids* **2014**, *3*, e182. [CrossRef] [PubMed]

23. Gilboa, E.; Berezhnoy, A.; Schrand, B. Reducing toxicity of immune therapy using aptamer-targeted drug delivery. *Cancer Immunol. Res.* **2015**, *3*, 1195–1200. [CrossRef] [PubMed]

24. Jiang, F.; Liu, B.; Lu, J.; Li, F.; Li, D.; Liang, C.; Dang, L.; Liu, J.; He, B.; Badshah, S.A.; et al. Progress and challenges in developing aptamer-functionalized targeted drug delivery systems. *Int. J. Mol. Sci.* **2015**, *16*, 23784–23822. [CrossRef] [PubMed]

25. Kruspe, S.; Giangrande, P.H. Aptamer-siRNA chimeras: Discovery, progress, and future prospects. *Biomedicines* **2017**, *5*, 45. [CrossRef] [PubMed]

26. Kruspe, S.; Giangrande, P.H. Design and preparation of aptamer-siRNA chimeras (AsiCs) for targeted cancer therapy. *Methods Mol. Biol.* **2017**, *1632*, 175–186. [PubMed]

27. Kruspe, S.; Hahn, U. An aptamer intrinsically comprising 5-fluoro-2′-deoxyuridine for targeted chemotherapy. *Angew. Chem. Int. Ed. Engl.* **2014**, *53*, 10541–10544. [CrossRef] [PubMed]

28. Fujimoto, K.; Ida, H.; Hirota, Y.; Ishigai, M.; Amano, J.; Tanaka, Y. Intracellular dynamics and fate of a humanized anti-interleukin-6 receptor monoclonal antibody, tocilizumab. *Mol. Pharmacol.* **2015**, *88*, 660–675. [CrossRef] [PubMed]

29. Blattner, F.R.; Williams, B.G.; Blechl, A.E.; Denniston-Thompson, K.; Faber, H.E.; Furlong, L.; Grunwald, D.J.; Kiefer, D.O.; Moore, D.D.; Schumm, J.W.; et al. Charon phages: Safer derivatives of bacteriophage lambda for DNA cloning. *Science* **1977**, *196*, 161–169. [CrossRef] [PubMed]

30. Li, Y.; Yu, Y.; Kang, L.; Lu, Y. Effects of chlorin e6-mediated photodynamic therapy on human colon cancer SW480 cells. *Int. J. Clin. Exp. Med.* **2014**, *7*, 4867–4876. [PubMed]

31. Yoon, I.; Li, J.Z.; Shim, Y.K. Advance in photosensitizers and light delivery for photodynamic therapy. *Clin. Endosc.* **2013**, *46*, 7–23. [CrossRef] [PubMed]

32. Agostinis, P.; Berg, K.; Cengel, K.A.; Foster, T.H.; Girotti, A.W.; Gollnick, S.O.; Hahn, S.M.; Hamblin, M.R.; Juzeniene, A.; Kessel, D.; et al. Photodynamic therapy of cancer: An update. *CA Cancer J. Clin.* **2011**, *61*, 250–281. [CrossRef] [PubMed]

33. Choudhary, S.; Nouri, K.; Elsaie, M.L. Photodynamic therapy in dermatology: A review. *Lasers Med. Sci.* **2009**, *24*, 971–980. [CrossRef] [PubMed]

34. Kruspe, S.; Meyer, C.; Hahn, U. Chlorin e6 conjugated interleukin-6 receptor aptamers selectively kill target cells upon irradiation. *Mol. Ther. Nucleic Acids* **2014**, *3*, e143. [CrossRef] [PubMed]

35. Salk, R.S.; Grogan, K.A.; Chang, T.J. Topical 5% 5-fluorouracil cream in the treatment of plantar warts: A prospective, randomized, and controlled clinical study. *J. Drugs Dermatol.* **2006**, *5*, 418–424. [PubMed]

36. Heidelberger, C.; Chaudhuri, N.K.; Danneberg, P.; Mooren, D.; Griesbach, L.; Duschinsky, R.; Schnitzer, R.J.; Pleven, E.; Scheiner, J. Fluorinated pyrimidines, a new class of tumour-inhibitory compounds. *Nature* **1957**, *179*, 663–666. [CrossRef] [PubMed]

37. Longley, D.B.; Harkin, D.P.; Johnston, P.G. 5-fluorouracil: Mechanisms of action and clinical strategies. *Nat. Rev. Cancer* **2003**, *3*, 330–338. [CrossRef] [PubMed]

38. Goette, D.K. Topical chemotherapy with 5-fluorouracil. A review. *J. Am. Acad Dermatol.* **1981**, *4*, 633–649. [CrossRef]

39. Zuker, M. Mfold web server for nucleic acid folding and hybridization prediction. *Nucleic Acids Res.* **2003**, *31*, 3406–3415. [CrossRef] [PubMed]

40. Szameit, K.; Berg, K.; Kruspe, S.; Valentini, E.; Magbanua, E.; Kwiatkowski, M.; Chauvot de Beauchene, I.; Krichel, B.; Schamoni, K.; Uetrecht, C.; et al. Structure and target interaction of a G-quadruplex RNA-aptamer. *RNA Biol.* **2016**, *13*, 973–987. [CrossRef] [PubMed]

41. Berg, K.; Lange, T.; Mittelberger, F.; Schumacher, U.; Hahn, U. Selection and characterization of an α6β4 Integrin blocking DNA Aptamer. *Mol. Ther. Nucleic Acids* **2016**, *5*, e294. [CrossRef] [PubMed]

42. Guo, W.; Giancotti, F.G. Integrin signalling during tumour progression. *Nat. Rev. Mol. Cell Biol.* **2004**, *5*, 816–826. [CrossRef] [PubMed]

43. Nikolopoulos, S.N.; Blaikie, P.; Yoshioka, T.; Guo, W.; Giancotti, F.G. Integrin β4 signaling promotes tumor angiogenesis. *Cancer Cell* **2004**, *6*, 471–483. [CrossRef] [PubMed]

44. Zhou, J.; Rossi, J.J. Cell-specific aptamer-mediated targeted drug delivery. *Oligonucleotides* **2011**, *21*, 1–10. [CrossRef] [PubMed]

45. Chen, Z.; Tai, Z.; Gu, F.; Hu, C.; Zhu, Q.; Gao, S. Aptamer-mediated delivery of docetaxel to prostate cancer through polymeric nanoparticles for enhancement of antitumor efficacy. *Eur. J. Pharm. Biopharm.* **2016**, *107*, 130–141. [CrossRef] [PubMed]

46. Prisner, L.; Bohn, N.; Hahn, U.; Mews, A. Size dependent targeted delivery of gold nanoparticles modified with the IL-6R-specific aptamer AIR-3A to IL-6R-carrying cells. *Nanoscale* **2017**, *9*, 14486–14498. [CrossRef] [PubMed]

47. Wang, R.E.; Wu, H.; Niu, Y.; Cai, J. Improving the stability of aptamers by chemical modification. *Curr. Med. Chem.* **2011**, *18*, 4126–4138. [CrossRef] [PubMed]

48. Tolle, F.; Brandle, G.M.; Matzner, D.; Mayer, G. A Versatile approach towards nucleobase-modified aptamers. *Angew. Chem. Int. Ed. Engl.* **2015**, *54*, 10971–10974. [CrossRef] [PubMed]

49. Karl, D.M.; Bailiff, M.D. The measurement and distribution of dissolved nucleic acids in aquatic environments. *Limnol. Oceanogr.* **1989**, *34*, 543–558. [CrossRef]

5

Unraveling Prion Protein Interactions with Aptamers and Other PrP-Binding Nucleic Acids

Bruno Macedo * and Yraima Cordeiro *

Faculty of Pharmacy, Federal University of Rio de Janeiro (UFRJ), Av. Carlos Chagas Filho 373, Bloco B, Subsolo, Sala 17, Rio de Janeiro, RJ 21941-902, Brazil
* Correspondence: brunomacedo@ufrj.br (B.M.); yraima@pharma.ufrj.br (Y.C.)

Academic Editors: Julian Alexander Tanner, Andrew Brian Kinghorn and Yee-Wai Cheung

Abstract: Transmissible spongiform encephalopathies (TSEs) are a group of neurodegenerative disorders that affect humans and other mammals. The etiologic agents common to these diseases are misfolded conformations of the prion protein (PrP). The molecular mechanisms that trigger the structural conversion of the normal cellular PrP (PrP^C) into the pathogenic conformer (PrP^{Sc}) are still poorly understood. It is proposed that a molecular cofactor would act as a catalyst, lowering the activation energy of the conversion process, therefore favoring the transition of PrP^C to PrP^{Sc}. Several in vitro studies have described physical interactions between PrP and different classes of molecules, which might play a role in either PrP physiology or pathology. Among these molecules, nucleic acids (NAs) are highlighted as potential PrP molecular partners. In this context, the SELEX (Systematic Evolution of Ligands by Exponential Enrichment) methodology has proven extremely valuable to investigate PrP–NA interactions, due to its ability to select small nucleic acids, also termed aptamers, that bind PrP with high affinity and specificity. Aptamers are single-stranded DNA or RNA oligonucleotides that can be folded into a wide range of structures (from harpins to G-quadruplexes). They are selected from a nucleic acid pool containing a large number (10^{14}–10^{16}) of random sequences of the same size (~20–100 bases). Aptamers stand out because of their potential ability to bind with different affinities to distinct conformations of the same protein target. Therefore, the identification of high-affinity and selective PrP ligands may aid the development of new therapies and diagnostic tools for TSEs. This review will focus on the selection of aptamers targeted against either full-length or truncated forms of PrP, discussing the implications that result from interactions of PrP with NAs, and their potential advances in the studies of prions. We will also provide a critical evaluation, assuming the advantages and drawbacks of the SELEX (Systematic Evolution of Ligands by Exponential Enrichment) technique in the general field of amyloidogenic proteins.

Keywords: prion protein; nucleic acids; SELEX (Systematic Evolution of Ligands by Exponential Enrichment); aptamers

---

1. Introduction

Aberrant prion proteins (PrPs) responsible for the transmissible spongiform encephalopathies (TSEs) are misfolded conformations of the natively expressed prion protein, the innocuous cellular PrP (PrP^C) [1]. The misfolded conformers, termed scrapie PrP (PrP^{Sc}), have the ability to self-perpetuate and to become infectious entities [1]. Therefore, they are the primary culprit of TSEs, which form a group of fatal neurodegenerative disorders that affect humans and other mammals [1]. Currently, the "prion" term has emerged as a new phenomenon in molecular biology, describing proteins with the ability to undergo autoconversion, autopropagation, and dissemination between cells [2]. Remarkably, pathogenic PrPs can be transmitted not only between cells but also among organisms of the same species and this can ultimately lead to epidemic outbreaks [3]. To date, only the prion protein fulfills the infectious characteristics of true prions. There is, apparently, a lack of conformational properties in other prion-like proteins to define them as bona fide prions.

PrP^C is a constitutive cell-surface glycoprotein, highly conserved among species, expressed in several cell types, mainly in the central nervous system (CNS) [1]. High-resolution studies have revealed two structurally distinct domains: the flexible N-terminal region (residues 23–~120) and the globular C-terminal domain (residues ~120–231), the latter composed of three α-helices and a small antiparallel β-sheet [4,5]. Is it still not known how the drastic conformational changes occur in the PrP^C structure—even without any mutations in the *PRNP* gene—to give rise to the abnormal PrP^{Sc}. However, once formed, PrP^{Sc} can propagate in an autocatalytic manner, recruiting more PrP^C to fold into new PrP^{Sc}, leading to its accumulation in tissues with severe cellular damage and further neurodegeneration [6]. In contrast to PrP^C, PrP^{Sc} is a β-structure-rich protein, insoluble, and resistant to proteolysis. It can form toxic oligomers and aggregates either with an amyloid-like architecture or with an amorphous disposition [6]. Besides prion diseases, protein aggregation is the central event of many other neurodegenerative disorders, including Alzheimer's (AD) and Parkinson's (PD) diseases [7]. In each scenario, the misfolding of a specific protein, that is, the amyloid β-protein (Aβ) for AD, α-synuclein (α-syn) for PD, and the prion protein itself (PrP) for TSEs, can lead to its aggregation and cell-to-cell transfer, forming insoluble deposits or plaques in different regions of the brain (depending on the particular protein under discussion) [8]. To date, there is no available treatment to halt or to delay the neurodegeneration process triggered by one or more of these misfolded and aggregated proteins in the CNS; therefore, these diseases are still invariably fatal. Understanding the molecular basis of protein misfolding and conformational conversion are major priorities in the search for therapeutic strategies that could block or modulate the aggregation process from its very beginning.

The mechanisms that lead a soluble and natively folded protein to adopt an aberrant conformation with a higher tendency to form aggregates depend on the different intermediate structures formed during the folding process, the free energy of these intermediates, the energy barrier between them, and the exposition of hydrophobic surfaces that should be normally buried and solvent-excluded in a functional conformation [9]. Misfolded forms are normally degraded by cell protein quality control mechanisms, but during aging these mechanisms begin to fail, losing or reducing their ability to prevent protein accumulation [10]. Mutations, posttranslational modifications, environmental variations, or interactions with external agents are also factors that can drive protein misfolding and aggregation [11]. PrP is also known as a "promiscuous" protein that can bind to different classes of molecules, including metallic ions, glycosaminoglycans, lipids, and nucleic acids. The biological relevance of most of these interactions is still not clear, but these ligands may participate in the PrP structural conversion and, consequently, in disease progression [12–17].

Nowadays, the cofactor hypothesis has gained more visibility. It postulates that the presence of an adjuvant factor that interacts with PrP favors its interconversion, aggregation, and infectivity [13,16,18–20]. Such a cofactor may act as a catalyst in PrP conversion, lowering the high-energy barrier that prevents the spontaneous conversion of PrP^C into PrP^{Sc} (Figure 1). In this context, nucleic acid (NAs) molecules have been ascribed an important role. PrP has been shown to interact with DNAs and RNAs both in vitro and in vivo [21–26], indicating their suitable involvement in PrP pathophysiology. Many studies have evaluated the effects of NAs as molecular cofactors for PrP conversion into PrP^{Sc}-like species. The in vitro-methodology called SELEX (Systematic Evolution of Ligands by Exponential Enrichment) [27,28] is an interesting tool that has been used to identify and select small oligonucleotides, known as "aptamers" that bind with high affinity and high specificity to the wild-type (full-length) PrP and/or its different domains.

In this review, we will focus on published studies about PrP–NA interactions, the SELEX methodology, the knowledge these bring to the prion field, and the new avenues they offer for the therapy and diagnosis of such devastating diseases. Besides PrP, several other amyloid-forming proteins related to conformational diseases can also bind nucleic acids [25]; therefore, we will also present an overall critical assessment of the aptamer literature in the general field of amyloids, reviewing some relevant SELEX studies against other amyloidogenic proteins, focusing also on the possible drawbacks of this approach regarding aptamer specificity and selectivity against the monomeric or fibrillar forms of these proteins.

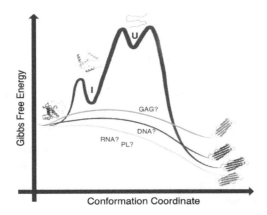

Figure 1. Free energy diagram representing the role of cofactors in prion protein (PrP) conformational conversion. DNA, RNA, phospholipid (PL), and glycosaminoglycan (GAG) candidates may interact with PrP^C, lowering the energy barrier that prevents its spontaneous conversion to the PrP^{Sc}. Different cofactor molecules may stimulate the conversion to the different PrP pathogenic forms and may result in the generation of PrP^{Sc} with varying conformations, providing a possible explanation for the existence of various prion strains. I: intermediate state; U: unfolded state. Reproduced from [9].

2. PrP and Nucleic Acids Interactions

The crosstalk between PrP and NAs has captured the attention of the prion research community for the last twenty years. The first study was conducted by Pradip Nandi in 1997 with the human-derived neurotoxic prion peptide (PrP$^{106-126}$) and showed, through fluorescence measurements, the ability of this peptide to bind to a small single-stranded DNA (ssDNA) sequence with micromolar affinity and that this interaction induced a structural change in the DNA molecule [29]. In subsequent publications, Nandi showed that PrP$^{106-126}$ polymerizes into amyloid aggregates in the presence of DNA, either in its circular or in its linearized forms, under experimental conditions where the peptide alone did not polymerize [30]. Wild-type murine recombinant PrP (rPrP) also underwent polymerization in a nucleic acid aqueous solution [31]. In 2001, our group was the first to show the dual role of NAs in changing PrP conformation and aggregation [21]. While PrP interaction with double-stranded DNA (dsDNA) induced the conversion of the full-length recombinant PrP (rPrP) to β-sheet-rich structures and led to rPrP aggregation as revealed by spectroscopic techniques, the same dsDNA oligonucleotides inhibited the aggregation of a PrP hydrophobic domain, the PrP$^{109-149}$ [21]. PrP$^{109-149}$ undergoes prompt aggregation when diluted from a denaturing condition into an aqueous solution; however, the aggregation is completely inhibited in the presence of DNA in a concentration-dependent manner, as verified by light scattering (LS) measurements and through transmission electron microscopy [21,24]. It was also reported that an anti-DNA antibody (OCD4), as well as the gene 5 protein, a DNA-binding protein, is able to catch PrP only from the brain material of prion-infected humans or animals, but they do not capture PrP from non-infected brains [26]. OCD4 seems to present immunoreaction with DNA-associated molecules and this antibody can form a complex with PrP in prion diseases [26]. Moreover, OCD4 detects PrP^{Sc} over ten times more efficiently than an antibody against PrP [26] supporting the proposal that nucleic acids are associated with PrP^{Sc} in vivo. Collectively, these results reinforce the proposal by our group that DNA can participate in PrP misfolding, shifting the equilibrium between PrP^C and PrP^{Sc} by reducing protein mobility and favoring protein–protein interactions [21,32].

Following these initial observations, many groups started evaluating the interaction of NAs with both PrP^C and PrP^{Sc}, unraveling many aspects of this crosstalk. One important area of exploration was to characterize the DNA-binding site on PrP. Studies with different rPrP constructs, mainly using nuclear magnetic resonance (NMR) and small angle X-ray spectroscopy (SAXS) measurements, identified at least two DNA-binding sites in rPrP; one of them in the C-terminal globular domain and the other in the flexible N-terminal region [33–35]. In 2012, our group showed that different small dsDNA sequences can individually bind to rPrP, inducing protein aggregation in a supramolecular structure resembling less-ordered amyloid fibrils [24]. We have observed different effects on the structure, stability, and aggregation of rPrP upon interaction with different DNA sequences [24]. The resultant PrP–DNA complex was toxic to murine neuroblastoma (N2a) cell lines, depending on the DNA sequence, but caused no toxicity to human kidney (HK-2) cell lines [24]. Our results suggested that the DNA GC-content is important to dictate the aggregation pattern and the formation of toxic species; in addition, the PrP expression level or some specific factors from the cellular lineage also appeared to be important to mediate PrP toxicity [24]. In 2013, Cavaliere et al. showed that G-quadruplex forming DNA can bind to different forms of PrP with nanomolar affinity and, in accordance with our previous studies, the PrP–DNA interaction led to loss of the secondary structure of both the PrP and the DNA molecule, indicating that there are reciprocal structural changes after DNA binds to PrP [24,36].

PrP–RNA interactions have also been described. The work of the Darlix group showed that PrP has nucleic acid chaperoning activities, similar to nucleocapsid retroviral proteins, indicating that PrP might participate in nucleic acid metabolism (both RNA and DNA) [37,38]. Indeed, rPrP binds different RNAs with high affinity in vitro and in vivo. This interaction promotes the formation of PrP aggregates where PrP becomes resistant to proteinase K (PK) digestion and the RNAs bound to the complex are resistant to ribonuclease (RNase) attack [22,23,39,40]. This interaction is normally abolished when the PrP construct has its N-terminal region truncated (residues 23–~121), as shown by different groups, suggesting that the flexible PrP N-terminal region is important to establish the interaction with RNA [33,40]. In 2003, Deleault et al. described the role of RNA molecules in stimulating prion protein conversion in vitro [23]. The amplification of a protease-resistant PrP^{Sc}-like molecule, termed PrP^{Res} (from the PK-resistance property), was evaluated by the in vitro conversion assay based on the protein-misfolding cyclic amplification (PMCA) method [41,42]. PMCA uses diluted prion-infected brain homogenate as a seed to trigger the conversion of PrP^{C} in healthy brain homogenates; the final amplified PrP^{Res} shares many specific characteristics with the scrapie prion propagated in vivo; therefore, it is widely used in prion conversion studies [41]. It was found that RNase inhibits PrP^{Res} amplification in a dose-dependent manner, evidencing that RNA is required for the efficient formation and accumulation of scrapie-like PrP in vitro. Moreover, only the addition of specific RNAs (isolated from mammalian brains) was able to stimulate this conversion reaction [23]. Subsequent work, using only purified and synthetic molecules, revealed that PrP^{C}, PrP^{Sc}, co-purified lipids, and poly-A RNA can form the minimal set of components necessary to amplify the PrP^{Res} conformation in vitro with the ability to infect normal wild-type hamsters in vivo [18]. The requirement of a negatively charged accessory molecule for the efficient production of infectious prions in vitro (synthetic prions) is in good agreement with the proposed cofactor hypothesis, where endogenous or extracellular factors may participate in prion propagation in vivo. Nevertheless, more studies are required to determine what the exact molecular characteristics of PrP conversion catalysts are and to establish whether one or more cofactors could be considered 'ideal' for forming true prions in vitro or to participate in prion pathogenesis in vivo. Our group showed, through several biophysical approaches, that depending on the RNA source—whether from mammalian, yeast, or bacterial cells—the interaction with murine rPrP led to aggregation with different extents. rPrP–RNA interaction led to secondary structural changes in both rPrP, which loses α-helical content, and in the RNA molecule [40]. Finally, only the aggregated species formed upon incubation with RNA extracted from N2a cells were highly toxic to N2a cells in culture [40]. RNA-binding to ovine PrP was also investigated, and the results revealed a likewise PrP conformational shift to a higher β-sheet content, as well as the neurotoxicity of this complex [39]. In accordance with previous work, the PrP N-terminal region seems to be essential to mediate these effects [40,43].

Although PrP^C is typically localized anchored at the plasmatic membrane, it has been reported that it can be found in the nucleus of neuronal and endocrine cells and can interact with chromatin [44]. The translocation and deposition of misfolded PrPs in the nucleus of infected cells, where the misfolded PrP was able to interact with chromatin components, has also been shown [45]. Although converging experimental evidence indicates that the endocytic pathway is the principal site of prion conversion [46,47], one might speculate that an abnormal nuclear compartmentalization of PrP may contribute to its encounter with non-native partners that could be involved in prion pathogenesis. Nevertheless, PrP and nucleic acids could crosstalk even along the endocytic pathway, as would be the case of endocytosis of exogenous (or from the membrane) PrP bound to nucleic acid. It would also be possible that cytosolic forms of PrP [46] encounter small NAs in the cytoplasm, triggering conversion. In fact, cytosolic PrP has been shown to induce the formation of large ribonucleoprotein organelles in the N2a cell line [48]. Moreover, PrP^C to PrP^{Sc} conversion can also occur on the plasma membrane, being the primary site of conversion when the host is infected with scrapie from external sources [49,50]. In this latter case, a nucleic acid released from a cell or from an exogenous source could encounter PrP^C/PrP^{Sc} at the membrane.

Altogether, the evidence compiled here concerning PrP–NA interactions strongly suggest that these molecules can be partners in vivo. Both of them can trigger PrP misfolding, leading to its aggregation in vitro, and they can also stimulate PrP^{Sc} conversion and propagation in vivo. Although they are not identical, the misfolded PrPs formed can be toxic to cultured cells depending on the nucleic acid sequence evaluated. We strongly believe that the sequence and structure adopted by the NAs are essential to dictate those effects. More studies about this partnership may be fundamental not only to understand prion function or dysfunction but also for the development of effective therapeutic approaches.

3. SELEX Technique and the Aptamer Discovery

The SELEX technique consists of finding NA ligands with high affinity and specificity against a given target [27,28]. In this review, our target is the PrP. Generally, the core of the selection process consists of the following essential steps: (i) incubation of a randomly synthesized DNA or RNA library (containing 10^{14}–10^{16} different oligonucleotides sequences) with the selected target to allow binding; (ii) separation of bound from non-bound species (unbound oligonucleotides are removed by several stringent washing steps of the binding complexes); (iii) elution of the target-bound oligonucleotides with higher salt concentrations; and (iv) amplification of the oligonucleotide bound species by a polymerase chain reaction (PCR). A new enriched pool of selected oligonucleotides is generated by purification of ssDNAs from the PCR products (DNA SELEX) or by in vitro transcription (RNA SELEX). Then, this selected NA pool is used for the next selection round. This process can be repeated several times to enhance the affinity and specificity of the isolated NA sequences. The final selected NA sequences are called aptamers; they have to be cloned and individual aptamers have to be sequenced and validated against its target (Figure 2). The stoichiometry of the target and the NAs can be altered as well as the number of washes, and competitive inhibitors can also be added to the binding buffer to enhance the stringency of the SELEX conditions [51]. A counter-SELEX procedure can also be performed to exclude sequences recognizing other non-interested targets by using similar structural targets, therefore increasing the selectivity of the aptamers [52]. Over the years, many modifications and improvements have been introduced to the classical SELEX methodology in order to decrease the selection time and enhance the binding affinity, which include capillary electrophoresis (CE)-SELEX, automated SELEX, and whole cell SELEX [52]. The cell–SELEX technique is fast, straightforward, and very promising because it can be performed with normal living cells, thus guaranteeing that target proteins on the cell maintain their native conformation and function along the selection procedures [52]. Some aptamers have already been discovered to work against different cancer cells by this method; for example, one of them could specifically recognize leukemia cells [53]. To date, there is only one federally approved aptamer, the Pegaptanib drug, selected against vascular endothelial growth factor

(VEGF) to treat the age-related macular degeneration, although there are more than 10 aptamers under different stages of clinical trials for treatment of coagulation, inflammation, cancer, etc. [54]. However, various crucial aspects have delayed the clinical translation of therapeutic aptamers, including their intrinsic physicochemical properties and the lack of safety data.

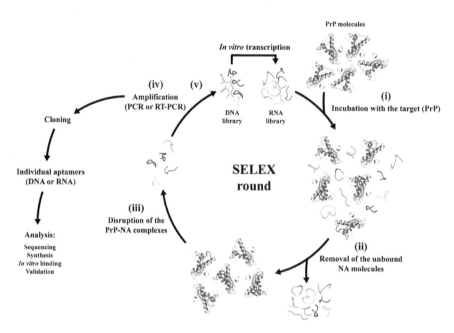

Figure 2. General scheme of the SELEX method using recombinant PrP as the target. A SELEX round consists of the following essential steps: (i) binding after the incubation of a randomly synthesized DNA or RNA library (containing 10^{14}–10^{16} different sequences) with the molecular target (full-length recombinant PrP or other PrP constructions); (ii) removal of the non-bound NA species; (iii) elution of NA sequences from the immobilized PrP (either in-column, in ELISA dishes, or other); (iv) amplification of the eluted NA sequences; (v) back to Step (i). This process can be repeated several times to enhance the affinity and specificity of the isolated NA sequences. The final selected NA pool contains the aptamers that have to be further cloned, and individual aptamers have to be sequenced and validated for binding against its target, PrP.

Aptamers are small single-stranded DNA or RNA nucleotides with a length varying from 20 to 100 bases, flanked by two constant sequences that contain the primer-binding sites [48]. This name comes from the Latin "aptus," which means "fit". Single-stranded NAs can fold into a variety of loops, stems, hairpins, quadruplexes, and bugles, among other shapes, to generate a vast range of secondary and tertiary NAs structures [52]. A well-defined three-dimensional structure of these oligonucleotides can specifically recognize and interact with several target molecules, including amino acids [55,56], proteins [57], antibodies [58], or even whole cells [59]. The interaction affinity occurs in the nanomolar to femtomolar range and is established by various intermolecular forces such as hydrogen bonding, van der Waals forces, and base stacking. To date, thousands of aptamers have already been selected against a wide range of target ligands [52]. They are compared to antibodies but have more advantages over them. They are smaller in size, which allows them to reach the cellular target more easily than classical antibodies; they might have more affinity to the specific target; they are easier to synthesize; and they are non-toxic and non-immunogenic [52]. Aptamers are also more thermally stable and can restore their original structure easily and quickly if denatured, while antibodies cannot. These characteristics support the evaluation of aptamers as great therapeutic candidates. Their clinical limitations—such as low stability, because of nucleases' action in blood, and fast clearance, due to their smaller size—can be easily overcome by further chemical modifications [60]. Site-specific modifications are difficult to make in antibodies, in contrast to aptamers, where chemical modifications can be easily introduced at any desired position in the nucleotide sequence. Because of their ability to bind to proteins and to block their functions, they are also being investigated in the interruption or prevention of misfolded protein accumulation, which is related to many diseases, as discussed here. Moreover, aptamers might differentiate even between isoforms of a given protein, which would aid the understanding of prion function and/or dysfunction, in addition to the diagnostic potential.

4. Nucleic Acids Aptamers against PrP

Considering PrP interaction with a large variety of NAs, this protein was proposed as a NA chaperone, although many doubts still persist concerning either its functional or pathological role. Pursuing the main objectives of dissecting PrP–NA interaction, elucidating PrP conformational conversion, identifying specific NAs with which PrPs from different species may interact, and looking for therapeutic and/or diagnostic methods, several studies have applied the SELEX method [61]. With regard to therapeutic perspectives, aptamer(s) binding to PrPs could prevent their conversion and accumulation by stabilizing PrP^C or PrP^{Sc}, inhibiting PrP^C-PrP^{Sc} interaction, or blocking PrP^C binding to some pathological stimulating cofactor. Specific aptamers for PrPs could also set up new diagnostic tools, since early diagnosis is crucial and will definitely improve the efficacy of any attempt at therapy.

The first aptamer for PrP was selected by Weiss et al. in 1997 using the SELEX method directed against recombinant Syrian hamster full-length prion protein ($rPrP^{23-231}$). The isolated unmodified RNA aptamer did not recognize the C-terminal domain construct ($rPrP^{90-231}$) that lacks almost all the N-terminal residues [33]. They mapped the RNA aptamer-binding site between amino acid residues 23–52 at the PrP N-terminal region. They also suggested that these RNA aptamers may fold into guanine(G)-quartet-containing structural elements that seem to be essential for PrP recognition since the G replacement for uridines (U) in the aptamer sequence abolished its binding to PrP [33]. These individual RNA aptamers interact specially with PrP^C from brain homogenates of healthy mice, hamsters, and cattle but did not recognize PrP^{Res} in brain homogenates from prion-infected mice; moreover, the interaction was not observed in PrP knockout mice [33]. The conservation of the specially PrP^C–RNA interaction in different species provided the first landmark to the development of new diagnostic assays for prion diseases using aptamers.

The same group in 2001 was also the first to show the therapeutic potential of aptamers against prion diseases by selecting the 2′-amino-2′-deoxypyrimidine-modified RNA aptamer (DP7) that was able to reduce PrP^{Sc} accumulation in prion-infected neuroblastoma cells [62]. The RNA chemical modification was made—after using SELEX—as a strategy to enhance the RNA resistance to nucleases. DP7 was highly specific to human PrP^{90-129}, a region involved in PrP pathological conversion, and its binding was sustained even for full-length PrP from different species, including humans, mice, and hamsters [62]. One might speculate that blocking this region could supply promise for controlling the conversion. In 2003, Rhie et al. were the first to characterize RNA aptamers that bind preferentially to the infection-related conformations of PrP [63]. The 2′fluoro-RNA aptamer against scrapie-associated fibrils (SAF-93) had a tenfold higher affinity for PrP^{Sc} than for PrP^C and inhibited prion propagation in an in vitro conversion assay, highlighting its therapeutic and diagnostic potential [63].

Following these studies, many authors have used the SELEX methodology either against PrP^{23-231} from different species, the PrP C-terminal domains, PrP peptides, or pathogenic-related conformations (Table 1). It has been revealed that aptamer interaction with PrP is sustained even after immobilization and aptamer chemical modification [64]. Briefly, the first DNA aptamers were selected against recombinant human cellular PrP^{23-231} and interacted specially with mammalian PrPs from normal brain homogenates of sheep, calf, piglets, and deer as well as with PrP^C expressed in N2a cells [65]. None of them bound to PK-digested prion-infected ScN2a cells, suggesting that these aptamers hold specificity for PrP^C [65]. Their binding affinity appeared to be both aptamer sequence- and structure-dependent, with dissociation constants in the micromolar to nanomolar range, in accordance with our work with nucleic-acid ligands selected individually [65]. The work from King et al. selected aptamers against the globular domain (residues 90–231) of hamster PrP, folded into an α-helical-rich native conformation, identifying a thioaptamer (with phosphorothioate modification to enhance their stability against nucleases) with an affinity of 0.58 (+/−0.1) nM for hamster PrP [66]. Lower affinities for bovine (Bo) and human (Hu) PrP were found, suggesting some specificity where the interaction is dependent on the primary structure of PrP [66]. A control oligonucleotide with the same length and a scrambled consensus sequence could not differentiate among the three PrP sequences, and control oligonucleotides encompassing non-selected sequences bound to PrP at a sequence-independent DNA-binding site with much lower affinities [66]. The results confirm that the high-affinity binding of thioaptamers to PrP depends on backbone modifications, oligonucleotide sequence, and PrP sequence [66].

As well as RNA aptamers, the ssDNA thioaptamers designed by Kocisko et al., were able to bind PrP^C on live cells, to be cell-internalized and potently inhibit PrP^{Res} accumulation in infected-cultured cells [67]. Interestingly enough, prophylactic treatments with these modified oligonucleotides tripled scrapie survival periods in mice [67]. A prolonged survival time was also observed when these phosphorothioate aptamers were previously mixed with the infectious brain inoculum [67]. The potent anti-scrapie activity of these modified nucleic acids represents a new class of drugs that hold promises for the treatment of prion diseases [67,68]. Mashima et al. in 2009 provided the first report, showing the high-resolution structure of an RNA aptamer (R12) against isolated domains of the bovine PrP by NMR. The GGAGGAGGAGGA sequence from R12 aptamer forms an intramolecular parallel G-quadruplex structure [69]. G-quadruplexes are formed by G-rich sequences and are built around tetrads of hydrogen-bonded guanine bases (Hoogsteen base pair). Two or more G-tetrads can stack on top of each other to form the structure, and the quadruplex is stabilized by a cation, especially potassium [70]. Two R12 quadruplexes form a dimer through intermolecular hexad–hexad stacking [69]. Most of the RNA aptamers obtained by this group contain GGA tandem repeats and bind both rPrP^C and the beta-form of PrP with high affinities [71]. The DNA counterpart aptamer (D12) can also bind to PrP, but the affinity is weaker for both cellular PrP and its β-form [36,69]. The GGA tandem repeats form peculiar quadruplex structures that appear to be critical for the higher affinities and recognition of PrP. This tight binding is expected to stabilize PrP^C or block its interconversion and to thereby prevent the onset of prion diseases.

NMR measurements also provide the first high-resolution 3D-structure of the complex formed with N-terminal PrP peptides (P1 and P16) and the R12 aptamer [72]. The G-quadruplex structured RNA is preserved even after interaction with PrP. The RNA forms a dimer where each monomer simultaneously binds to two portions of the PrP^C N-terminal region, which can explain the strong binding, where electrostatic and stacking interactions drive the affinity of each portion [72]. Additionally, the authors demonstrate that the driving force for the binding between R12 and P16 (a PrP peptide) is a robust gain of water entropy, and the energy decrease driven by attractive interactions between R12 and P16 is compensated by the energetic dehydration effect after binding or vice-versa. The interaction of the complex occurs via stacking of flat moieties, via electrostatic interactions, including specific hydrogen bonding, and through molecular geometry complementarity [73].

One should not forget that an appropriate geometrical correspondence of hydrogen bond donors and acceptors may allow more stable complexes to be formed, but it is mainly due to stacking interactions that significant stabilization occurs [74]. Moreover, R12 was shown to reduce the accumulation of PrP^{Sc} levels in scrapie-infected neuronal cells, demonstrating its great therapeutic potential [65]. Remarkably, G-quadruplex forming NAs were shown to change the PrP^C structure after binding, so that they may actually lower the free energy barrier between the two conformers and therefore prompt the conversion process [36]. As one of the strongest PrP binders, more attention should be given to these peculiar NA structures. In fact, many NA ligands directed against PrP discussed here might form G-quadruplex structures, as they contain at least four GG repeats in their sequence that could form quadruplexes after dimerization. The PrP messenger RNA (mRNA) itself has sequences with the propensity to form G-quadruplex depending on environmental conditions such as fluctuations in potassium levels [75]. One cannot rule out the possibility that PrP interaction with its own mRNA might be involved in their physiological or pathological pathways.

Table 1. Binding characteristics of mammalian PrPs and nucleic acids.

Author, Year (Ref.)	Nucleic Acid Type	$K_D{}^1$ (nM)	Binding Assay	PrP SELEX Target	PrPs Recognized	PrP Binding Region(s)
Weiss, 1997 [33]	RNA-aptamer	ND	Gel-shift of labeled aptamer	Hamster $rPrP^{23-231}$	Mouse, hamster, cow (PrP in brain homogenates)	(23-36)
Nandi, 1997 [29-31]	Plasmid DNA	250	Fluorescent dye displacement	NS	Human $rPrP^{106-126}$ and $rPrP^{23-231}$	ND
Cordeiro, 2001 [21]	Short dsDNAs	25	Fluorescence polarization	NS	Murine $rPrP^{23-231}$	N-terminal and C-terminal domains
Gabus, 2001 [76]	HIV-1 LTR DNA (1000 bp)	ND	Gel-shift assay	NS	Human $rPrP^{23-231}$ or 23-144	N-terminal
Gabus, 2001 [37]	HIV-1 5'-leader RNA (415 nt)	ND	Gel shift assay	NS	Human $rPrP^{23-231}$, Ovine $rPrP^{25-234}$	N-terminal
Proske, 2002 [62]	RNA-aptamer	100	Filter-binding assay	Human PrP^{90-129}	Hamster, mouse or human rPrP	(90-129)
Adler, 2003 [22]	Small, highly structured RNAs	3.8	Gel shift, filter-binding assay	NS	Human rPrP, PrP from brain homogenates of mouse, rat and hamster	N-terminal domain
Rhie, 2003 [63]	RNA-aptamer	16	Homologous competition binding assay	SAF material from infected brain homogenates	Bovine rPrP in b-oligomeric or a-helical form, PK-untreated SAF, PK-treated SAF	N-terminal and SAF conformation-specific site in (110-230)
Sayer, 2004 [77]	RNA-aptamer	6.8	Equilibrium binding	Bovine $rPrP^{23-230}$	Bovine rPrP	ND
Sekiya, 2005 [78]	RNA-aptamer	ND	ND	Murine $rPrP^{23-231}$ and murine SAF infected material	Murine $rPrP^{23-231}$ and mouse SAF	(23-108) of Murine rPrP and mouse SAF
Sekiya, 2006 [79]	RNA-aptamer	5.6	Filter-binding assay	Murine $rPrP^{23-231}$ with competitive selection	Murine $rPrP^{23-231}$, Bovine rPrP, Mouse PrP in brain homogenate	(23-108) and (23-88)
Mercey, 2006 [35]	RNA-aptamer	15	Surface plasmon resonance, filter-binding assay	Ovine PrP^{23-231} with mutations associated with disease	Ovine rPrP(ARR, VRQ, AHQ, ARQ), Murine rPrP, Bovine rPrP	(25-34) and (101-110)
Lima, 2006 [34]	Short dsDNAs	90	Fluorescence polarization and SAXS	NS	Murine $rPrP^{23-231}$	N-terminal and C-terminal
Takemura, 2006 [65]	DNA-aptamer	16	End-point titration method in microplate, gel-shift, and dot-blot assays	Human $rPrP^{23-231}$	Murine $rPrP^{23-231}$, PrP from brain homogenates of sheep, calves, pigs, deer, PK-untreated PrP from ScN2a cells	(23-89)

Table 1. *Cont.*

Author, Year (Ref.)	Nucleic Acid Type	K_D [1] (nM)	Binding Assay	PrP SELEX Target	PrPs Recognized	PrP Binding Region(s)
Ogasawara, 2007 [80]	DNA-aptamer	100	Surface plasmon resonance, dot-blot, competitive selection and fluorescence measurements	Murine $rPrP^{23-231}$	Murine $rPrP^{23-231}$	ND
Murakami, 2008 [71]	RNA-aptamer	31	Surface plasmon resonance	Bovine PrP^{23-231}	Bovine $rPrP^{23-231}$	(125–231)
Bibby, 2008 [81]	DNA-aptamer	18	Saturation binding using PrP-coated Ni-NTA beads	Ni-NTA beads coated Murine PrP^{90-231}	Murine $rPrP^{90-231}$, Ovine rPrP and Human $rPrP^{90-231}$	(90–230)
Mashima, 2009 [69]	G4 RNA-aptamer	8.5	Northwestern blotting assay	Bovine PrP^{23-231}	Bovine PrP^C	(25–34) and (110–118)
	G4 RNA-aptamer	280			Amyloidogenic bovine PrP-β	ND
	G4 DNA-aptamer	85			Bovine PrP^C	(25–34) and (110–118)
	G4 DNA-aptamer	>280			Amyloidogenic Bovine PrP-β	ND
Wang, 2011 [82]	DNA-aptamer biosensor immobilized	22	Surface plasmon resonance	PrP^{Sc} from brain tissues of scrapie-infected animals with counter-selection with PrP^C	Pathological isoforms of PrP from distinct species	ND
Macedo, 2012 [24]	Small dsDNAs	ND	Fluorescence measurements	NS	murine $rPrP^{23-231}$ and $rPrP^{109-149}$	N-terminal and C-terminal domains
Cavaliere, 2013 [36]	G4 DNA-aptamer	62	Surface plasmon resonance and Isothermal Titration Calorimetry (ITC)	Ovine rPrP-23-231	Ovine $rPrP^{23-24}$	23-134
	G4 RNA-aptamer	75			Ovine $rPrP^{23-24}$	23-134
	G4 DNA-aptamer	300			Amyloidogenic Ovine PrP-β	ND
	G4 RNA-aptamer	400			Amyloidogenic Ovine PrP-β	ND

We chose the KD (dissociation constant) value of the best interaction when several aptamers were described by the same study. When many types of PrP were investigated in binding assays, the PrP species or fragment with the lowest KD value is the first shown. NS: non-SELEX (NA sequences found individually); ND: non-determined.

5. Aptamers against Other Amyloidogenic Proteins

The prion protein has been understood as the etiological agent of TSEs amyloidosis, but nowadays the term "prion" has evolved to describe a phenomenon in molecular biology that is more ubiquitous than previously thought and that is shared by other prion-like proteins, especially the ones that can aggregate into amyloid fibrils through a highly ordered mechanism. Unrelated proteins in their sequence or structure can form amyloid aggregates that possess a common cross-architecture with a β-sheet enriched core; it appears that the amyloid fibril formation is the result of an intrinsic, conservative, and generic process of proteins, and amyloids formed by the same protein sequence can still be found with different structural and phenotypic properties [7]. Evidence indicates that amyloid fibrils share specific structural characteristics and aggregate morphologies; however, some structural polymorphisms between amyloids can be found either in vivo or in vitro [83]. Although our review focuses on PrP interactions with nucleic acids, we find it useful to also provide an overall critical consideration regarding aptamer interactions with other prion-like amyloid proteins.

In this context, RNA aptamers were selected by SELEX against the wild-type bovine PrP (bPrP); the recognition was shown to occur mainly through the PrP N-terminal (25–131) region, as expected and confirmed by other researchers [40,71]. However, and interestingly, those same aptamers also bound to bPrP in the β-conformation (bPrP-β), which resembles the amyloid core, but with a tenfold lower interaction affinity [71]. Thus, it led the group to conclude that the selected aptamers bind with high affinity to both native PrP and the amyloid-like PrP conformation. The transition from the native α-form to a β-form was achieved, and it occurs briefly in the presence of phospholipid micelle solutions at pH 5.0 [71]. One question raised is this: How did an aptamer selected against a non-amyloid prion form recognize the β-conformation? The same bPrP-β formation protocol was applicable to human, cow, elk, pig, dog, and mouse PrP, even in wild-type or truncated PrP forms, and it showed that part of the flexible domain encompassing the 105–120 region must be present for the generation of bPrP-β [71]. Although that region is normally unfolded in the native PrP, it can also undergo dynamic structural shifts. This region is normally positively charged but also contains hydrophobic amino acids, both important for NA-binding. One might speculate that the epitope recognition motif found in the 105–120 PrP region can be found both in the native or in the amyloid PrP conformation; the aptamer interaction is established with high affinity for the two forms (bPrP and bPrP-β), but with a significant difference between them, probably because, for the latter, the new structure potentially adopted by the 105–120 region might change the dissociation constant value. Besides, the results showed that the high affinity for bPrP-β by the selected aptamer is abolished at a high salt concentration (1M NaCl); this behavior is not observed for the native bPrP, which maintains approximately the same interaction affinity for the same selected aptamer in equal experimental conditions [71]. This suggests that this aptamer interaction with the amyloid PrP form occurs mainly through electrostatic interactions, but for the native PrP, electrostatic interactions may bring the two partners in proximity, allowing them to establish more specific intermolecular forces (hydrophobic and base stacking), resulting in a higher affinity for the native PrP. In addition, when minimizing the aptamers' length, they lose their affinity for the bPrP-β (30-fold less) in comparison to bPrP, demonstrating that a differential aptamer specificity indeed exists for the two PrP conformations [71]. We thus suggest the use of non-aggregated PrP forms as competitors to improve selection and binding-ability to β-forms of PrP.

Besides PrP, we will also discuss briefly some studies about other amyloidogenic proteins related to diseases, such as Aβ protein and β-2-microgobulin (β2m), regarding their interaction with aptamers. RNA aptamers were selected against Alzheimer's amyloid Aβ peptide (1–40), and, apparently, the selection targeting a non-amyloid Aβ conformation led to the selection of aptamers that recognized the β-sheet-rich fibrils of Aβ [84]. Aβ (1–40) presents a hydrophobic domain that aggregates easily, especially in amyloid fibrils, but the experimental condition of Aβ (1–40) immobilization for the SELEX process in this particular study did not guarantee that the monomeric conformation of Aβ (1–40), that is, the SELEX target, was maintained in-column [84]. Aβ (1–40) could possibly aggregate in-column and/or will also be found in the trimeric or tetrameric form. It cannot be ruled out that those oligomeric forms present a β-sheet core enriched enough to be considered a pre-amyloid aggregate, which could explain the positive selection of these aptamers to the mature amyloid fibrils as well. Preformed aggregates induce fast aggregation of amyloidogenic proteins, resulting in poor experimental reproducibility [85], and are not desirable in aptamer selection for non-aggregated, non-fibrillar forms of the respective proteins. The interaction of Aβ (1–40) fibrils with these particular RNA aptamers can be easily detected through electron microscopy; the aptamer was labeled with colloidal gold that stained as black dots along the amyloid-fibrils, showing that the interaction occurs in specific regions of the fibril with an apparently specific distribution pattern [84]. Again, the addition of experimental controls like other amyloid fibrils or completely non-aggregated proteins is of interest to refine the executed assay.

Subsequently, the study performed by Rahimi et al. provided a step forward, exploiting aptamers' interaction with amyloid fibrils formed by distinct proteins [86]. The results showed that an aptamer selection targeting a non-fibrillar Aβ preparation led to a selection of aptamers that recognized fibrils of Aβ and fibrils of other amyloidogenic proteins [86]. Although we consider this work strongly relevant, some considerations can be explored for relevant discussion and speculation of other possibilities regarding the experimental evidences and conditions; some of them were also raised by the authors [87]. One of the aptamer selection targets was the cross-linked trimeric form of the Aβ protein using the filter-binding SELEX assay. The authors selected the Aβ trimeric conformation by direct purification from the SDS-PAGE and stored it for approximately twenty-four hours in a solution containing traces of SDS before the final procedure, which consisted of long-term dialysis to remove impurities until the sample was ready for the SELEX's first round [88]. These procedures suggest that traces of SDS may have been present in the sample and/or the extensive dialysis duration might have accelerated Aβ self-aggregation, enhancing the β-sheet content and/or favoring protein–protein interaction and aggregation, as verified for other proteins [89,90]. In addition, a nitrocellulose filter-binding assay is not the most suitable way to retain low-molecular-weight proteins such as Aβ (1–40), due to the poor retention of the peptide on the filter. This approach might result in the apparent unexpected lack of interaction between the aptamers selected against this specific Aβ assembly (in the trimeric expected conformation), with the same expected assembly verified through a filter-binding assay [86]. The aptamers were probably not selected against a homogeneous non-aggregated Aβ-form; we therefore believe the SELEX target was a common cross-beta structure present both in the trimeric form, oligomers and in the amyloid fibrils. Moreover, the evidence that two aptamers have different binding affinities between fibrils formed with the Aβ (1–40) and Aβ (1–42) reveals that there must be some specificity governing the aptamer interaction [84]. The reactivity of aptamers against Aβ (1–42) fibrils was somewhat lower than their reactivity with Aβ (1–40) fibrils, again suggesting moderate specificity for Aβ (1–40) [84]. Given that observation, there are minor differences between Aβ (1–40) and Aβ (1–42) that come from a dissimilar enzymatic cleavage site of the amyloid precursor protein (APP), and there are still significant differences upon binding to the selected RNA aptamers, strongly suggesting some specificity, which we are still looking for. Remarkably, these same aptamers recognize fibrils of other amyloidogenic proteins, including insulin, islet amyloid polypeptide, calcitonin, lysozyme, and PrP[106–126], but with significant differences between some fibrils that might correlate with residual specificity [86].

Similar results were obtained in other work, where aptamers were selected against fibrils of β2-microglobulin (β2-m) or against monomeric β2-m at low pH [91]. The aptamers bind with high affinity to β2-m fibrils with different morphologies formed under different conditions in vitro, as well as to amyloid fibrils isolated from tissues of β2-m-related amyloidosis patients, demonstrating that they can detect conserved epitopes between different fibrillar assemblies of β2-m, including those formed in vivo. At this time, the group's data demonstrate that the selections generated aptamers able to bind with high affinity to all three forms of β2-m, including two distinct fibrils and the low pH monomeric form [91]. The validation of the interaction was performed through surface plasmon resonance (SPR), a more suitable and refined method for this approach than dot-blot only [91]. These results suggest that the β2-m fibrils share at least one epitope in common with the monomeric β2-m, and the higher affinity for the fibrillar form might be due to more epitopes being available and the greater ease of interacting with these aptamers in the macromolecular structure of the fibril. The aptamers also reacted with some (but not all) other amyloid fibrils, either generated in vitro or isolated from ex vivo sources; but for these other proteins, none of the aptamers were able to bind to native monomers, confirming that the epitopes being recognized are fibril specific [91]. Thus, the native folded species seem not to share epitopes in common with the fibrillar and pre-fibrillar states. SPR measurements also showed that there are large signal differences between the naive SELEX RNA pool and the final selected individual aptamers against the target, confirming that the SPR responses seen are due to specific binding and not to inherent affinity of the oligonucleotides for fibrillar amyloid structures [91]. The same behavior was reported for certain antibodies, which could recognize conformational epitopes in Aβ assemblies and interact with similar assemblies of other amyloid-forming proteins [92–95]; some antibodies were raised against oligomers but reacted with both oligomeric and fibrillar assemblies [96–98].

One must conclude, based on these observations, that, although amyloid fibrils have many common structural properties, they also have features that are unique to individual fibril types. Some amyloids may hinder more structural differences than others, and because of these structural polymorphisms, aptamers selected against a specific amyloidogenic protein can interact with other unrelated amyloid-forming proteins with similar or significantly different affinities, depending on the amyloid protein aggregated structure. There were also cases where these aptamers have not at all recognized amyloid fibrils from other amyloidogenic proteins, such as apomyoglobin, Aβ (1–40), or transthyretin, but significant binding was observed to fibrils formed from lysozyme [91]. The aptamer binding to lysozyme fibrils cannot be the cause of nonspecific interactions, as evidenced by the inability of the aptamer to bind to native monomeric lysozymes together with the observation that the naive RNA SELEX pool binds relatively weakly to the lysozyme fibrils under these conditions [91]. Although these aptamers recognize an epitope present in different amyloid fibrils, the epitope for each aptamer must be either distinct or differentially accessible between different amyloids. Understanding the structural molecular basis of why some aptamers and antibodies raised against monomeric proteins can recognize either the oligomeric forms or the amyloid polymeric architecture requires an ongoing investigation, but will definitely improve the discussion about their potential in the pharmaceutical and biotechnology fields.

6. Conclusions and Perspectives

Over the last twenty years, NAs have been proposed as potential cofactors that can bind to different disease-related proteins and can trigger their misfolding and aggregation processes [9,99]. Protein interactions with NAs are governed by several molecular forces, including hydrogen bonding mediated by aqueous solvent, electrostatic, hydrophobic, and stacking interactions [52]. Because of the many structural motifs existing both in proteins and NAs, as well as the variations in the nucleotide sequences, it is very difficult to characterize a single model for protein–NA interactions. Hydrophobic interactions seem to be more efficient than charge effects for driving protein aggregation [100]. However, both factors (hydrophobic and charge effects) can be critical and determinant for protein aggregation and need to be considered for understanding the process in vivo and the role of amino acid composition, sequence, and substitutions in protein misfolding diseases and protein design [100,101].

The knowledge acquired from the PrP studies discussed here permitted us to map the NA-binding sites on PrP. This interaction involves at least three different binding sites: two of them localized at the N-terminal flexible region and the other in the structured globular C-terminal domain. Apparently, the two lysine clusters in the N-terminal domain, encompassing residues 23–52 and 101–110, are involved in all non-specific NA interactions, since this is a positively charged region with enough flexibility to bind DNAs, RNAs, and even heparin molecules, mainly through electrostatic interactions with the sugar-phosphate backbone [35,63]. Through NMR studies, residues encompassing the lysine cluster were shown to mediate the interaction with DNA [34] or RNA [72]. Although contributions from hydrophobic interactions appear to be more important than those involving charge interactions, the influence of charge factors on protein aggregation must not be underestimated [100]. Structural data show that the PrP globular domain in the normal conformation can interact with DNA but not with RNA [35,40]. However, conformational changes in the C-terminal domain, especially in its transition to beta forms, can expose structural or sequence motifs that are able to bind even RNA aptamers through more specific interactions than those established with the N-terminal region [63].

Although many efforts have been made to find the sequence specificity governing PrP-NA interactions, no consensus has yet been found. Partial consensus sequences are clearly present, confirming that selection had occurred, but there has been no obviously dominant epitope-binding consensus. Comparison of the aptamers sequences reported elsewhere for either anti-PrP or anti-Aβ (1–40) or anti-β2-m aptamers did not show significant sequence motif matches, suggesting that the aptamers raised are specific to their selection targets. Some structural features can be highlighted, such as the G-forming quadruplex in many DNA and RNA ligands that provides tight bonding; however, other structural motifs, such as hairpins and the double helix, have also been described and proven to have a high affinity for PrPs [24,36]. Through NMR studies, it appears that the geometric characteristics (overall shapes, sizes, and detailed polyatomic structures) of the molecules are the most important factors governing PrP recognition [73]. The strong diversity between PrP nucleic acid ligands should not rule out the existence of some specificity, once many modifications in the NA molecule can alter its PrP-binding affinity and can trigger different changes on PrP properties [24,33].

NA molecules might play a dual role in prion biology, either by triggering PrP conversion and aggregation or by preventing them. Understanding this intriguing partnership could be critical to explaining how prion diseases arise, and to developing effective diagnostic and therapeutic methodologies. In terms of the pathological aspect, NA-binding to PrP could lead to reciprocal conformational changes, altering both the PrP and NA structure and promoting distinct modes of polymerization depending on the NA source. Possibly the charge neutralization of the positively charged PrP N-terminal domain after NA binding favors the association of PrP molecules, which might explain the immediate NA-induced PrP aggregation [102].

What remains to be elucidated is whether these interactions are relevant in vivo, regarding either the pathology or biology of prions. Although RNA molecules were found to be associated with plaques in the brains of AD-diseased patients [103–105], no direct evidence of in vivo association of specific nucleic acid sequences with PrP scrapie in affected humans has been found yet. Nevertheless, the work of Manuelidis' group showed that circular DNAs could be co-purified along with infectivity in 22L-infected cell lines, in hamster 263K scrapie-infected brain samples, and in FU-CJD infected mouse brain [106]. Additionally, the same group showed that prion infectivity was retained when PrP was digested; in contrast, when different prion strains were treated with nucleases, infectivity (prion titer) was substantially reduced [107]. In addition, PrP-NA interaction was shown to be fundamental in generating synthetic scrapie prions; free small RNAs, extracted from highly infectious scrapie-associated fibrils (SAFs) and incubated with PrP^C, were shown to promote PrP^C–PrP^{Sc} conversion with the acquisition of infectivity, inducing prion disease in wild-type healthy Syrian hamsters [108]. These results indicate that nucleic acids are essential for prion infectivity and might be involved in prion pathogenesis.

Altogether, these studies show that NAs are potential PrP cofactors able to catalyze the formation of PrP^{Sc} in vivo. The "NA cofactor hypothesis" initially proposed by our group and reinforced by other contemporaneous studies does not necessarily refute the commonly accepted "protein-only hypothesis", where PrP is the solely proteinaceous infectious agent. In fact, we otherwise add to this vision the suitable participation of molecules that could facilitate protein aggregation and the formation of infectious conformations that could, even alone, template the conversion of normal PrPs into abnormal conformations, thus leading to prion disease progression.

Protein folding and protein aggregation are dynamic and competitive events constantly fighting inside the cell and driven by the same molecular forces, which explains why these processes are well controlled and balanced by the protein quality control of the cell machinery [109]. The process is so complex that an increasing number of proteins with the same amino acid sequence were shown to adopt, under native conditions, various folded conformations that coexist in dynamic equilibrium [109]. Especially for the amyloid aggregation pathway, there are many precursor species along the way to fibril maturation: amyloid seeds, oligomeric forms, prefibrillar, and fibrillar states [7]. Therefore, amyloid aggregation is also a dynamic and potentially reversible process where different species may be present even after fibril formation. Regarding aptamer selection against amyloid prions, the characterization of aptamers is particularly important, because the natural affinity of oligonucleotides for fibrillar amyloid structures potentially hinders the development of aptamers that are specific for non-fibrillar amyloid proteins under physiological conditions [86]. Particular studies discussed in this review describe the selection of nucleic acids that inherently bind fibrillar or β-sheet-rich structures of amyloid proteins. This tendency must be more deeply explored. In addition, it is important to determine the size distribution profile of the amyloid aggregates as well as the morphology of each species that can coexist in the aggregation protocols used in those studies, making it more difficult to guarantee which exact species are exposing the exact epitope that led to the aptamer selection. In general, aptamers selected against amyloidogenic proteins recognize a structural motif, probably the backbone of the proteins in a cross-β structure that is common to the fibrillar state of these proteins. Nucleic acid reactivity clearly depends on the protein assembly state and to some extent on the protein sequence. Based on the idea that the amyloid fold is ancient and may have co-evolved with RNAs [110,111], it is plausible to propose that such proteins present a general nucleic acid binding property resulting from this evolution process. NA-binding can thus result in ribonucleoprotein complexes that possess important cellular functions, for instance, being related to functional amyloids [112] or to amyloidogenic diseases [99]. Accordingly, it is expected that amyloids and amyloid-forming proteins will present promiscuous RNA- (or even DNA)-binding characteristics.

Developing effective therapies against prion diseases and other amyloidosis remains a hard challenge. Together with the therapeutic potential of aptamers against PrP, ligands able to bind and stabilize the native state of an amyloidogenic protein provide one such potential strategy for controlling protein accumulation and the disease progression of many neurodegenerative disorders including Alzheimer's and Parkinson's diseases [51]. Efforts to generate aptamers that would specifically recognize oligomeric pre-amyloid species have also been a challenge, likely due to the dynamic nature of the oligomers preventing long-lasting NA–oligomer interactions. This inherent, apparently sequence-independent, affinity of oligonucleotides may have led to the generation of fibril-cross-reactive aptamers in studies aiming to generate aptamers for non-fibrillar amyloidogenic proteins. Recently, Takahashi et al. have selected aptamers against an oligomeric model of Aβ (1–40) and demonstrated an interaction with monomeric Aβ with micromolar affinity [112]. However, the cross-reactivity of these aptamers with fibrillar Aβ (1–40) or with other fibrillar amyloidogenic proteins was not determined [113]. In addition, data in the literature indicate that aptamers can also be used to detect early β-sheet formation more sensitively than the common thioflavin-T (ThT) amyloid dye [86]. Thus, these aptamers could be highly efficient detection tools of β-sheet formation in histopathological and in biophysical studies in vitro.

Overall, if aptamers are to be obtained for diagnostic and therapeutic approaches in amyloid diseases, the use of such a selective powerful tool is yet to be achieved in this field. Additional experiments to generate devoted and specific aptamers for prefibrillar assemblies (including monomers and oligomers) will have to deal with the apparent inherent affinity of oligonucleotides for fibrillar structures. Nevertheless, small differences in specificity and affinity of aptamers for amyloid and monomeric proteins may indeed allow their application in diagnosis or therapy.

In the context of the biology and pathology of prion proteins as well as in other protein-misfolding diseases, it would be valuable for those who would like to continue researching aptamers and their applications to find the ideal aptamer against therapeutic targets of the future, specific enough to warrant their use as recognition tools or therapeutics. Maybe the literature is being too optimistic in this regard, but we cannot forget those are still invariably fatal diseases where researchers are avidly waiting for a new drug discovery to treat or cure illnesses for both humans and animals. Several nucleic acids, especially the aptamers for PrP, have been shown to bind to PrP^C or PrP^{Sc} and to interfere with PrP^{Sc} biogenesis, providing a new class of promising molecules that could be used for the treatment of prion diseases. Even if they bind monomeric PrP^C to some extent, the benefit of preventing conversion into PrP^{Sc} would surpass the drawback of lack of specificity. Some of the investigated nucleic acids have shown therapeutic efficacy in infected mice models by tripling their survival time [67], but to our knowledge none of them have proceeded to clinical studies so far. Alternative attempts based on antibody therapy also have potential [114]; however, the stimulation of the autoimmune system presents challenges to further developments in this area [114].

Unfortunately, there is no therapy to treat or prevent prion diseases. Most of the lead compounds found in the drug screening for anti-scrapie activity lack efficacy (possibly due to prion strain specificity), and have poor pharmacokinetic profiles, such as high toxicity and/or an inability to efficiently cross the blood–brain barrier (BBB) [115]. In fact, the aptamer pharmacokinetic profile is especially relevant for neurodegenerative disorders, pushing the development of strategies towards crossing the BBB, as it is unlikely that they can easily enter the brain. Nevertheless, aptamers may surpass this barrier via pinocytosis, transcytosis, channel, and/or receptors to their uptake [116]. Additionally, quadruplex-structured aptamers may cross the BBB through binding to nucleolin via micropinocytosis [117]. Using a nicely executed in vivo selection protocol, Cheng and collaborators selected aptamers that permeated the brain after peripheral injection of the library in wild-type mice [118]. Fortunately, aptamers are molecules that can be easily modified to overcome their clinical limitations: for example, nanoparticle-encapsulated aptamers were reported to cross the BBB, and delivery of liposome-based aptamers was well tolerated in clinical trials [119]. We still do not have the safety profile of these molecules, although they are expected to be non-toxic and non-immunogenic [52].

In conclusion, NA aptamers can distinguish normal and abnormal conformations of PrP, representing the first reagents able to identify PrP pathological conformations from multiple host species. They can even differentiate prion strains and can be used to detect infectious prions in blood samples, which cannot be accomplished using conventional diagnostic tools that rely on antibody-based detection methods. The hard challenge of prion disease diagnosis before the symptomatic stage is how to discriminate and detect the minute quantity of disease-associated prion protein isoform (PrP^{Res}) sensitively and selectivity in complex biological samples, from plasma to brain homogenate. The development of a dual-aptamer strategy for diagnostic tools began with an investigation of the advantages of aptamers, the great separation ability of magnetic microparticles (MMPs), and the high fluorescence emission features of quantum dots (QDs) [120]. Two aptamers were coupled to the surfaces of MMPs and QDs, respectively, which then could be co-associated through the specific interaction of the two aptamers with their two corresponding different PrP epitopes, forming an aptasensor platform [120]. Moreover, aptamers can enrich a target, for example, the PrP molecule, from biological fluids; in this context, RNA aptamers have been successfully utilized for the concentration of PrP^C and PrP^{res} taken from serum, urine, and brain homogenate [121]. There is also an interesting proposal for PrP^{Sc}-enrichment, using PrP^C-specific aptamers to capture normal prions from biological samples, which could be used as a diagnostic tool in double ligand assay systems and other aptasensors [65]. There is an urgent necessity to develop more sensitive and more efficient assays to detect the pathological forms of PrP in pre-symptomatic screening of tissue, blood, or other body fluids. Based on these promising studies, NA aptamers appear to be good candidates to reach this goal. Although many aptamers have been identified against PrP, with great potential for use in diagnostic tools, the community is still relying on antibody-based detection methods. Among the limiting factors that make aptamers especially promising is the sensitivity of detection. Thus, many efforts are now being made to build aptasensing platforms based on electrochemical or dual-signal systems to develop highly sensitive prion assays [122,123]. This new class of molecules thus has great potential.

Acknowledgments: We thank the Fundação de Amparo a Pesquisa do Estado do Rio de Janeiro (FAPERJ) and INCT-INBEB from CNPq (process #465395/2014-7) for financial support. We are thankful to Lucas M. Ascari for the art design of Figure 2 and to Professors Julia R. Clarke, Luís M. T. R. Lima, Monica S. Freitas from UFRJ and Sotiris Missailidis from FIOCRUZ for critical revision of the manuscript.

Author Contributions: Bruno Macedo wrote the manuscript, and Yraima Cordeiro revised it critically for important intellectual content.

References

1. Prusiner, S.B. Nobel Prize Lecture: Prions. *Proc. Natl. Acad. Sci. USA* **1998**, *95*, 13363–13383. [CrossRef] [PubMed]
2. Aguzzi, A.; Lakkaraju, A.K.K. Cell Biology of Prions and Prionoids: A Status Report. *Trends Cell Biol.* **2016**, *26*, 40–51. [CrossRef] [PubMed]
3. Collinge, J. Prion diseases of humans and animals: Their causes and molecular basis. *Annu. Rev. Neurosci.* **2001**, *24*, 519–550. [CrossRef] [PubMed]
4. Riek, R.; Hornemann, S.; Wider, G.; Billeter, M.; Glockshuber, R.; Wüthrich, K. NMR structure of the mouse prion protein domain $PrP^{121-231}$. *Nature* **1996**, *382*, 180–182. [CrossRef] [PubMed]
5. Knaus, K.J.; Morillas, M.; Swietnicki, W.; Malone, M.; Surewicz, W.K.; Yee, V.C. Crystal structure of the human prion protein reveals a mechanism for oligomerization. *Nat. Struct. Biol.* **2001**, *8*, 770–774. [CrossRef] [PubMed]
6. Caughey, B.; Baron, G.S.; Chesebro, B.; Jeffrey, M. Getting a grip on prions: Oligomers, amyloids, and pathological membrane interactions. *Annu. Rev. Biochem.* **2009**, *78*, 177–204. [CrossRef] [PubMed]
7. Knowles, T.P.J.; Vendruscolo, M.; Dobson, C.M. The amyloid state and its association with protein misfolding diseases. *Nat. Rev. Mol. Cell Biol.* **2014**, *15*, 384–396. [CrossRef] [PubMed]
8. Guo, J.L.; Lee, V.M.Y. Cell-to-cell transmission of pathogenic proteins in neurodegenerative diseases. *Nat. Med.* **2014**, *20*, 130–138. [CrossRef] [PubMed]
9. Silva, J.L.; Cordeiro, Y. The "Jekyll and Hyde" Actions of Nucleic Acids on the Prion-like Aggregation of Proteins. *J. Biol. Chem.* **2016**, *291*, 15482–15490. [CrossRef] [PubMed]
10. Chiti, F.; Dobson, C.M. Protein Misfolding, Functional Amyloid, and Human Disease. *Annu. Rev. Biochem.* **2006**, *75*, 333–366. [CrossRef] [PubMed]
11. Uversky, V.N.; Dunker, A.K. Understanding protein non-folding. *Biochim. Biophys. Acta Proteins Proteom.* **2010**, *1804*, 1231–1264. [CrossRef] [PubMed]
12. Yen, C.-F.; Harischandra, D.S.; Kanthasamy, A.; Sivasankar, S. Copper-induced structural conversion templates prion protein oligomerization and neurotoxicity. *Sci. Adv.* **2016**, *2*, e1600014. [CrossRef] [PubMed]
13. Silva, J.L.; Gomes, M.P.; Vieira, T.C.; Cordeiro, Y. PrP interactions with nucleic acids and glycosaminoglycans in function and disease. *Front. Biosci.* **2010**, *15*, 132–150. [CrossRef]
14. Vieira, T.C.R.G.; Reynaldo, D.P.; Gomes, M.P.B.; Almeida, M.S.; Cordeiro, Y.; Silva, J.L. Heparin binding by murine recombinant prion protein leads to transient aggregation and formation of rna-resistant species. *J. Am. Chem. Soc.* **2011**, *133*, 334–344. [CrossRef] [PubMed]
15. Liu, C.; Zhang, Y. Nucleic acid-mediated protein aggregation and assembly. *Adv. Protein Chem. Struct. Biol.* **2011**, *84*, 1–40. [PubMed]
16. Supattapone, S. Elucidating the role of cofactors in mammalian prion propagation. *Prion* **2014**, *8*, 100–105. [CrossRef] [PubMed]
17. Critchley, P.; Kazlauskaite, J.; Eason, R.; Pinheiro, T.J.T. Binding of prion proteins to lipid membranes. *Biochem. Biophys. Res. Commun.* **2004**, *313*, 559–567. [CrossRef] [PubMed]
18. Deleault, N.R.; Harris, B.T.; Rees, J.R.; Supattapone, S. Formation of native prions from minimal components in vitro. *Proc. Natl. Acad. Sci. USA* **2007**, *104*, 9741–9746. [CrossRef] [PubMed]
19. Miller, M.B.; Wang, D.W.; Wang, F.; Noble, G.P.; Ma, J.; Woods, V.L.; Li, S.; Supattapone, S. Cofactor molecules induce structural transformation during infectious prion formation. *Structure* **2013**, *21*, 2061–2068. [CrossRef] [PubMed]
20. Soto, C. Prion hypothesis: The end of the controversy? *Trends Biochem. Sci.* **2011**, *36*, 151–158. [CrossRef] [PubMed]

21. Cordeiro, Y.; Machado, F.; Juliano, L.; Juliano, M.A.; Brentani, R.R.; Foguel, D.; Silva, J.L. DNA Converts Cellular Prion Protein into the β-Sheet Conformation and Inhibits Prion Peptide Aggregation. *J. Biol. Chem.* **2001**, *276*, 49400–49409. [CrossRef] [PubMed]

22. Adler, V.; Zeiler, B.; Kryukov, V.; Kascsak, R.; Rubenstein, R.; Grossman, A. Small, highly structured RNAs participate in the conversion of human recombinant PrPSen to PrPRes in vitro. *J. Mol. Biol.* **2003**, *332*, 47–57. [CrossRef]

23. Deleault, N.R.; Lucassen, R.W.; Supattapone, S. RNA molecules stimulate prion protein conversion. *Nature* **2003**, *425*, 717–720. [CrossRef] [PubMed]

24. Macedo, B.; Millen, T.A.; Braga, C.A.C.A.; Gomes, M.P.B.; Ferreira, P.S.; Kraineva, J.; Winter, R.; Silva, J.L.; Cordeiro, Y. Nonspecific prion protein-nucleic acid interactions lead to different aggregates and cytotoxic species. *Biochemistry* **2012**, *51*, 5402–5413. [CrossRef] [PubMed]

25. Chaves, J.A.P.; Sanchez-López, C.; Gomes, M.P.B.; Sisnande, T.; Macedo, B.; de Oliveira, V.E.; Braga, C.A.C.; Rangel, L.P.; Silva, J.L.; Quintanar, L.; et al. Biophysical and morphological studies on the dual interaction of non-octarepeat prion protein peptides with copper and nucleic acids. *J. Biol. Inorg. Chem.* **2014**, *19*, 839–851. [CrossRef] [PubMed]

26. Zou, W.-Q.; Zheng, J.; Gray, D.M.; Gambetti, P.; Chen, S.G. Antibody to DNA detects scrapie but not normal prion protein. *Proc. Natl. Acad. Sci. USA* **2004**, *101*, 1380–1385. [CrossRef] [PubMed]

27. Tuerk, C.; Gold, L. Systematic evolution of ligands by exponential enrichment: RNA ligands to bacteriophage T4 DNA polymerase. *Science* **1990**, *249*, 505–510. [CrossRef] [PubMed]

28. Ellington, A.D.; Szostak, J.W. In vitro selection of RNA molecules that bind specific ligands. *Nature* **1990**, *346*, 818–822. [CrossRef] [PubMed]

29. Nandi, P.K. Interaction of prion peptide HuPrP106–126 with nucleic acid: Brief report. *Arch. Virol.* **1997**, *142*, 2537–2545. [CrossRef] [PubMed]

30. Nandi, P.K. Polymerization of human prion peptide HuPrP 106–126 to amyloid in nucleic acid solution. *Arch. Virol.* **1998**, *143*, 1251–1263. [CrossRef] [PubMed]

31. Nandi, P.K.; Leclerc, E. Polymerization of murine recombinant prion protein in nucleic acid solution. *Arch. Virol.* **1999**, *144*, 1751–1763. [CrossRef] [PubMed]

32. Cordeiro, Y.; Silva, J.L. The hypothesis of the catalytic action of nucleic acid on the conversion of prion protein. *Protein Pept. Lett.* **2005**, *12*, 251–255. [CrossRef] [PubMed]

33. Weiss, S.; Proske, D.; Neumann, M.; Groschup, M.H.; Kretzschmar, H.A.; Famulok, M.; Winnacker, E.L. RNA aptamers specifically interact with the prion protein PrP. *J. Virol.* **1997**, *71*, 8790–8797. [PubMed]

34. Lima, L.M.T.R.; Cordeiro, Y.; Tinoco, L.W.; Marques, A.F.; Oliveira, C.L.P.; Sampath, S.; Kodali, R.; Choi, G.; Foguel, D.; Torriani, I.; et al. Structural insights into the interaction between prion protein and nucleic acid. *Biochemistry* **2006**, *45*, 9180–9187. [CrossRef] [PubMed]

35. Mercey, R.; Lantier, I.; Maurel, M.C.; Grosclaude, J.; Lantier, F.; Marc, D. Fast, reversible interaction of prion protein with RNA aptamers containing specific sequence patterns. *Arch. Virol.* **2006**, *151*, 2197–2214. [CrossRef] [PubMed]

36. Cavaliere, P.; Pagano, B.; Granata, V.; Prigent, S.; Rezaei, H.; Giancola, C.; Zagari, A. Cross-talk between prion protein and quadruplex-forming nucleic acids: A dynamic complex formation. *Nucleic Acids Res.* **2013**, *41*, 327–339. [CrossRef] [PubMed]

37. Gabus, C.; Derrington, E.; Leblanc, P.; Chnaiderman, J.; Dormont, D.; Swietnicki, W.; Morillas, M.; Surewicz, W.K.; Marc, D.; Nandi, P.; et al. The Prion Protein Has RNA Binding and Chaperoning Properties Characteristic of Nucleocapsid Protein NCp7 of HIV-1. *J. Biol. Chem.* **2001**, *276*, 19301–19309. [CrossRef] [PubMed]

38. Guichard, C.; Ivanyi-Nagy, R.; Sharma, K.K.; Gabus, C.; Marc, D.; Mély, Y.; Darlix, J.L. Analysis of nucleic acid chaperoning by the prion protein and its inhibition by oligonucleotides. *Nucleic Acids Res.* **2011**, *39*, 8544–8558. [CrossRef] [PubMed]

39. Liu, M.; Yu, S.; Yang, J.; Yin, X.; Zhao, D. RNA and CuCl2 induced conformational changes of the recombinant ovine prion protein. *Mol. Cell. Biochem.* **2007**, *294*, 197–203. [CrossRef] [PubMed]

40. Gomes, M.P.B.; Millen, T.A.; Ferreira, P.S.; Cunha E Silva, N.L.; Vieira, T.C.R.G.; Almeida, M.S.; Silva, J.L.; Cordeiro, Y. Prion protein complexed to N2a cellular RNAs through its N-terminal domain forms aggregates and is toxic to murine neuroblastoma cells. *J. Biol. Chem.* **2008**, *283*, 19616–19625. [CrossRef] [PubMed]

41. Saborio, G.P.; Permanne, B.; Soto, C. Sensitive detection of pathological prion protein by cyclic amplification of protein misfolding. *Nature* **2001**, *411*, 810–813. [CrossRef] [PubMed]

42. Lucassen, R.; Nishina, K.; Supattapone, S. In vitro amplification of protease-resistant prion protein requires free sulfhydryl groups. *Biochemistry* **2003**, *42*, 4127–4135. [CrossRef] [PubMed]

43. Cordeiro, Y.; Kraineva, J.; Gomes, M.P.B.; Lopes, M.H.; Martins, V.R.; Lima, L.M.T.R.; Foguel, D.; Winter, R.; Silva, J.L. The amino-terminal PrP domain is crucial to modulate prion misfolding and aggregation. *Biophys. J.* **2005**, *89*, 2667–2676. [CrossRef] [PubMed]

44. Strom, A.; Wang, G.S.; Picketts, D.J.; Reimer, R.; Stuke, A.W.; Scott, F.W. Cellular prion protein localizes to the nucleus of endocrine and neuronal cells and interacts with structural chromatin components. *Eur. J. Cell. Biol.* **2011**, *90*, 414–419. [CrossRef] [PubMed]

45. Mangé, A.; Crozet, C.; Lehmann, S.; Béranger, F. Scrapie-like prion protein is translocated to the nuclei of infected cells independently of proteasome inhibition and interacts with chromatin. *J. Cell. Sci.* **2004**, *117*, 2411–2416. [CrossRef] [PubMed]

46. Marijanovic, Z.; Caputo, A.; Campana, V.; Zurzolo, C. Identification of an intracellular site of prion conversion. *PLoS Pathog.* **2009**, *5*, e1000426. [CrossRef] [PubMed]

47. Yim, Y.-I.; Park, B.-C.; Yadavalli, R.; Zhao, X.; Eisenberg, E.; Greene, L.E. The multivesicular body is the major internal site of prion conversion. *J. Cell. Sci.* **2015**, 1434–1443. [CrossRef] [PubMed]

48. Beaudoin, S.; Vanderperre, B.; Grenier, C.; Tremblay, I.; Leduc, F.; Roucou, X. A large ribonucleoprotein particle induced by cytoplasmic PrP shares striking similarities with the chromatoid body, an RNA granule predicted to function in posttranscriptional gene regulation. *Biochim. Biophys. Acta Mol. Cell. Res.* **2009**, *1793*, 335–345. [CrossRef] [PubMed]

49. Baron, G.S.; Magalhães, A.C.; Prado, M.A.M.; Caughey, B. Mouse-adapted scrapie infection of SN56 cells: Greater efficiency with microsome-associated versus purified PrP-res. *J. Virol.* **2006**, *80*, 2106–2117. [CrossRef] [PubMed]

50. Rouvinski, A.; Karniely, S.; Kounin, M.; Moussa, S.; Goldberg, M.D.; Warburg, G.; Lyakhovetsky, R.; Papy-Garcia, D.; Kutzsche, J.; Korth, C.; et al. Live imaging of prions reveals nascent PrP^{Sc} in cellsurface, raft-associated amyloid strings and webs. *J. Cell. Biol.* **2014**, *204*, 423–441. [CrossRef] [PubMed]

51. Qu, J.; Yu, S.; Zheng, Y.; Zheng, Y.; Yang, H.; Zhang, J. Aptamer and its applications in neurodegenerative diseases. *Cell. Mol. Life Sci.* **2017**, *74*, 683–695. [CrossRef] [PubMed]

52. Zhou, J.; Rossi, J. Aptamers as targeted therapeutics: Current potential and challenges. *Nat. Rev. Drug Discov.* **2017**, *16*, 181–202. [CrossRef] [PubMed]

53. Shangguan, D.; Li, Y.; Tang, Z.; Cao, Z.C.; Chen, H.W.; Mallikaratchy, P.; Sefah, K.; Yang, C.J.; Tan, W. Aptamers evolved from live cells as effective molecular probes for cancer study. *Proc. Natl. Acad. Sci. USA* **2006**, *103*, 11838–11843. [CrossRef] [PubMed]

54. Gogtay, N.J.; Sridharan, K. Therapeutic Nucleic Acids: Current clinical status. *Br. J. Clin. Pharmacol.* **2016**, *82*, 659–672.

55. Famulok, M. Molecular Recognition of Amino Acids by RNA-Aptamers: An L-Citrulline Binding RNA Motif and Its Evolution into an L-Arginine Binder. *J. Am. Chem. Soc.* **1994**, *116*, 1698–1706. [CrossRef]

56. Lupold, S.E.; Hicke, B.J.; Lin, Y.; Coffey, D.S. Identification and characterization of nuclease-stabilized RNA molecules that bind human prostate cancer cells via the prostate-specific membrane antigen. *Cancer Res.* **2002**, *62*, 4029–4033. [PubMed]

57. Tabarzad, M.; Jafari, M. Trends in the Design and Development of Specific Aptamers Against Peptides and Proteins. *Protein J.* **2016**, *35*, 81–99. [CrossRef] [PubMed]

58. Williams, K.P.; Liu, X.H.; Schumacher, T.N.; Lin, H.Y.; Ausiello, D.A.; Kim, P.S.; Bartel, D.P. Bioactive and nuclease-resistant L-DNA ligand of vasopressin. *Proc. Natl. Acad. Sci. USA* **1997**, *94*, 11285–11290. [CrossRef] [PubMed]

59. Tang, Z.; Shangguan, D.; Wang, K.; Shi, H.; Sefah, K.; Mallikratchy, P.; Chen, H.W.; Li, Y.; Tan, W. Selection of aptamers for molecular recognition and characterization of cancer cells. *Anal. Chem.* **2007**, *79*, 4900–4907. [CrossRef] [PubMed]

60. Yu, Y.; Liang, C.; Lv, Q.; Li, D.; Xu, X.; Lui, B.; Lu, A.; Zhang, G. Molecular Selection, Modification and Development of Therapeutic Oligonucleotide Aptamers. *Int. J. Mol. Sci.* **2016**, *17*, 358. [CrossRef] [PubMed]

61. Marc, D. Aptamers to explore prion protein interactions with nucleic acids. *Front. Biosci.* **2010**, *15*, 550–563. [CrossRef]

62. Proske, D.; Gilch, S.; Wopfner, F.; Schatzl, H.M.; Winnacker, E.L.; Famulok, M. Prion-protein-specific aptamer reduces PrPSc formation. *ChemBioChem* **2002**, *3*, 717–725. [CrossRef]

63. Rhie, A.; Kirby, L.; Sayer, N.; Wellesley, R.; Disterer, P.; Sylvester, I.; Gill, A.; Hope, J.; James, W.; Tahiri-Alaoui, A. Characterization of 2'-fluoro-RNA aptamers that bind preferentially to disease-associated conformations of prion protein and inhibit conversion. *J. Biol. Chem.* **2003**, *278*, 39697–39705. [CrossRef] [PubMed]

64. Kouassi, G.K.; Wang, P.; Sreevatan, S.; Irudayaraj, J. Aptamer-mediated magnetic and gold-coated magnetic nanoparticles as detection assay for prion protein assessment. *Biotechnol. Prog.* **2007**, *23*, 1239–1244. [CrossRef] [PubMed]

65. Takemura, K.; Wang, P.; Vorberg, I.; Surewicz, W.; Priola, S.A.; Kanthasamy, A.; Pottathil, R.; Chen, S.G.; Sreevatsan, S. DNA aptamers that bind to PrP^C and not PrP^{Sc} show sequence and structure specificity. *Exp. Biol. Med.* **2006**, *231*, 204–214.

66. King, D.J.; Safar, J.G.; Legname, G.; Prusiner, S.B. Thioaptamer Interactions with Prion Proteins: Sequence-specific and Non-specific Binding Sites. *J. Mol. Biol.* **2007**, *369*, 1001–1014. [CrossRef] [PubMed]

67. Kocisko, D.A.; Vaillant, A.; Lee, K.S.; Arnold, K.M.; Bertholet, N.; Race, R.E.; Olsen, E.A.; Juteau, J.M.; Caughey, B. Potent antiscrapie activities of degenerate phosphorothioate oligonucleotides. *Antimicrob. Agents Chemother.* **2006**, *50*, 1034–1044. [CrossRef] [PubMed]

68. Karpuj, M.V.; Giles, K.; Gelibter-Niv, S.; Scott, M.R.; Lingappa, V.R.; Szoka, F.C.; Peretz, D.; Denetclaw, W.; Prusiner, S.B. Phosphorothioate oligonucleotides reduce PrP levels and prion infectivity in cultured cells. *Mol. Med.* **2007**, *13*, 190–198. [CrossRef] [PubMed]

69. Mashima, T.; Matsugami, A.; Nishikawa, F.; Nishikawa, S.; Katahira, M. Unique quadruplex structure and interaction of an RNA aptamer against bovine prion protein. *Nucleic Acids Res.* **2009**, *37*, 6249–6258. [CrossRef] [PubMed]

70. Burge, S.; Parkinson, G.N.; Hazel, P.; Todd, A.K.; Neidle, S. Quadruplex DNA: Sequence, topology and structure. *Nucleic Acids Res.* **2006**, *34*, 5402–5415. [CrossRef] [PubMed]

71. Murakami, K.; Nishikawa, F.; Noda, K.; Yokoyama, T.; Nishikawa, S. Anti-bovine prion protein RNA aptamer containing tandem GGA repeat interacts both with recombinant bovine prion protein and its beta isoform with high affinity. *Prion* **2008**, *2*, 73–80. [CrossRef] [PubMed]

72. Mashima, T.; Nishikawa, F.; Kamatari, Y.O.; Fujiwara, H.; Saimura, M.; Nagata, T.; Kodaki, T.; Nishikawa, S.; Kuwata, K.; Katahira, M. Anti-prion activity of an RNA aptamer and its structural basis. *Nucleic Acids Res.* **2013**, *41*, 1355–1362. [CrossRef] [PubMed]

73. Hayashi, T.; Oshima, H.; Mashima, T.; Nagata, T.; Katahira, M.; Kinoshita, M. Binding of an RNA aptamer and a partial peptide of a prion protein: Crucial importance of water entropy in molecular recognition. *Nucleic Acids Res.* **2014**, *42*, 6861–6875. [CrossRef] [PubMed]

74. Yakovchuk, P.; Protozanova, E.; Frank-Kamenetskii, M.D. Base-stacking and base-pairing contributions into thermal stability of the DNA double helix. *Nucleic Acids Res.* **2006**, *34*, 564–574. [CrossRef] [PubMed]

75. Olsthoorn, R.C.L. G-quadruplexes within prion mRNA: The missing link in prion disease? *Nucleic Acids Res.* **2014**, *42*, 9327–9333. [CrossRef] [PubMed]

76. Gabus, C.; Auxilien, S.; Péchoux, C.; Dormont, D.; Swietnicki, W.; Morillas, M.; Surewicz, W.; Nandi, P.; Darlix, J.L. The prion protein has DNA strand transfer properties similar to retroviral nucleocapsid protein. *J. Mol. Biol.* **2001**, *307*, 1011–1021. [CrossRef] [PubMed]

77. Sayer, N.M.; Cubin, M.; Rhie, A.; Bullock, M.; Tahiri-Alaoui, A.; James, W. Structural Determinants of Conformationally Selective, Prion-binding Aptamers. *J. Biol. Chem.* **2004**, *279*, 13102–13109. [CrossRef] [PubMed]

78. Sekiya, S.; Nishikawa, F.; Noda, K.; Kumar, P.K.R.; Yokoyama, T.; Nishikawa, S. In vitro selection of RNA aptamers against cellular and abnormal isoform of mouse prion protein. *Nucleic Acids Symp. Ser.* **2005**, 361–362. [CrossRef] [PubMed]

79. Sekiya, S.; Noda, K.; Nishikawa, F.; Yokoyama, T.; Kumar, P.K.R.; Nishikawa, S. Characterization and application of a novel RNA aptamer against the mouse prion protein. *J. Biochem.* **2006**, *139*, 383–390. [CrossRef] [PubMed]

80. Ogasawara, D.; Hasegawa, H.; Kaneko, K.; Sode, K.; Ikebukuro, K. Screening of DNA aptamer against mouse prion protein by competitive selection. *Prion* **2007**, *1*, 248–254. [CrossRef] [PubMed]

81. Bibby, D.F.; Gill, A.C.; Kirby, L.; Farquhar, C.F.; Bruce, M.E.; Garson, J.A. Application of a novel in vitro selection technique to isolate and characterise high affinity DNA aptamers binding mammalian prion proteins. *J. Virol. Methods* **2008**, *151*, 107–115. [CrossRef] [PubMed]

82. Wang, P.; Hatcher, K.L.; Bartz, J.C.; Chen, S.G.; Skinner, P.; Richt, J.; Liu, H.; Sreevatsan, S. Selection and characterization of DNA aptamers against PrP^{Sc}. *Exp. Biol. Med.* **2011**, *236*, 466–476. [CrossRef] [PubMed]

83. Eisenberg, D.; Jucker, M. The amyloid state of proteins in human diseases. *Cell* **2012**, *148*, 1188–1203. [CrossRef] [PubMed]

84. Ylera, F.; Lurz, R.; Erdmann, V.A.; Fürste, J.P. Selection of RNA aptamers to the Alzheimer's disease amyloid peptide. *Biochem. Biophys. Res. Commun.* **2002**, *290*, 1583–1588. [CrossRef] [PubMed]

85. Jucker, M.; Walker, L.C. Self-propagation of pathogenic protein aggregates in neurodegenerative diseases. *Nature* **2013**, *501*, 45–51. [CrossRef] [PubMed]

86. Rahimi, F.; Murakami, K.; Summers, J.L.; Chen, C.H.B.; Bitan, G. RNA aptamers generated against oligomeric Aβ40 recognize common amyloid aptatopes with low specificity but high sensitivity. *PLoS ONE* **2009**, *4*, e7694. [CrossRef] [PubMed]

87. Rahimi, F.; Bitan, G. Selection of aptamers for amyloid β-protein, the causative agent of Alzheimer's disease. *J. Vis. Exp.* **2010**, *13*, 1–7. [CrossRef] [PubMed]

88. Bitan, G.; Teplow, B. Preparation of Aggregate-Free, Low Molecular Weight Amyloid beta for Assembly and Toxicity Assays. *Methods Mol. Biol. Protein* **2005**, *299*, 3–9.

89. Pesarrodona, M.; Unzueta, U.; Vázquez, E. Dialysis: A characterization method of aggregation tendency. In *Insoluble Proteins: Methods and Protocols*; Springer: Berlin, Germany, 2014; pp. 321–330.

90. Hamada, H.; Arakawa, T.; Shiraki, K. Effect of additives on protein aggregation. *Curr. Pharm. Biotechnol.* **2009**, *10*, 400–407. [CrossRef] [PubMed]

91. Bunka, D.H.J.; Mantle, B.J.; Morten, I.J.; Tennent, G.A.; Radford, S.E.; Stockley, P.G. Production and characterization of RNA aptamers specific for amyloid fibril epitopes. *J. Biol. Chem.* **2007**, *282*, 34500–34509. [CrossRef] [PubMed]

92. Kayed, R.; Head, E.; Sarsoza, F.; Saing, T.; Cotman, C.W.; Necula, M.; Margol, L.; Wu, J.; Breydo, L.; Thompson, J.L.; et al. Fibril specific, conformation dependent antibodies recognize a generic epitope common to amyloid fibrils and fibrillar oligomers that is absent in prefibrillar oligomers. *Mol. Neurodegener.* **2007**, *2*, 18. [CrossRef] [PubMed]

93. Kayed, R.; Glabe, C.G. Conformation-Dependent Anti-Amyloid Oligomer Antibodies. *Methods Enzymol.* **2006**, *413*, 326–344. [PubMed]

94. O'Nuallain, B.; Wetzel, R. Conformational Abs recognizing a generic amyloid fibril epitope. *Proc. Natl. Acad. Sci. USA* **2002**, *99*, 1485–1490. [CrossRef] [PubMed]

95. Kayed, R.; Head, E.; Thompson, J.L.; McIntire, T.M.; Milton, S.C.; Cotman, C.W.; Glabe, C.G. Common structure of soluble amyloid oligomers implies common mechanism of pathogenesis. *Science* **2003**, *300*, 486–489. [CrossRef] [PubMed]

96. Lambert, M.P.; Velasco, P.T.; Chang, L.; Viola, K.L.; Fernandez, S.; Lacor, P.N.; Khuon, D.; Gong, Y.; Bigio, E.H.; Shaw, P.; et al. Monoclonal antibodies that target pathological assemblies of Aβ. *J. Neurochem.* **2007**, *100*, 23–35. [CrossRef] [PubMed]

97. Lacor, P.N.; Buniel, M.C.; Chang, L.; Fernandez, S.J.; Gong, Y.; Viola, K.L.; Lambert, M.P.; Velasco, P.T.; Bigio, E.H.; Finch, C.E.; et al. Synaptic targeting by Alzheimer's-related amyloid β oligomers. *J. Neurosci.* **2004**, *24*, 10191–10200. [CrossRef] [PubMed]

98. Lee, E.B.; Leng, L.Z.; Zhang, B.; Kwong, L.; Trojanowski, J.Q.; Abel, T.; Lee, V.M.Y. Targeting amyloid-β peptide (Aβ) oligomers by passive immunization with a conformation-selective monoclonal antibody improves learning and memory in Aβ precursor protein (APP) transgenic mice. *J. Biol. Chem.* **2006**, *281*, 4292–4299. [CrossRef] [PubMed]

99. Cordeiro, Y.; Macedo, B.; Silva, J.L.; Gomes, M.P.B. Pathological implications of nucleic acid interactions with proteins associated with neurodegenerative diseases. *Biophys. Rev.* **2014**, *6*, 97–110. [CrossRef]

100. Calamai, M.; Taddei, N.; Stefani, M.; Ramponi, G.; Chiti, F. Relative Influence of Hydrophobicity and Net Charge in the Aggregation of Two Homologous Proteins. *Biochemistry* **2003**, *42*, 15078–15083. [CrossRef] [PubMed]

101. Chiti, F. Relative Importance of Hydrophobicity, Net Charge, and Secondary Structure Propensities in Protein Aggregation. *Analysis* **2004**, *4*, 43–59.

102. Groveman, B.R.; Kraus, A.; Raymond, L.D.; Dolan, M.A.; Anson, K.J.; Dorward, D.W.; Caughey, B. Charge neutralization of the central lysine cluster in prion protein (PrP) promotes PrP^{Sc}-Like folding of recombinant PrP amyloids. *J. Biol. Chem.* **2015**, *290*, 1119–1128. [CrossRef] [PubMed]

103. Ginsberg, S.D.; Galvin, J.E.; Chiu, T.S.; Lee, V.M.Y.; Masliah, E.; Trojanowski, J.Q. RNA sequestration to pathological lesions of neurodegenerative diseases. *Acta Neuropathol.* **1998**, *96*, 487–494. [CrossRef] [PubMed]

104. Ginsberg, S.D.; Crino, P.B.; Hemby, S.E.; Weingarten, J.A.; Lee, V.M.Y.; Eberwine, J.H.; Trojanowski, J.Q. Predominance of neuronal mRNAs in individual Alzheimer's disease senile plaques. *Ann. Neurol.* **1999**, *45*, 174–181. [CrossRef]

105. Marcinkiewicz, M. BetaAPP and furin mRNA concentrates in immature senile plaques in the brain of Alzheimer patients. *J. Neuropathol. Exp. Neurol.* **2002**, *61*, 815–829. [CrossRef] [PubMed]

106. Manuelidis, L. Nuclease resistant circular DNAs copurify with infectivity in scrapie and CJD. *J. Neurovirol.* **2011**, *17*, 131–145. [CrossRef] [PubMed]

107. Botsios, S.; Manuelidis, L. CJD and Scrapie Require Agent-Associated Nucleic Acids for Infection. *J. Cell. Biochem.* **2016**, *117*, 1947–1958. [CrossRef] [PubMed]

108. Simoneau, S.; Thomzig, A.; Ruchoux, M.-M.; Vignier, N.; Daus, M.L.; Poleggi, A.; Lebon, P.; Freire, S.; Durand, V.; Graziano, S.; et al. Synthetic Scrapie Infectivity: Interaction between Recombinant PrP and Scrapie Brain-Derived RNA. *Virulence* **2015**, *6*, 132–144. [CrossRef] [PubMed]

109. Hartl, F.U.; Bracher, A.; Hayer-Hartl, M. Molecular chaperones in protein folding and proteostasis. *Nature* **2011**, *475*, 324–332. [CrossRef] [PubMed]

110. Maury, C.P.J. Self-Propagating beta-sheet polypeptide structures as prebiotic informational molecular entities: The amyloid world. *Orig. Life Evol. Biosph.* **2009**, *39*, 141–150. [CrossRef] [PubMed]

111. Si, K. Prions: What Are They Good For? *Annu. Rev. Cell. Dev. Biol* **2015**, *31*, 149–169. [CrossRef] [PubMed]

112. Maury, C.P.J. The emerging concept of functional amyloid. *J. Intern. Med.* **2009**, *265*, 329–334. [PubMed]

113. Takahashi, T.; Tada, K.; Mihara, H. RNA aptamers selected against amyloid β-peptide (Aβ) inhibit the aggregation of Abeta. *Mol. Biosyst.* **2009**, *5*, 986–991. [PubMed]

114. Burchell, J.T.; Panegyres, P.K. Prion diseases: Immunotargets and therapy. *ImmunoTargets Ther.* **2016**, *5*, 57–68. [PubMed]

115. Cordeiro, Y.; Ferreira, N.C. New approaches for the selection and evaluation of anti-prion organic compounds. *Mini Rev. Med. Chem.* **2015**, *15*, 84–92. [PubMed]

116. Hanss, B.; Leal-Pinto, E.; Bruggeman, L.; Copeland, T.; Klotman, P. Identification and characterization of a cell membrane nucleic acid channel. *Proc. Natl. Acad. Sci. USA* **1998**, *95*, 1921–1926. [PubMed]

117. Reyes-Reyes, E.M.; Teng, Y.; Bates, P.J. A new paradigm for aptamer therapeutic AS1411 action: Uptake by macropinocytosis and its stimulation by a nucleolin-dependent mechanism. *Cancer Res.* **2010**, *70*, 8617–8629. [PubMed]

118. Cheng, C.; Chen, Y.H.; Lennox, K.A.; Behlke, M.A.; Davidson, B.L. In vivo SELEX for Identification of Brain-penetrating Aptamers. *Mol. Ther. Nucleic Acids* **2013**, *2*, E67. [PubMed]

119. Kim, Y.; Wu, Q.; Hamerlik, P.; Hitomi, M.; Sloan, A.E.; Barnett, G.H.; Weil, R.J.; Leahy, P.; Hjelmeland, A.B.; Rich, J.N. Aptamer identification of brain tumor-initiating cells. *Cancer Res.* **2013**, *73*, 4923–4936. [PubMed]

120. Xiao, S.J.; Hu, P.P.; Wu, X.D.; Zou, Y.L.; Chen, L.Q.; Peng, L.; Ling, J.; Zhen, S.J.; Zhan, L.; Li, Y.F.; et al. Sensitive discrimination and detection of prion disease-associated isoform with a dual-aptamer strategy by developing a sandwich structure of magnetic microparticles and quantum dots. *Anal. Chem.* **2010**, *82*, 9736–9742. [PubMed]

121. Zeiler, B.; Adler, V.; Kryukov, V.; Grossman, A. Concentration and removal of prion proteins from biological solutions. *Biotechnol. Appl. Biochem.* **2003**, *37*, 173–182. [PubMed]

122. Yu, P.; Zhang, X.; Xiong, E.; Zhou, J.; Li, X.; Chen, J. A label-free and cascaded dual-signaling amplified electrochemical aptasensing platform for sensitive prion assay. *Biosens. Bioelectron.* **2016**, *85*, 471–478. [CrossRef] [PubMed]

123. Zhou, J.; Battig, M.R.; Wang, Y. Aptamer-based molecular recognition for biosensor development. *Anal. Bioanal. Chem.* **2010**, *398*, 2471–2480. [CrossRef] [PubMed]

6

Selection and Characterization of a DNA Aptamer Specifically Targeting Human HECT Ubiquitin Ligase WWP1

Wesley O. Tucker, Andrew B. Kinghorn, Lewis A. Fraser, Yee-Wai Cheung and Julian A. Tanner *

School of Biomedical Sciences, Li Ka Shing Faculty of Medicine, The University of Hong Kong, 21 Sassoon Road, Hong Kong, China; wtucker@hku.hk (W.O.T.); kinghorn@hku.hk (A.B.K.); lewis-fraser@hku.hk (L.A.F.); cheungw@hku.hk (Y.-W.C.)
* Correspondence: jatanner@hku.hk

Abstract: Nucleic acid aptamers hold promise as therapeutic tools for specific, tailored inhibition of protein targets with several advantages when compared to small molecules or antibodies. Nuclear WW domain containing E3 ubiquitin ligase 1 (WWP1) ubiquitin ligase poly-ubiquitinates Runt-related transcription factor 2 (Runx2), a key transcription factor associated with osteoblast differentiation. Since WWP1 and an adapter known as Schnurri-3 are negative regulators of osteoblast function, the disruption of this complex has the potential to increase bone deposition for osteoporosis therapy. Here, we develop new DNA aptamers that bind and inhibit WWP1 then investigate efficacy in an osteoblastic cell culture. DNA aptamers were selected against three different truncations of the HECT domain of WWP1. Aptamers which bind specifically to a C-lobe HECT domain truncation were observed to enrich during the selection procedure. One particular DNA aptamer termed C3A was further evaluated for its ability to bind WWP1 and inhibit its ubiquitination activity. C3A showed a low μM binding affinity to WWP1 and was observed to be a non-competitive inhibitor of WWP1 HECT ubiquitin ligase activity. When SaOS-2 osteoblastic cells were treated with C3A, partial localization to the nucleus was observed. The C3A aptamer was also demonstrated to specifically promote extracellular mineralization in cell culture experiments. The C3A aptamer has potential for further development as a novel osteoporosis therapeutic strategy. Our results demonstrate that aptamer-mediated inhibition of protein ubiquitination can be a novel therapeutic strategy.

Keywords: aptamer; SELEX; ubiquitin ligase; WWP1; targeted drug delivery

1. Introduction

In mammalian bone, the osteoblast is exclusively responsible for the deposition of bone matrix in the form of hydroxyapatite crystals together with various structural proteins [1]. The osteoblast lineage is largely controlled by Runx2 (Runt–related transcription factor 2/CBFA1/PEBP2αA), often considered the "master regulator" of osteoblast differentiation [2]. The importance of Runx2 is illustrated by observations that Runx2 null mice cannot produce mineralized bone due to a lack of early stage osteoblasts [3]. The Nedd4 family HECT ('Homologous to the E6-AP Carboxyl Terminus') domain E3 ligase WWP1, in complex with an adapter Schnurri-3 (Shn3), ubiquitinates Runx2 thus leading to its degradation and reduction of osteoblast-mediated bone matrix synthesis [2]. WWP1 and Shn3 proteins negatively regulate Runx2 at the protein level, hence are promising targets to stimulate osteoblast differentiation, and by extension, higher bone mass in osteoporosis patients [4,5]. Human WWP1 was discovered in 1997 and contains an independently active HECT domain (Homologous to the E6 AP carboxyl terminus) which has its crystal structure solved [6–8]. WWP1 is structurally and functionally characterized, is readily "expressible" in *E. coli*, and inhibition of WWP1 to maintain Runx2 levels is a promising therapeutic strategy to stimulate osteoblast function.

Nucleic acid aptamers are short, single-stranded nucleic acid chains evolved to bind specifically to a given target. Riboswitches can be considered as natural non-coding RNA aptamers which regulate gene expression upon binding a metabolic target [9]. Since 1990, aptamers have been created in the laboratory using a process known as Systematic Evolution of Ligands by Exponential Enrichment (SELEX) [10,11]. First, a large pool of random sequences ($\sim10^{15}$) is exposed to a target, washed to exclude non-binding species, and eluted to recover binding species. Second, the pool is PCR amplified by having included 3' and 5' primer binding regions. Finally, the sequences are enriched by repeating the process successively while introducing increased stringency (i.e., more rigorous washing conditions and counter-selections of related targets). The SELEX iterations are continued until enriched pools of homologous sequences are revealed, which then hold the potential to be tightly binding and specific [10,11]. Aptamers often bind to their targets with K_D values in the µM to nM range with aptamer affinity maturation developed to increase affinity and specificity [12]. Aptamers have secondary structure including hairpins, loops, pseudoknots, triplexes, and quadruplexes [13–15]. SELEX can be applied to a wide variety of targets ranging from small molecules to cells [16,17]. Due to these complexities and features, aptamers are sometimes compared and contrasted with both antibodies and small molecules for use as therapeutics [18,19], diagnostics [20–22], and laboratory tools [23]. Importantly, recent work has demonstrated methods that lower the cost of single stranded oligonucleotide synthesis by several orders of magnitude [24].

Aptamers have traditionally focused on extracellular targets. However, polynucleotides such as aptamers can enter the intracellular space by either: (1) extraneous means such as liposomes, polymers, viruses, microinjection, electroporation, particle bombardment, calcium phosphate precipitation and ultrasound, or by (2) inherent mechanisms such as endocytosis or pinocytosis [25–28]. Using cationic reagents, aptamers can be transfected as is commonly practiced with plasmids. Naked DNA, without the company of reagents, does enter the cell to some degree in spite of electrostatic repulsion, as was demonstrated when internalized plasmids were found to express proteins [29]. Uptake depends both on temperature and oligonucleotide length [30]. It has also been found that two putative cell surface receptors are specific to oligonucleotides generally [31]. More recently, aptamers have been conjugated to moieties to bind an internalizing receptor, or may be selected to bind the internalizing receptor themselves [26,32]. Approximately 19,000 DNA molecules can be internalized by one cell once cationic lipoplexes have fused with the anionic plasma membrane [33]. Aptamers can also be transcribed directly in the cell, this type of aptamer has been coined an "intramer" [34–36].

Previously, our group selected aptamers which bind and inhibit the activity of the extracellular protein sclerostin, which is a known negative regulator of bone formation [19]. Here, we investigate an alternative intracellular strategy where the intracellular protein WWP1 is targeted with a DNA aptamer. Herein we select and characterize DNA aptamers against the HECT domain of WWP1. We investigate aptamer binding to its target, specific inhibition of its function, observe localization, and evaluate efficacy in SaOS-2 osteoblastic cells.

2. Results

2.1. Strategy to Select DNA Aptamers against HECT Domain of WWP1

As it is not possible to express the full length WWP1 protein in *E. coli*, truncations of the smaller yet active HECT domain were used as the targets for aptamer selection. The HECT domain has two lobes: the N-lobe is involved in E2 binding and the C-lobe is important in ubiquitination [8]. Rotation around a hinge between the two lobes is critical to WWP1 ubiquitin ligation activity [8], hence the HECT domain is central to WWP1 function.

As shown in Figure 1A, we expressed and purified three WWP1 truncations for use as targets in our aptamer selections: the entire HECT domain, the N-lobe of the HECT domain and the C-lobe of the HECT domain.

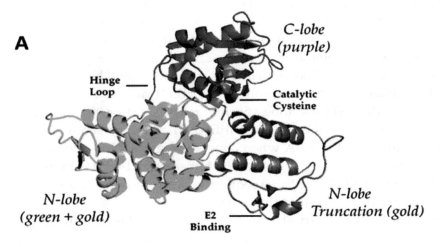

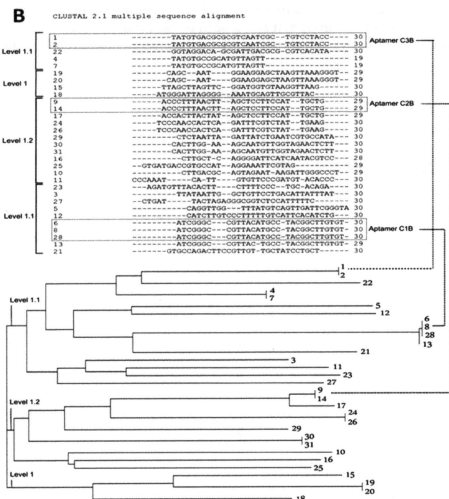

Figure 1. Structure of WWP1 HECT domain target and selected pool. (**A**) WWP1 HECT structure as generated by pyMOL with functionally important regions labeled; (**B**) Variable regions of the enriched pool for C-lobe with the three most abundant groups of identical sequences are boxed (C1–C3 where B indicates without constant regions), and phylogenetic tree of sequenced aptamers showing inter familial relationships.

To begin, we used the entire HECT domain for the first 8 rounds of DNA aptamer selection. Important to its inhibition, HECT domain requires a flexible hinge loop, the binding of an E2 adapter, and a catalytic cysteine residue for activity (Figure 1A). We then proceeded to lobe-specific selections to target particular functional regions more specifically. To further increase specificity as well as stringency, C-lobe selections included N-lobe counter selections, N-lobe selections included C-lobe counter selections, and HECT selections included ubiquitin as a general protein counter selection. During the sequencing of pools throughout the process, the C-lobe pools proved to enrich promisingly, while HECT and N-lobe selections did not enrich. This indicated that DNA aptamers had been selected in the initial rounds that likely bind to the C-lobe of the HECT domain.

We performed 4 further SELEX lobe-specific rounds (after the 8 round HECT domain pre-enrichment). Cloning and sequencing identified 31 aptamer sequences against the C-lobe truncation (Figure 1B), annotated as C#A when constant region is included and C#B without. The relationships between the variable regions is somewhat scattered—C3 is more closely related to C1 while C2 is completely unrelated. These three sequences were chosen to move forward in the project due to their higher copy number in sequencing data.

To compare the predicted secondary structures of the three aptamers, mFold software (version 2.3, The RNA Institute, State University of New York at Albany, Albany, NY, USA) was used (Figure 2). The structure of the variable region mirrors that of its full-length counterpart for aptamer C1, but not for C2 and C3; possibly indicating that full length is required for C2 and C3 binding (Figure 2). Experiments were therefore carried out subsequently with constant regions included (indicated by the A suffix).

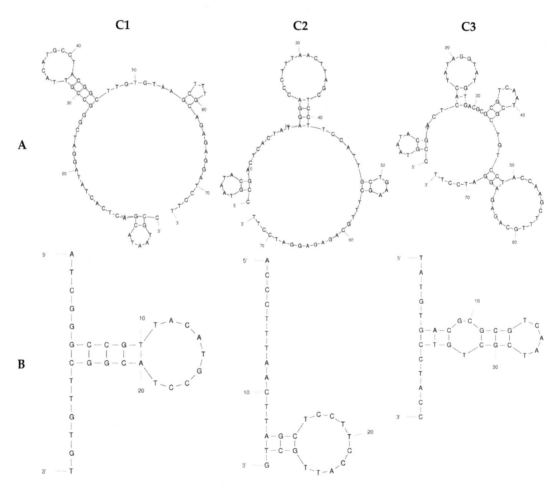

Figure 2. Predicted 2D secondary structures for C-lobe aptamers using mFold software comparing full length (**A**) with variable region (**B**) simulated at 150 mM NaCl and 25 °C.

2.2. Determination of Binding Affinity of DNA Aptamers Binding to HECT Domain

We next investigated the relative affinity of the DNA aptamers for the protein target relative to control. To demonstrate qualitative binding of the C-lobe aptamers to our recombinant HECT, Electrophoretic Mobility Shift Assay (EMSA) was employed (Figure 3). EMSA is a hallmark method for characterization of protein-nucleic acid binding. However, EMSA has shortcomings such as complexes differing in stability depending on the gel medium and secondary binding confusing the magnitude of the binding of interest. Thus, we view EMSA here as a relative and semi-quantitative characterization of binding. The full-length aptamers C1A–C3A (the A indicating inclusion of the constant flanking region) were incubated with a range of HECT concentrations then analyzed by PAGE electrophoresis (Figure 3A), and the resultant band intensities were plotted (Figure 3B). Aptamer bound to protein remains at the top of the gel, while free aptamer migrates normally. Of the three C-lobe aptamers, C3A showed the strongest binding affinity. We observed that C3A shows a multimeric band above its monomer band also, which may imply a multimer in equilibrium with the monomer (74 bp). A K_D roughly in the low µM range was estimated for aptamer C3A, while other C-lobe aptamers and controls did not produce an inflection point with which to base a determination but with clearly weaker binding. Therefore, in later experiments we focused on aptamer C3A as the most promising of the three aptamer candidates.

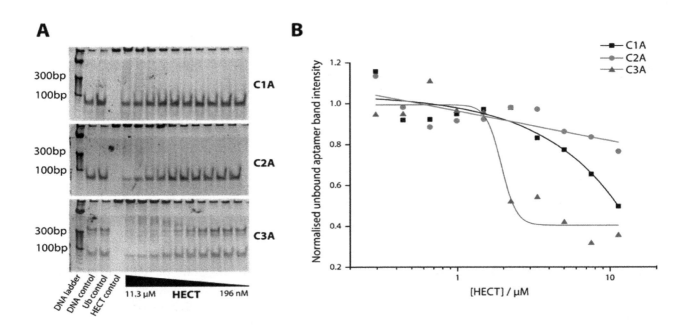

Figure 3. Determination of K_D for aptamers using electrophoretic mobility shift assay (EMSA) for aptamers C1A, C2A, and C3A. (**A**) 10% PAGE gels showing bound (upper bands) versus unbound (lower bands) aptamer; (**B**) Plot of HECT concentration versus normalized band intensity (unbound) with a one-site binding fit from representative data in Figure 3A.

2.3. Determination of C-Lobe DNA Aptamer-Mediated Inhibition of Ubiquitination Activity of HECT Domain

The HECT domain's ability to self-ubiquitinate with the requirement of an E1 and E2 protein in vitro provides an approach to observe the kinetics of HECT domain activity. When the HECT domain is ubiquitinated, the HECT domain-ubiquitin conjugate is observed by SDS-PAGE appearing several kilodaltons above HECT and can be used for quantitation (Figure 4C). We used this assay to generate Michaelis-Menten curves between aptamers C1A–C3A which implicates aptamer C3A as the strongest inhibitor of the three (Figure 4A).

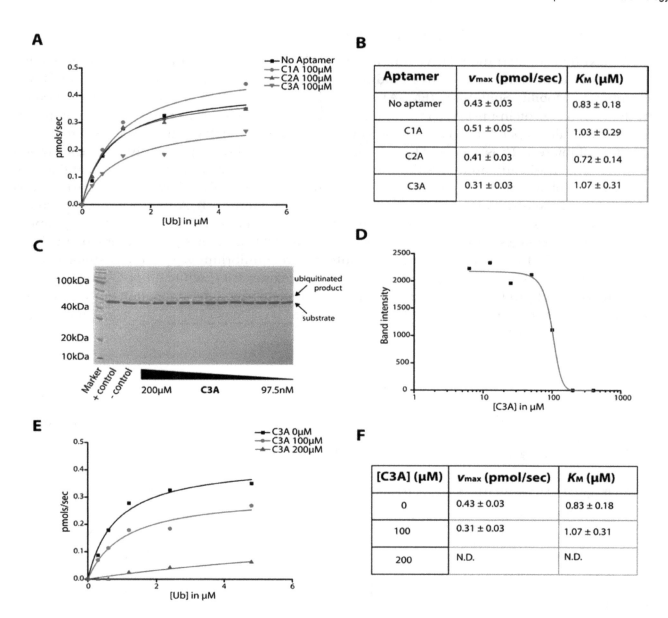

Figure 4. Aptamer mediated inhibition of HECT ubiquitination. (**A**) Rate of reaction product formation plotted against ubiquitin concentration (Michaelis-Menten) for aptamers C1A–C3A to compare inhibition rates between aptamers from representative gel data. (**B**) Resultant V_{max} and K_M values from the Michaelis-Menten plot. (**C**) 12% SDS-PAGE gel showing the reaction product over increasing C3A concentrations; (**D**) Logarithmic plot of C3A aptamer concentration to band intensity as measured to determine IC_{50} from representative gel data (**E**) Michaelis-Menten plot of two different C3A concentrations for determination of mode of inhibition from representative gel data. (**F**) Summary of V_{max} and K_M values tabulated for the two C3A concentrations.

The resulting v_{max} and K_M are tabulated in Figure 4B and reconfirms aptamer C3A as the strongest inhibitor. This result is also consistent with our observations of aptamer binding in Figure 3. Figure 4C shows the ubiquitination assay on an example gel over a series of aptamer C3A concentrations, where substrate and reaction product are labeled (residual ubiquitin (~10 kD) and E2 (~10 kD) can also be seen). The plot of C3A concentration to band intensity of reaction product (Figure 4D) determines an IC_{50} of approximately 100 μM. To determine mode of inhibition, two different C3A concentrations were tested (100 μM and 200 μM) while varying the concentration of ubiquitin substrate to determine K_M and v_{max} as shown in Figure 4E,F. The presence of the inhibitor caused a significant reduction in the apparent v_{max} while the apparent K_M remained similar. These results imply that the aptamer acts as a non-competitive inhibitor of ubiquitination.

2.4. Cellular Localization, Runx2 Levels, and Bone Mineralization in Saos-2 Osteoblastic Cells

Fixed and stained SaOS-2 cells were imaged after the transfection process to investigate differences in localization. These stained images were used to compare aptamer C3A and control, both with and without transfection reagent as shown in Figure 5A. Although the images are overexposed, we were able to gain information about the relative abundance of aptamer in the cytoplasm versus nucleus. In this representative example, aptamer C3A without transfection (CN) appears to enter the cell, but appears mostly proximal to the nucleus, whereas aptamer C3A with transfection (CT) seems to localize both in the nucleus and cytosol.

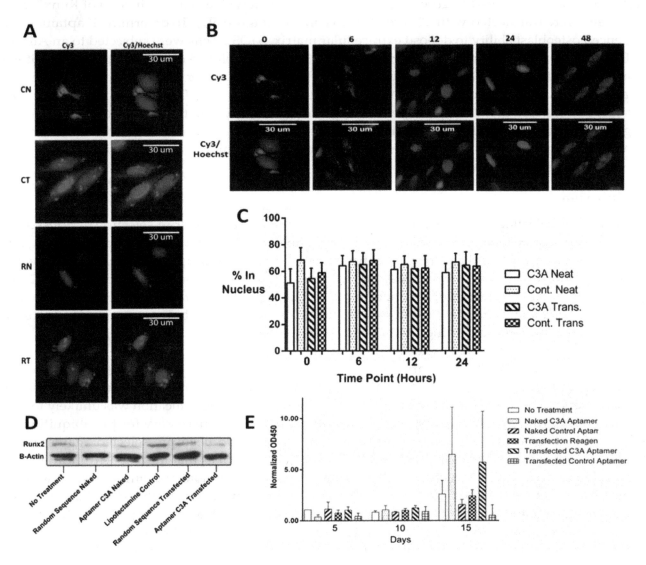

Figure 5. Aptamer C3A localization, and effect on Runx2 levels and bone mineralization in SaOS-2 cells. (**A**) Aptamer (Cy3) nucleus (Hoechst) overlays where: RN—random sequence control without transfection, RT—random sequence control with transfection, CN—aptamer C3A without transfection, and CT—aptamer C3A with transfection; (**B**) Aptamer and nucleus overlays over time points in hours; (**C**) Quantitative assessment of nuclear localization using MetaMorph™ software for transfected or naked (labeled as "NEAT") aptamer C3A and random sequence control; (**D**) Western Blot of Runx2 performed on SaOS-2 cell extracts; (**E**) Quantitation of extracellular matrix deposition over time using Alizarin Red assay.

After determining that aptamer enters the cells without transfection reagent, we compared localization over time by incubating aptamer C3Awith SaOS-2 cells without transfection reagent over 5 time points (Figure 5B). All aptamer samples appear to enter the cell and the images depict migration towards the nucleus over time. Percent nuclear localization over time was quantified by image intensity where hundreds of cells were batch processed using MetaMorph™ software (64-bit version, Molecular Devices Corp., Sunnyvale, CA, USA) for aptamer C3A and control, with and without transfection (Figure 5C). Whether or not transfection reagent was used, aptamers appear to enter the cell and localize partially in both the cytoplasm and nucleus to some degree. There are examples where all aptamers are exclusively in the cytoplasm for a period of time, but these results show that for most cells, a proportion (>50%) of the aptamers end up in the nucleus quickly after exposure. To see if treatment with aptamer C3A had an effect on Runx2 protein expression levels in SaOS-2 cell extracts, a western blot was performed (Figure 5D). Data were consistent with a slight reduction of Runx2 levels when cells were transfected with C3A relative to controls (Figure 5D). To determine if aptamer C3A influenced osteoblast ability to deposit extracellular matrix, SaOS-2 cells were subjected to an Alizarin Red assay to quantify calcific deposition (Figure 5E). At a timepoint 15 days after transfection it was clear that both the naked C3A aptamer and the transfected C3A aptamer were able to increase the rate of calcific deposition of the cells relative to control aptamers (Figure 5E). This data would be consistent with an observation that the C3A aptamer is able to enter the cell even in the absence of the transfection reagent and increase calcific extracellular matrix deposition. We also investigated the influence of aptamer C3A on apoptosis and observed that aptamer C3A promoted apoptosis significantly more than a random sequence control (Figure S1).

3. Discussion

Aptamer selections were begun with an entire HECT domain 'pre-selection'. Subsequent selections against N-lobe, C-lobe and HECT pool showed enrichment of Aptamers only against C-lobe. This indicated that the C-lobe likely had selected tight-binding aptamers early in the selection process. We observed that it was necessary to retain the flanking constant regions in the C-lobe aptamers. Two classic examples of aptamers for which the PCR constant regions were not necessary for binding are pegaptanib and the anti-thrombin aptamer, but others have been reported where constant regions were required for binding [37,38]. We surmise that the necessity of flanking regions in an aptamer varies from selection to selection.

Concerning our WWP1 truncation targets themselves (Figure 1), some structural features are relevant. First, the C-lobe portion of the related E3 ubiquitin ligase UbR5 has its crystal structure solved and was found to be active [39]. Nevertheless, we assumed our C-lobe truncation was unlikely to have activity considering previous reports had shown that the N-lobe is necessary for the ubiquitination reaction [8]. Our best C-lobe targeting aptamer C3A, which had been pre-selected against HECT domain, was shown to specifically bind HECT (Figure 3). This implies that the C-lobe truncation most likely folded into a similar structure as that in the context of the whole protein, consistent with the demonstrated activity of UbR5 [39]. Many ubiquitin ligases have been identified as disease targets. This is the first attempt to inhibit HECT ubiquitin ligase activity with an aptamer and we therefore do not have a direct comparison of binding affinity for this specific family of targets [40]. Aptamers often bind to targets with nanomolar affinities, but the K_D of our C3A could be estimated in the low μM range. We were not able to obtain estimations from our other aptamers and control because they did not bind sufficiently at the tested range of concentrations, but one may conclude their binding was far weaker (Figure 3).

Ubiquitination assays to determine inhibition are well-established [41]. HECT ligases in particular are commonly studied from many angles with such assays [41]. Other groups have detected ubiquitination by anti-ubiquitin western blotting, but we found this step unnecessary given the quantitative Coomassie stain from the reaction product on the gel, which we confirmed to be ubiquitin conjugated HECT with mass spectrometry [40]. Nevertheless, IC$_{50}$ values for aptamers are rarely reported because enzymes are typically targeted by small molecules. There are some examples, however, such as G-quadruplex aptamers against Shp2 phosphatase which inhibited Shp2 activity at 29 nM, while small molecules inhibited in the µM range [42]. Conversely, anti-sclerostin antibodies inhibit in the nM range, while aptamers for the same protein inhibit at 15 µM [19,43]. Evidence of the mode of inhibition could be seen with our Michaelis-Menten curves and apparent v_{max} and K_M values, which were consistent with non-competitive inhibition. Overall these pieces of evidence support the possibility that aptamer C3A binds specifically to HECT domain and inhibits its activity in a non-competitive manner (Figures 3 and 4).

There are 3 known mechanisms for DNA to enter a eukaryotic cell without transfection or a virus: pinocytosis, absorptive endocytosis, or receptor mediated endocytosis [31,44]. In our assessment of aptamer C3A localization in SaOS-2 cells, we found qualitatively that the aptamer appeared to be able to enter the cell in the presence of absence of transfection reagent (Figure 5A). Here, we can only assume that the DNA in the non-transfected samples entered by the abovementioned mechanisms. Recently, aptamers for the intracellular and membrane target nucleolin were demonstrated to enter via macropinocytosis [45]. In addition, non-specific DNA internalizing receptors exist in other cell types but are not known in osteoblasts [27,28]. Both transfected and non-transfected samples were treated with DNA for 5 h, a time frame where aptamers could conceivably diffuse passively through the nuclear pore complex (Figure 5B), which could be compared to eukaryotic cells expressing proteins from plasmids which must enter the nucleus to be transcribed [29,46]. The percentage of signal appearing in the nucleus region for the hundreds of cells we analyzed with batch processing averaged around 60% (Figure 5C). One possible reason for this could be positively charged nuclear proteins such as histones, trapping a certain proportion of the aptamers.

Regarding aptamer C3A effect on phenotype, Western blot results were not clear-cut but were consistent with a slight reduction in Runx2 levels relative to controls (Figure 5D). The extracellular matrix deposition assay was performed under the same treatment conditions as the localization experiments and increased extracellular matrix was observed in the C3A treated cells relative to controls for both transfected and non-transfected cells.

Overall these experiments show that a DNA aptamer can bind specifically to WWP1 C-lobe and inhibit its target. Observations were consistent with the DNA aptamer entering the SaOS-2 cell nucleus, inhibiting WWP1 ubiquitination of Runx2 then increasing extracellular matrix deposition. WWP1 has also recently been implicated as an oncogene indicating that further work on WWP1 inhibition is warranted [47]. Finally, a recent paper demonstrated that interference of WWP1 led to the induction of apoptosis in osteosarcoma cells [47], which is also consistent with apoptosis experiments which we performed at the characterization stage (Figure S1). Further cell-based and animal-based experiments will be required to better understand the applicability of the aptamer to promote bone mineralization.

4. Materials and Methods

4.1. HECT, C-Lobe, N-Lobe and E2 Cloning, Expression and Purification

HECT, C-lobe and N-lobe were amplified per Platinum® Pfx Polymerase (Invitrogen, Waltham, MA, USA) guidelines from a human liver cDNA library. The following sequences were those for which the expression inserts were designed: HECT Domain (bp 1916–3047 NCBI Reference Sequence: NM_007013.3), C-lobe (bp 2687–3047 NCBI Reference Sequence: NM_007013.3), N-lobe (bp 2342–2587 in NCBI Reference Sequence: NM_007013.3). The resulting sequences, once inserted into pET28a(+) (Novagen, Madison, WI, USA) vector, gave an N-terminal hexahistidine tags. Subcloning was initiated by 1% agarose gel purification of the PCR product with Platinum® Pfx Polymerase (Invitrogen) of each insert (approx 20 cycles) using QIAquick® gel extraction kit (Qiagen, Hilden, Germany). Inserts were then digested with EcoRI and NdeI (New England Biolabs, Ipswich, MA, USA), 1% agarose gel purified and extracted, and the concentration of the insert and vector was then determined using absorbance at OD260/280 nm. Approximately 20 ng of insert and 10 ng of vector were put into ligation reaction with 1 µL ligase in total of 10 µL per instructions of T4 DNA ligase (New England Biolabs). A frozen eppendorf of XL-1 Blue *E. coli* (Stratagene, San Diego, CA, USA) provided in house was made competent and transformed by roughly following the pET System manual (Novagen). Colonies were selected on 50 µg/mL kanamycin agar plates, grown in 1 mL cultures for exponential growth phase, DNA was purified by QIAprep (Qiagen), and sequenced by Tech Dragon Limited™, Hong Kong, China.

The pET28a(+) (Novagen) vectors containing in-frame and un-mutated insert for all three truncations of HECT and pET-15b (Novagen) containing UbE2D2 (Addgene, Cambridge, MA, USA) were transformed into *E. coli* BL21(DE3) using the vendor transformation protocol (Novagen®). Overnight starter cultures of 30 µg/mL kanamycin containing LB media seeded 2–3 L cultures of the same media. These large cultures were grown at 37 °C under shaking at 200 rpm for approximately 6 h at 37 °C and induced to express with 500 µM IPTG, and shaken at room temperature for 4–5 h (C-lobe, N-lobe, and UbE2D2) and 4 °C overnight (HECT domain). Cell pellet was harvested by centrifugation at 4000 rpm, for 25 min, at 4 °C. After careful decanting, cell pellet was stored at −20 °C until purification. Column load was prepared by resuspending cell pellets in 20 mL of sonication buffer (Tris/500 mM NaCl/20 mM imidazole/0.1% (*v/v*) TritonX (pH 7.4)) for every 500 mL of culture and supplemented with Complete EDTA-free EASYpack protease inhibitor cocktail (Roche, Mannheim, Germany) followed by sonication for 10 min on ice at 30% amplitude. Cell lysate was centrifuged at 13,000 rpm for 25 min at 4 °C. Supernatant was filtered and loaded onto His-trap HP Ni^{2+} affinity column (GE Healthcare, Chicago, IL, USA) using a peristaltic pump. After loading, the column was transferred to Vision™ Workstation (Applied Biosystems, Waltham, MA, USA). Real time UV trace was analyzed with the accompanying Vision™ Software (SDS v. 2.3, Applied Biosystems, Foster City, CA, USA). A gradient was run from 0% mobile phase A (50 mM Tris/500 mM NaCl/20 mM Imidazole (pH 7.4)) → 100% mobile phase B (50 mM Tris/500 mM NaCl/500 mM Imidazole (pH 7.4)) over 20 min at 3.5 mL/min and ~2 mL fractions were taken by hand throughout the gradient. Fractions were stored at −20 °C. Overall, the proteins were of correct size, were derived from the correct sequence, had purities of around 95% (excluding UbE2D2), and maintained a solubility of around 50% of the starting fraction pool after one week at 4 °C. All proteins expressed in high yield (concentrations of 2–4 mg/mL) and signal during imidazole gradients generally coincided with predicted isoelectric points.

4.2. SELEX Procedures

Basic methodologies and tools for aptamer selection were adapted from our labs previous methods. A random DNA library was obtained from Tech Dragon Limited, Hong Kong containing 6.9 nanomoles of single stranded DNA of the following design: Aptamers→ 3'-CTAATACGACTC ACTATAGG(N30)AAGCTTTGCAGAGAGGATCCTT-5', Primers→ 3'-GATTATGCTGAGTGATATCC, TTCGAAACGTCTCTCCTAGGAA-5'-Biotin. DNA was reconstituted in MilliQ H_2O and contained a total of 4.2×10^{15} DNA molecules of 1.15×10^{18} possible sequences from a variable region of $N = 30$. Determination of the amount of protein target to be used was based on the Ni-NTA Agarose Beads Handbook (Qiagen, Hilden, Germany) estimate that approximately 3 µg of protein bind to 10 µL of bead solution. The amount of target protein necessary to saturate 10 µL of bead solution was determined by exposing the beads to an increasing amount of protein, washing with selection buffer (50 mM Tris/0.05% Tween-20/0 → 1.0 M NaCl (pH 7.3)), eluting with 1.0 M imidazole in selection buffer, and visualizing on SDS-PAGE. After washing beads with protein buffer (50 mM Tris/400 mM NaCl/0.05% Tween-20 (pH 7.3)), appropriate amounts of protein were added and beads, and washed 3 times again with protein buffer. A 25 µL aliquot of library was then diluted with 200 µL of selection buffer and added to the beads, gently mixed, and set for 1 min. The beads were then washed with 1 mL of selection buffer 6–12 times depending on the round of selection. Protein bound to DNA was eluted with 50 µL elution buffer (1.0 M imidazole in TBS, pH 7.5) and 5 µL was carried on to PCR amplification with biotinylated reverse primer for ~10 cycles. The resulting double stranded amplification product from each consecutive round was separated by washing 50 µL of Dynabeads

M-280 Streptavidin magnetic beads (Invitrogen, Waltham, MA, USA) with 1 mL separation buffer (50 mM Tris/0.05% Tween-20 (pH 7.3)), binding the entire 50 µL PCR product and 900 µL separation buffer with the washed beads, washing three times with separation buffer, and eluting with 50 µL of 100 mM NaOH. DNA containing NaOH was diluted with 150 µL TBS for neutralization. Progression of the selection process was monitored by PCR and PAGE. Pools of double stranded DNA from final rounds of selections were blunt end cloned into vectors using Zero Blunt® TOPO® PCR Cloning Kit (Invitrogen) and transfection of XL-1 Blue *E. coli* (Stratagene, San Diego, CA, USA) was performed as previously mentioned. 50 colonies from 50 µg/mL kanamycin selective plates were grown in 2.5 mL cultures for approximately 10 h, and said cultures were prepared for sequencing using the QIAprep® Spin Miniprep Kit (Qiagen, Hilden, Germany). Sequencing of plasmids was performed by Tech Dragon Limited, Hong Kong, China.

4.3. HECT Ubiquitin Ligase Activity Assay

A HECT ubiquitination assay was devised based on the general guidelines of the Boston Biochem® Company which specializes in ubiquitin assays. Active HECT domain and the human E2 protein (UbE2D2) were expressed in *E. coli*, purified as described previously and were kept at 100 ng/µL and 160 ng/µL, respectively, at −20 °C in ~25% glycerol. His$_6$-tagged human recombinant Ubiquitin and recombinant Human His$_6$-Ubiquitin E1 Enzyme (UBE1) (Boston Biochem®) were kept at 5 µg/µL and 250 ng/µL, respectively, at −20 °C. The reaction was optimized by starting at 250 ng E1, 300 ng, HECT, 320 ng E2, and 5 µg Ub, and then systematically altering assay parameters such as reaction volume, time, reagent amounts, and temperature. Finally, the reaction contained 2.6 µM HECT, 45.5 nM UBE1, 507.8 nM UBE2D2, and varying from 300 nM to 4.8 µM Ubiquitin depending on the experiment. The total reaction volume was 25 µL, was run for 1 h at 37 °C in assay buffer (50 mM Tris/50 mM NaCl/5 mM MgCl$_2$/5 mM KCl/25 mM DTT/5 mM ATP (pH 7.5)), and stopped by addition of 25 µL SDS-PAGE loading buffer (100 mM Tris/40% *v/v* glycerol/8% *w/v* SDS/5% *w/v* beta-mercaptoethanol/0.04% *w/v* bromophenol blue (pH 7.5)) with a final gel load of 20 µL. Identity of the reaction product band (HECT~Ub conjugate) was confirmed by MS/MS (Genome Research Center, Hong Kong, China). Gels were stained with Coomassie or Silver depending on need for quantitation and analyzed with ImageJ (NIH, Bethesda, MD, USA) and Prism® (version 7, GraphPad, San Diego, CA, USA) softwares.

4.4. Electrophoretic Mobility Shift Assay (EMSA)

Briefly, 15 µL samples containing 11.3 µM → 293.9 nM HECT domain, 55.6 nM aptamer, and 4 µL of EMSA sample buffer (50 mM Tris/10% (*v/v*) Glycerol/0.02% (*w/v*) bromophenol blue (pH 6.8)) were created. The mixture was set for 60 min at 4 °C while gel was pre-run in TAE buffer at 100 V for equilibration. Samples were then loaded and run on 12% PAGE Gels which were prepared using vendor Midi Protean Cell guidelines (BioRad, Hercules, CA, USA) at 100 V for approximately 3 h until tracking dye reached bottom of the gel. Gels were stained with 3.5 µL of SYBR Gold Nucleic Acid Gel Stain (Molecular Probes, Waltham, MA, USA) per 50 mL of TAE buffer and imaged in the UV Transilluminator Imaging System (UVP, Hercules, CA, USA) and analyzed with ImageJ and Prism®. 6.5 pmol of aptamer was mixed with a series of concentrations of HECT, set for 1 h at 4 °C, run on 10% PAGE for 1.5 h at 100 V in 4 °C with minimal glycerol for gel loading, and stained with SYBR gold™.

4.5. Cell Culturing

SaOS-2 Cells (ATCC® number HTB-85®), a human osteosarcoma line, were obtained in house labeled at passage 5. Cells were thawed, seeded, passaged according to the general guidelines of the Cell Culture Basics guide (Thermo Fisher Scientific, Hercules, CA, USA) in a Class II bio safety cabinet with McCoy's 5a Media (Sigma-Aldrich, St. Louis, MO, USA) supplemented with Penicillin/Streptomycin and 15% FBS, and centrifugation was performed on bench top centrifuge. Cells were stored at a controlled environment of 37 °C with 5% CO$_2$ and viewed with an Eclipse

TS100 (Nikon, Tokyo, Japan) light microscope at 4× and 10× magnifications daily. Samples to be prepared for fixed cell imaging, western blot, apoptosis, bone mineralization, and alkaline phosphatase activity were initiated by growing cells to ~80% confluency on either 6, 24, or 96 well cell culture plates (Corning Inc., Corning, NY, USA). Cells were transfected with 100 ng of aptamer and control for 4 h per Lipofectamine 2000 (Invitrogen) guidelines using McCoy's 5a Media (Sigma-Aldrich) without Penicillin/Streptomycin or FBS or treated with aptamer and control without transfection reagent for the same amount of time. Cells were then replaced with normal supplemented media after transfection, which took place once (T0) for fixed cell imaging, western blot, and apoptosis assays and three times (T0 then every 3 days) for mineralization and alkaline phosphatase assays. The cells remained alive for the course of the experiments.

4.6. Fixed Cell Imaging

Cover slip containing Costar® 6 well flat bottom plates (Corning Inc., Corning, NY, USA) were prepared with sterile technique by placing cover slips into the wells with forceps in the BSC hood, rinsing with cold methanol, and rinsing 3× with ice cold sterile PBS. Cells were seeded 1:4 and grown to ~90% confluency and then treated with aptamer and controls, with and without Lipofectamine™ 2000. At each time point, wells were washed with ice cold PBS, gently fixed with ice cold methanol at −20 °C for one hour and washed with ice cold PBS. After damping dry, Acrytol® mounting media (Leica, Wetzlar, Germany) was used to mount slides, which were then dried overnight in the dark. Images were taken on an BX51 Fluorescence Microscope (Olympus, Tokyo, Japan) using brightfield and the appropriate filters for Cy3 (red) and Hoechst (blue) at 40×. MetaMorph® (Molecular Devices, San Jose, CA, USA) batch processing software was used to analyze at least 85 cells per sample.

4.7. Runx2 Western Blot

Cells were first washed with ice cold PBS, then 200 µL of ice cold RIPA buffer (150 mM sodium chloride/1.0% Triton X-100/0.5% sodium deoxycholate/0.1% SDS (sodium dodecyl sulfate)/50 mM Tris (pH 8.0)) supplemented with complete EDTA-free EASYpack protease inhibitor cocktail was added and cells, scraped and recovered by pipette. Lysates were agitated at 4 °C for 30 min by light shaking and centrifuged for 20 min at 12,000 rpm on a X-15R bench top centrifuge (Beckman Coulter) at 4 °C before recovering supernatant. In this case, Rabbit derived Anti-Runx2 (Novus Biologicals, Littleton, CO, USA) and Goat derived Anti-Rabbit IgG (whole molecule) Peroxidase conjugate (Novus Biologicals®) were used at 1/500 and 1/80,000, respectively, while generally following the Western Blotting Beginners Guide (Abcam, Cambridge, UK). 1 µL of B-Actin (Cell Signaling Technology, Danvers, MA, USA) stock was included with each primary antibody incubation as a loading control. Blots were imaged with UV Transilluminator Imaging System (UVP, Hercules, CA, USA).

4.8. Apoptosis Assay

Apoptosis samples were prepared as described in the cell culture section and were performed according to HT TiterTACS™ Assay kit (Trevigen, Gaithersburg, MD, USA) guidelines, read at 450 nm on a 200 SpectraMax 340PC 38 (Molecular Devices) and analyzed with Soft Max Pro (Molecular Devices) and Prism® (GraphPad) Software.

4.9. Bone Mineralization Assay

Assays to detect the calcium of bone deposition generally followed the Osteoblast Differentiation and Mineralization Guide (PromoCell, Heidelberg, Germany). First, cells washed with ice cold PBS, fixed with ice cold methanol overnight, stained with 20 g/mL Alizarin Red S for 45 min at room temperature, washed 5 times with diH₂O, and PBS added before digital photographs were taken.

5. Conclusions

We successfully generated a DNA aptamer (C3A) which binds to the C-lobe of WWP1 with a binding affinity of 1.9 μM. This aptamer was demonstrated to inhibit the ubiquitination activity of WWP1 in a non-competitive manner. The aptamer could internalize into SaOS2 cells even in absence of transfection agent. The aptamer was shown to stimulate extracellular matrix deposition relative to controls. Future work can improve binding affinity by microarray maturation, employ bioinformatics methods to modify the aptamer for greater functionality [48] or generate a library that could be used alongside a microfluidic selection method [49]. By such approaches, novel therapeutic aptamers can be further developed for a variety of human disease.

Acknowledgments: This work was supported by the Hong Kong University Grants Council under General Research Fund Grant HKU777109M.

Author Contributions: Wesley Tucker conceived, designed, prepared, and performed the experiments, analyzed the data, and helped prepare the manuscript. Julian Tanner conceived the study, guided the research, and helped prepare the manuscript. Andrew Kinghorn, Lewis Fraser, and Yee-Wai Cheung helped analyze data and to prepare the manuscript.

References

1. Dallas, S.L.; Bonewald, L.F. Dynamics of the transition from osteoblast to osteocyte. *Ann. N. Y. Acad. Sci.* **2010**, *1192*, 437–443. [CrossRef] [PubMed]

2. Glimcher, L.H.; Jones, D.C.; Wein, M.N. Control of postnatal bone mass by the zinc finger adapter protein Schnurri-3. *Ann. N. Y. Acad. Sci.* **2007**, *1116*, 174–181. [CrossRef] [PubMed]

3. Komori, T.; Yagi, H.; Nomura, S.; Yamaguchi, A.; Sasaki, K.; Deguchi, K.; Shimizu, Y.; Bronson, R.T.; Gao, Y.H.; Inada, M.; et al. Targeted disruption of Cbfa1 results in a complete lack of bone formation owing to maturational arrest of osteoblasts. *Cell* **1997**, *89*, 755–764. [CrossRef]

4. Wein, M.N.; Jones, D.C.; Shim, J.H.; Aliprantis, A.O.; Sulyanto, R.; Lazarevic, V.; Poliachik, S.L.; Gross, T.S.; Glimcher, L.H. Control of bone resorption in mice by Schnurri-3. *Proc. Natl. Acad. Sci. USA* **2012**, *109*, 8173–8178. [CrossRef] [PubMed]

5. Jones, D.C.; Wein, M.N.; Oukka, M.; Hofstaetter, J.G.; Glimcher, M.J.; Glimcher, L.H. Regulation of adult bone mass by the zinc finger adapter protein Schnurri-3. *Science* **2006**, *312*, 1223–1227. [CrossRef] [PubMed]

6. Huang, K.; Johnson, K.D.; Petcherski, A.G.; Vandergon, T.; Mosser, E.A.; Copeland, N.G.; Jenkins, N.A.; Kimble, J.; Bresnick, E.H. A HECT domain ubiquitin ligase closely related to the mammalian protein WWP1 is essential for Caenorhabditis elegans embryogenesis. *Gene* **2000**, *252*, 137–145. [CrossRef]

7. Pirozzi, G.; McConnell, S.J.; Uveges, A.J.; Carter, J.M.; Sparks, A.B.; Kay, B.K.; Fowlkes, D.M. Identification of novel human WW domain-containing proteins by cloning of ligand targets. *J. Biol. Chem.* **1997**, *272*, 14611–14616. [CrossRef] [PubMed]

8. Verdecia, M.A.; Joazeiro, C.A.; Wells, N.J.; Ferrer, J.L.; Bowman, M.E.; Hunter, T.; Noel, J.P. Conformational flexibility underlies ubiquitin ligation mediated by the WWP1 HECT domain E3 ligase. *Mol. Cell* **2003**, *11*, 249–259. [CrossRef]

9. Roth, A.; Breaker, R.R. The structural and functional diversity of metabolite-binding riboswitches. *Annu. Rev. Biochem.* **2009**, *78*, 305–334. [CrossRef] [PubMed]

10. Tuerk, C.; Gold, L. Systematic evolution of ligands by exponential enrichment: RNA ligands to bacteriophage T4 DNA polymerase. *Science* **1990**, *249*, 505–510. [CrossRef] [PubMed]

11. Ellington, A.D.; Szostak, J.W. In vitro selection of RNA molecules that bind specific ligands. *Nature* **1990**, *346*, 818–822. [CrossRef] [PubMed]

12. Kinghorn, A.B.; Dirkzwager, R.M.; Liang, S.; Cheung, Y.W.; Fraser, L.A.; Shiu, S.C.; Tang, M.S.; Tanner, J.A. Aptamer affinity maturation by resampling and microarray selection. *Anal. Chem.* **2016**, *88*, 6981–6985. [CrossRef] [PubMed]

13. Sampson, T. Aptamers and SELEX: The technology. *World Pat. Inf.* **2003**, *25*, 123–129. [CrossRef]

14. Ulrich, H. DNA and RNA aptamers as modulators of protein function. *Med. Chem.* **2005**, *1*, 199–208. [CrossRef] [PubMed]

15. Cheung, Y.W.; Kwok, J.; Law, A.W.; Watt, R.M.; Kotaka, M.; Tanner, J.A. Structural basis for discriminatory

recognition of plasmodium lactate dehydrogenase by a DNA aptamer. *Proc. Natl. Acad. Sci. USA* **2013**, *110*, 15967–15972. [CrossRef] [PubMed]

16. Zou, J.; Huang, X.; Wu, L.; Chen, G.; Dong, J.; Cui, X.; Tang, Z. Selection of intracellularly functional RNA mimics of green fluorescent protein using fluorescence-activated cell sorting. *J. Mol. Evol.* **2015**, *81*, 172–178. [CrossRef] [PubMed]

17. Daniels, D.A.; Chen, H.; Hicke, B.J.; Swiderek, K.M.; Gold, L. A tenascin-C aptamer identified by tumor cell SELEX: Systematic evolution of ligands by exponential enrichment. *Proc. Natl. Acad. Sci. USA* **2003**, *100*, 15416–15421. [CrossRef] [PubMed]

18. Stein, C.A.; Castanotto, D. FDA-approved oligonucleotide therapies in 2017. *Mol. Ther.* **2017**, *25*, 1069–1075. [CrossRef] [PubMed]

19. Shum, K.T.; Chan, C.; Leung, C.M.; Tanner, J.A. Identification of a DNA aptamer that inhibits sclerostin's antagonistic effect on Wnt signalling. *Biochem. J.* **2011**, *434*, 493–501. [CrossRef] [PubMed]

20. Dirkzwager, R.M.; Liang, S.; Tanner, J.A. Development of aptamer-based point-of-care diagnostic device for malaria using 3D printing rapid prototyping. *ACS Sens.* **2016**, *1*, 420–426. [CrossRef]

21. Cheung, Y.W.; Dirkzwager, R.M.; Wong, W.C.; Cardoso, J.; Costa, J.D.; Tanner, J.A. Aptamer-mediated plasmodium-specific diagnosis of malaria. *Biochimie* **2018**, *145*, 131–136. [CrossRef] [PubMed]

22. Fraser, L.A.; Kinghorn, A.B.; Dirkzwager, R.M.; Liang, S.; Cheung, Y.W.; Lim, B.; Shiu, S.C.; Tang, M.S.L.; Andrew, D.; Manitta, J.; et al. A portable microfluidic Aptamer-Tethered Enzyme Capture (APTEC) biosensor for malaria diagnosis. *Biosens. Bioelectron.* **2018**, *100*, 591–596. [CrossRef] [PubMed]

23. Ouellet, J. RNA fluorescence with light-up aptamers. *Front. Chem.* **2016**, *4*, 29. [CrossRef] [PubMed]

24. Praetorius, F.; Kick, B.; Behler, K.L.; Honemann, M.N.; Weuster-Botz, D.; Dietz, H. Biotechnological mass production of DNA origami. *Nature* **2017**, *552*, 84–87. [CrossRef] [PubMed]

25. Maurisse, R.; De Semir, D.; Emamekhoo, H.; Bedayat, B.; Abdolmohammadi, A.; Parsi, H.; Gruenert, D.C. Comparative transfection of DNA into primary and transformed mammalian cells from different lineages. *BMC Biotechnol.* **2010**, *10*, 9. [CrossRef] [PubMed]

26. Yu, B.; Zhao, X.; Lee, L.J.; Lee, R.J. Targeted delivery systems for oligonucleotide therapeutics. *AAPS J.* **2009**, *11*, 195–203. [CrossRef] [PubMed]

27. Zamecnik, P.C.; Stephenson, M.L. Inhibition of Rous sarcoma virus replication and cell transformation by a specific oligodeoxynucleotide. *Proc. Natl. Acad. Sci. USA* **1978**, *75*, 280–284. [CrossRef] [PubMed]

28. Akhtar, S.; Juliano, R.L. Cellular uptake and intracellular fate of antisense oligonucleotides. *Trends Cell Biol.* **1992**, *2*, 139–144. [CrossRef]

29. Bennett, R.M. As nature intended? The uptake of DNA and oligonucleotides by eukaryotic cells. *Antisense Res. Dev.* **1993**, *3*, 235–241. [CrossRef] [PubMed]

30. Loke, S.L.; Stein, C.A.; Zhang, X.H.; Mori, K.; Nakanishi, M.; Subasinghe, C.; Cohen, J.S.; Neckers, L.M. Characterization of oligonucleotide transport into living cells. *Proc. Natl. Acad. Sci. USA* **1989**, *86*, 3474–3478. [CrossRef] [PubMed]

31. Yakubov, L.A.; Deeva, E.A.; Zarytova, V.F.; Ivanova, E.M.; Ryte, A.S.; Yurchenko, L.V.; Vlassov, V.V. Mechanism of oligonucleotide uptake by cells: Involvement of specific receptors? *Proc. Natl. Acad. Sci. USA* **1989**, *86*, 6454–6458. [CrossRef] [PubMed]

32. Zhou, J.; Rossi, J.J. Therapeutic potential of aptamer-siRNA conjugates for treatment of HIV-1. *BioDrugs* **2012**, *26*, 393–400. [CrossRef] [PubMed]

33. Legendre, J.Y.; Szoka, F.C., Jr. Delivery of plasmid DNA into mammalian cell lines using pH-sensitive liposomes: Comparison with cationic liposomes. *Pharm. Res.* **1992**, *9*, 1235–1242. [CrossRef] [PubMed]

34. De Fougerolles, A.R. Delivery vehicles for small interfering RNA in vivo. *Hum. Gene Ther.* **2008**, *19*, 125–132. [CrossRef] [PubMed]

35. Mayer, G. The chemical biology of aptamers. *Angew. Chem.* **2009**, *48*, 2672–2689. [CrossRef] [PubMed]

36. Auslander, D.; Wieland, M.; Auslander, S.; Tigges, M.; Fussenegger, M. Rational design of a small molecule-responsive intramer controlling transgene expression in mammalian cells. *Nucleic Acids Res.* **2011**, *39*, e155. [CrossRef] [PubMed]

37. Ruckman, J.; Green, L.S.; Beeson, J.; Waugh, S.; Gillette, W.L.; Henninger, D.D.; Claesson-Welsh, L.; Janjic, N. 2′-Fluoropyrimidine RNA-based aptamers to the 165-amino acid form of vascular endothelial growth factor (VEGF165). Inhibition of receptor binding and VEGF-induced vascular permeability through interactions

requiring the exon 7-encoded domain. *J. Biol. Chem.* **1998**, *273*, 20556–20567. [CrossRef] [PubMed]

38. Bock, L.C.; Griffin, L.C.; Latham, J.A.; Vermaas, E.H.; Toole, J.J. Selection of single-stranded DNA molecules that bind and inhibit human thrombin. *Nature* **1992**, *355*, 564–566. [CrossRef] [PubMed]

39. Matta-Camacho, E.; Kozlov, G.; Menade, M.; Gehring, K. Structure of the HECT C-lobe of the UBR5 E3 ubiquitin ligase. *Acta Crystallogr. Sect. F Struct. Biol. Cryst. Commun.* **2012**, *68*, 1158–1163. [CrossRef] [PubMed]

40. Sun, Y. Targeting E3 ubiquitin ligases for cancer therapy. *Cancer Biol. Ther.* **2003**, *2*, 623–629. [CrossRef] [PubMed]

41. Beaudenon, S.; Dastur, A.; Huibregtse, J.M. Expression and assay of HECT domain ligases. *Methods Enzymol.* **2005**, *398*, 112–125. [PubMed]

42. Hu, J.; Wu, J.; Li, C.; Zhu, L.; Zhang, W.Y.; Kong, G.; Lu, Z.; Yang, C.J. A G-quadruplex aptamer inhibits the phosphatase activity of oncogenic protein Shp2 in vitro. *Chembiochem* **2011**, *12*, 424–430. [CrossRef] [PubMed]

43. Li, X.; Ominsky, M.S.; Warmington, K.S.; Morony, S.; Gong, J.; Cao, J.; Gao, Y.; Shalhoub, V.; Tipton, B.; Haldankar, R.; et al. Sclerostin antibody treatment increases bone formation, bone mass, and bone strength in a rat model of postmenopausal osteoporosis. *J. Bone Miner. Res.* **2009**, *24*, 578–588. [CrossRef] [PubMed]

44. Patil, S.D.; Rhodes, D.G.; Burgess, D.J. DNA-based therapeutics and DNA delivery systems: A comprehensive review. *AAPS J.* **2005**, *7*, E61–E77. [CrossRef] [PubMed]

45. Reyes-Reyes, E.; Šalipur, F.R.; Shams, M.; Forsthoefel, M.K.; Bates, P.J. Mechanistic studies of anticancer aptamer AS1411 reveal a novel role for nucleolin in regulating Rac1 activation. *Mol. Oncol.* **2015**, *9*, 1392–1405. [CrossRef] [PubMed]

46. Wente, S.R.; Rout, M.P. The nuclear pore complex and nuclear transport. *Cold Spring Harb. Perspect. Biol.* **2010**, *2*, a000562. [CrossRef] [PubMed]

47. Zhang, L.; Wu, Z.; Ma, Z.; Liu, H.; Wu, Y.; Zhang, Q. WWP1 as a potential tumor oncogene regulates PTEN-Akt signaling pathway in human gastric carcinoma. *Tumour. Biol.* **2015**, *36*, 787–798. [CrossRef] [PubMed]

48. Kinghorn, A.B.; Fraser, L.A.; Lang, S.; Shiu, S.C.C.; Tanner, J.A. Aptamer bioinformatics. *Int. J. Mol. Sci.* **2017**, *18*, 2516. [CrossRef]

49. Fraser, L.A.; Kinghorn, A.B.; Tang, M.S.; Cheung, Y.-W.; Lim, B.; Liang, S.; Dirkzwager, R.M.; Tanner, J.A. Oligonucleotide functionalised microbeads: Indispensable tools for high-throughput aptamer selection. *Molecules* **2015**, *20*, 21298–21312. [CrossRef] [PubMed]

7

Key Aspects of Nucleic Acid Library Design for in Vitro Selection

Maria A. Vorobyeva [1,*], **Anna S. Davydova** [1], **Pavel E. Vorobjev** [1,2], **Dmitrii V. Pyshnyi** [1,2] **and Alya G. Venyaminova** [1]

[1]   Institute of Chemical Biology and Fundamental Medicine, Siberian Division of Russian Academy of Sciences, Lavrentiev Ave., 8, 630090 Novosibirsk, Russia; anna.davydova@niboch.nsc.ru (A.S.D.); vorobyev@niboch.nsc.ru (P.E.V.); pyshnyi@niboch.nsc.ru (D.V.P.); ven@niboch.nsc.ru (A.G.V.)

[2]   Department of Natural Sciences, Novosibirsk State University, Pirogova St., 2, 630090 Novosibirsk, Russia

*   Correspondence: maria.vorobjeva@gmail.com

Abstract: Nucleic acid aptamers capable of selectively recognizing their target molecules have nowadays been established as powerful and tunable tools for biospecific applications, be it therapeutics, drug delivery systems or biosensors. It is now generally acknowledged that in vitro selection enables one to generate aptamers to almost any target of interest. However, the success of selection and the affinity of the resulting aptamers depend to a large extent on the nature and design of an initial random nucleic acid library. In this review, we summarize and discuss the most important features of the design of nucleic acid libraries for in vitro selection such as the nature of the library (DNA, RNA or modified nucleotides), the length of a randomized region and the presence of fixed sequences. We also compare and contrast different randomization strategies and consider computer methods of library design and some other aspects.

Keywords: SELEX; aptamers; design of nucleic acid libraries

---

1. Introduction

Nucleic Acid (NA) aptamers [1] are a special class of nucleic acid molecules capable of tight and specific binding with certain molecular or supramolecular targets, thanks to characteristic spatial structures. The range of their targets is enormously wide. Nowadays, NA aptamers have been generated to metal ions (e.g., mercury [2] and lead [3]), small organic molecules (e.g., theophylline [4] and cocaine [5]), larger molecules (e.g., fluorophores [6,7] and porphyrins [8]), peptides and proteins (e.g., hormones [9,10], enzymes [11,12], antibodies [13] and cell surface proteins [14]) and liposomes [15]. These are just a few examples selected from a large diversity of NA aptamers. Nucleic acid aptamers were selected from the NA libraries by means of the method of Selective Evolution of Ligands by Exponential enrichment (SELEX) [16,17]. SELEX technology incorporates a variety of related methods for selecting functional nucleic acids with the desired properties, including also catalytic nucleic acids and riboswitches [18,19]. A selection process could also be aimed at finding genomic sequences or expressible NAs with an affinity to a specific molecule, e.g., to reveal the sequence specificity of NA-enzyme interactions [20,21]. In this review, we focus particularly on NA aptamers.

The main characteristics of NA aptamers are defined by their chemical nature. As nucleic acids, these molecules possess a significant negative charge and are susceptible to nuclease hydrolysis, and surrounding conditions (pH, ionic strength and the presence of certain ions) can influence the stability of their secondary structure. Binding with a target molecule, the aptamer can change the properties of the target, e.g., inhibit the enzymatic activity [11] or alter the characteristics of fluorescent dyes [22].

The molecular recognition function specifies the areas of possible applications of NA aptamers. An ability to inhibit pathogenic proteins affords an opportunity to employ aptamers as therapeutics [22–25]. Aptamers specific to certain cell-surface receptors, which are able to induce an internalization process, could be used as vehicles for cell-targeted drug delivery [26]. Aptamers are anticipated to compete with therapeutic monoclonal antibodies since the chemical synthesis of nucleic acids is far simpler and more cost-effective than obtaining humanized antibodies (although the SELEX process itself could become rather laborious). A set of chemical modifications is available to improve the nuclease resistance and pharmacokinetics of NA aptamers [27]. It is also worth noting that aptamers have the benefit of having a low immunogenicity typical for most oligonucleotides.

Bioanalytics represents probably the broadest application area of nucleic acid aptamers. In principle, every aptamer can be considered as a recognizing module for a certain molecule. It is no wonder that such a vast diversity of aptamer-based biosensors (also known as "aptasensors") has been created (see [28–31] for a review).

The main success criteria for any given aptamer include binding affinity, nuclease resistance and convenience of chemical synthesis. All these properties are largely defined by the particular nucleic acid library employed for SELEX. Therefore, the choice of library design has a great impact on the overall efficiency of the selection. When generating the initial library, a researcher should keep in mind the properties of the target (such as in capture SELEX for small molecules [32]) and the end use of an aptamer (whether nuclease resistance is necessary or not) [27,33]. The importance of covering a maximal sequence space (a multi-dimensional space of different sequences of a certain length), the necessity of introducing a particular sequence or structural element should also be taken into account. In some cases, additional effort is needed to obtain a library that enables the generation of aptamers to SELEX-inaccessible (somewhat similar to non-immunogenic) targets [34,35]. Thus, at the beginning of the study, one has to fill out a kind of checklist of the key issues to choose the most suitable library design (Figure 1). The main aspects regarding the design of the initial libraries for aptamer selection and the basic trends in library design will be reviewed and discussed below.

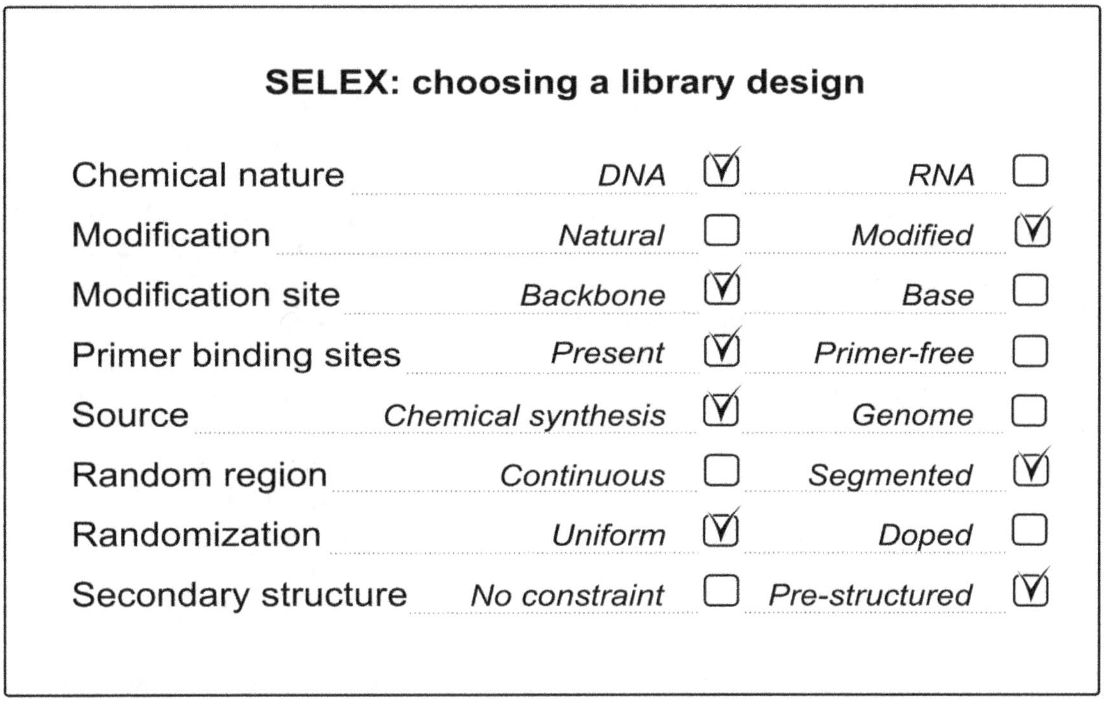

Figure 1. An example checklist for an NA library design with the key issues to be considered.

2. General Issues of Initial Library Design

2.1. DNA or RNA?

All SELEX studies can be generally divided into two groups. In the first group, the choice of the type of nucleic acid library is predetermined by the task of the study such as for the in vitro selection of ribozymes, riboswitches, DNAzymes or genomic SELEX studies. Experiments on the isolation of RNA aptamers or artificial riboswitches intended to be expressed in cells also relate to this group. The second group includes SELEX studies on aptamers that will be further employed for research, therapeutic or bioanalytical purposes. In this case, a researcher can deliberately choose the type of sugar-phosphate backbone.

The first decade in the development of SELEX technology was marked by a dominance of RNA aptamers [36,37]. This was possibly due to the common opinion that only RNA molecules could form functional motifs [38]. At the very beginning of the SELEX era, Ellington and Szostak demonstrated the ability of single-stranded DNA to fold into functional spatial structures [39]. Nevertheless, until 2007, about 70% of all experiments in the field related to RNA aptamers [36]. The distribution became quite the opposite in 2008–2013: DNA aptamers now occupy 70% of SELEX studies, and no significant differences were found in the distributions of the lowest K_D values [36]. DNA and RNA aptamers generated for a number of small-molecule targets have demonstrated similar affinities [40].

Thus, neither the RNA nor DNA libraries provide any systemic preferences for the isolation of affine aptamers [36]. Such preferences can clearly be attributed to some modified nucleic acids, e.g., Slow Off-rate Modified Aptamers (SOMAmers), which will be discussed below. The particular conditions of an aptamer's application also influence the choice of a sugar-phosphate backbone. An enhanced nuclease resistance could require the use of backbone chemical modifications, which will be briefly described in the next section. According to [37], the number of aptamers isolated from non-natural nucleic acid libraries increased significantly in 2011–2015.

2.2. Backbone Modifications of NA Libraries

A number of popular applications of in vitro selected aptamers—such as the design of new therapeutics or engineering of drug delivery systems and biosensors—assumes their use in biological media containing different nucleases. Both DNA and RNA aptamers are susceptible to nuclease degradation. To protect them, a large set of chemical modifications of the sugar-phosphate backbone has been developed. However, any post-selective chemical modification of individual aptamers can affect binding affinity, so the modification pattern should be optimized in every particular case, which is rather laborious and time-consuming. Therefore, it seems reasonable to introduce modified nucleotides into the initial library to select molecules that are both affine and nuclease-resistant. One of the most important criteria for such pre-SELEX modifications is the compatibility of modified nucleotides with all enzyme reactions involved in a selection protocol. A number of chemical modifications meeting this requirement are now available (see the reviews in [27,35,41–43]), including ribose ($2'$-NH$_2$, $2'$-F, $2'$-O-Me, $4'$-S-, LNA (locked nucleic acids), TNA (threose nucleic acid), FANA (fluoroarabino nucleic acid) and HNA (1,5-anhydro hexitol nucleic acid)) and internucleoside phosphate (boranophosphate or phosphorothioate) modifications (Figure 2). Among them, $2'$-modifications are clearly at the top of the list. The first SELEX-compatible $2'$-modification was the replacement of ribose $2'$-OH by an amino group [44]. However, this type of modification was then quite rarely used, owing to problems with the chemical synthesis of $2'$-NH$_2$-modified aptamers and the negative impact of the $2'$-amino group on the ribose conformation [35]. In contrast, the $2'$-F modification of pyrimidine nucleotides, which was proposed almost at the same time, gained outstanding popularity since it provided sufficient nuclease resistance, did not dramatically affect the RNA spatial structure and could be introduced by even using a non-modified T7 RNA polymerase under optimized conditions [45]. To apply any other

SELEX-compatible modifications as mentioned above, one should use mutant versions of polymerase enzymes (see [41,46,47] for reviews).

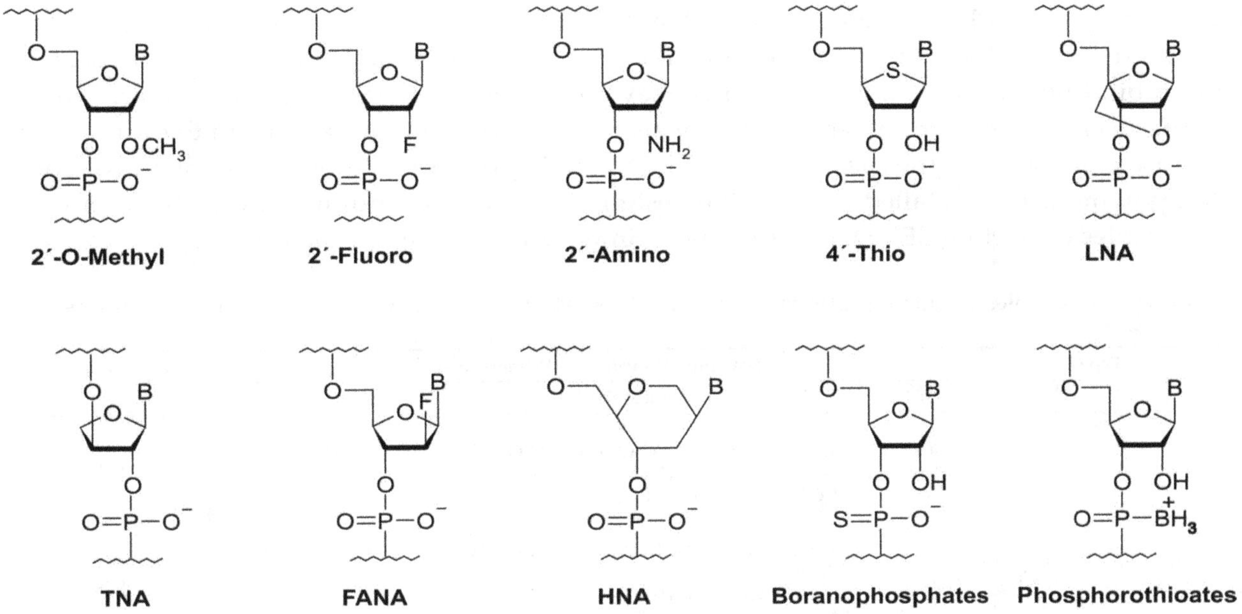

Figure 2. Sugar-phosphate backbone modifications compatible with a SELEX procedure. LNA: locked nucleic acids; TNA: threose nucleic acid; FANA: fluoroarabino nucleic acid; and HNA: 1,5-anhydro hexitol nucleic acid.

2.3. The Length of the Random Region

When choosing the length of the random region, a researcher should consider both the sequence space and structural diversity. In the general case, the maximum possible sequence space for a random sequence of N nucleotides comprises a total of 4^N possible sequences. Therefore, for those quantities of libraries that can be routinely obtained and handled, a maximal theoretical diversity can only be reached for random regions shorter than 28 nt (7×10^{16} sequences $\approx$ 0.1 μmol corresponds to a fully-represented library) [48]. Longer libraries are unable to extensively cover the sequence space. On the other hand, longer sequences can fold into more complex structures that may be needed to form a target-binding domain. Thus, a balance should be kept between the diversity of the sequences and the desired complexity of the spatial structures formed by these sequences. For in vitro selection of aptamers, 30–50-nt randomized regions are the most abundant [49].

With regard to the minimal sequence diversity to provide a sufficient selection, a value of 10^{11} is often used (see [50]), based on SELEX publications from the early 1990s [1,51,52]. It should be noted that all these works deal with RNA SELEX to small-molecule targets, so the question arises as to whether such estimation is applicable for all possible types of targets and libraries.

Aside from the theoretical considerations, from a practical point of view, the length of the library is governed by: (1) the convenience and cost of its chemical synthesis; (2) the possibility of PCR (polymerase chain reaction) artifact formation in the course of an amplification of long libraries; and (3) future applications of the selected aptamers. When an aptamer is further used for practical applications, a shorter length of the oligonucleotide chain is always better. To minimize the length of an individual aptamer, a series of its truncated variants has to be synthesized and tested to choose the minimal one retaining target binding affinity. To avoid this resource-consuming procedure, Thiel et al. [50] employed a short 51-nt library with a randomized region as short as 20 nt and demonstrated that this length was sufficient to generate high-affinity 2′-F-RNA aptamers to protein targets.

2.4. Primer-Binding Sites and Primer-Free SELEX

Traditional SELEX protocols, which are still prevalent today, imply the use of two fixed sequences flanking the randomized region for primer annealing during amplification (Figure 3a). As a rule, primer-binding sites (PBS) are about 20 nt in length. According to the statistical analysis performed in [49], their length does not correlate with the length of a randomized region. The sequences of primer-binding sites are designed to meet several general requirements, particularly to avoid PCR artifacts emerging from self-association or secondary structure formation and to ensure efficient polymerase extension. In the case of RNA SELEX, the 5′-primer contains a promoter sequence for T7 RNA polymerase. A detailed guide to the design of the primer-binding sites can be found in [53]. Some examples of starting SELEX libraries and primers are given in the Table 1.

Table 1. Examples of starting libraries for SELEX. SOMAmers, Slow Off-rate Modified Aptamers.

Type	Starting Libraries and Primers (5′->3′)	Ref.
	Classical SELEX	
DNA	Library: GGGAGACAAGAATAAACGCTCAA-N40-TTCGACAGGAGGCTCACAACAGGC 5′-primer: GGGAGACAAGAATAAACGCTCAA 3′-primer: GCCTGTTGTGAGCCTCCTGTCGAA	[45]
RNA, 2′-F-pyrimidine (Py) modified RNA, 2′-NH₂ Py modified RNA	Library: GGGAGACAAGAAUAAACGCUCAA-N40-UUCGACAGGAGGCUCACAACAGGC ssDNA template: GCCTGTTGTGAGCCTCCTTGTCGAA-N40-TTGAGCGTTTATTCTTGTCTCCC 5′-primer: <u>TAATACGACTCACTATAG</u>GGAGACAAGAATAAACGCTCAA [1] 3′-primer: GCCTGTTGTGAGCCTCCTGTCGAA	[45]
2′-O-Me RNA	Library: GGGAGAGAGGAACGUUCUCG-N30-GGAUCGUUACGACUAGCAUCGAUG ssDNA template: CATCGATGCTAGTCGTAACGATCC-N30-CGAGAACGTTCTCTCTCCCTATAGTGA GTCGTATTA 5′-primer: <u>TAATACGACTCACTATAG</u>GGAGAGGAGAGAAACGTTCTCG 3′-primer: CATCGATGCTAGTCGTAACGATCC	[54]
dRmY (2′-deoxy purine ribonucleotides, 2′-O-CH₃ Py ribonucleotides)	Library: GGGAGAGGAGAAGGUUCUAC-N30-GCGUGUCGAUCGAUCGAUCGAUG ssDNA template: CATCGATCGATCGATCGACAGCG-N30-GTAGAACGTTCTCTCCTCTCCCTATAGTGA GTCGTATTA 5′-primer: <u>TAATACGACTCACTATAG</u>GGAGAGGAGAGAACGTTCTAC 3′-primer: CATCGATCGATCGATCGACAGC	[55]
SOMAmers	Library: GATGTGAGTGTGTGACGAG-N40-CACAGAGAAGAAACAAGACC, random region containing 5-(N-benzylcarboxamide)-2′-deoxyuridine (Bn-dU) or 5-[N-(1-naphthylmethyl)carboxamide]-2′-deoxyuridine (Nap-dU) in place of dT 5′-primer: GATGTGAGTGTGTGACGAG 3′-primer: GGTCTTGTTTCTTCTCTGTG	[56]
	Capture SELEX	
DNA	Library: ATACCAGCTTATTCAATT-N10-TGAGGCTCGATC-N40-AGATAGTAAGTGCAATCT Capture oligonucleotide: Bio-GTC-(CH₂CH₂O)₆-GATCGAGCCTCA or GATCGAGCCTCA-(CH₂CH₂O)₆-GTC-Bio 5′-primer: ATACCAGCTTATTCAATT 3′-primer: AGATTGCACTTACTATCT	[57]
	Pre-structured libraries	
RNA	Library: GGAGGCGCCAACTGAATGAA-N26-CUGCUUCGGCAG-N26-UCCGUAACUAGUUCG CGUCAC ssDNA template: GTGACGCGACTAGTTACGGA-N26-CTGCCGAAGCAG-N26-TTCATTCAGTTGGCGCCT CCTATAGTGAGTCGTATTACAT 5′-primer: ATG<u>TAATACGACTCACTATAG</u>GAGGCGCCAACTGAATGAA 3′-primer: GTGACGCGACTAGTTACGGA	[58]

[1] Hereinafter in the table, the T7 promoter sequence is underlined.

Ideally, aptamer sequences generated by in vitro selection should bind their targets by means of spatial structures formed only by nucleotides from a random region. For most aptamers, this is indeed the case: the analysis of >2000 sequences from the Aptamer Database revealed that for a majority of aptamers, their secondary structure was independent of primer-binding sites [49]. However, there was a number of outliers (examples in [59–61]). Taking this into account, primer-binding sites cannot be simply cut off to minimize the length of the sequence during aptamer truncation, and additional minimization studies are needed. Moreover, during the SELEX, primer-binding sites could interact with sequences in the random region, hampering their target binding and/or amplification (for more details, see [62] and the references therein).

These problems stimulated a search for SELEX approaches that minimize the influence of primer-binding sites on the sequence and structure of selected aptamers (schematically depicted in Figure 3). For instance, Shtatland et al. [61] showed that fixed regions of a genomic RNA library (with *Escherichia coli* (*E. coli*) genome fragments as a random region) interacted with a random region, which resulted in a large number of experimental artifacts. After the traditional SELEX from this library on MS2 bacteriophage capsid protein, about 90% of the generated sequences were represented by artifacts (not found in the *E. coli* genome). The authors proposed two alternative selection strategies to neutralize the negative impact of constant regions: primer-annealing genomic SELEX and primer-switching genomic SELEX. In the primer-annealing genomic SELEX protocol, prior to selection, the RNA library was hybridized with two oligonucleotides complementary to the primer-binding sites (Figure 3b). This approach provided 60% of the artifacts in the obtained clones. During the course of primer-switching genomic SELEX, several rounds of classical SELEX were performed, followed by a replacement of primer-binding sites and subsequent classical or primer-annealing SELEX. To replace the flanking regions, the purified library was digested by the FokI restrictase (restriction sites were introduced 9–13 nt from the random region); the sticky ends were extended to blunt ends by a Klenow reaction; then, new primer-binding sites were ligated to the library. This approach enabled the authors to decrease the fraction of unwanted products down to 10%.

Ouellet et al. successfully adapted the primer-annealing SELEX protocol for completely random libraries [62,63]. Blocking oligonucleotides annealed with primer-binding sites eliminated their negative impact in several selections on therapeutically-important targets.

The approach proposed by Shtatland et al. was further developed for the genomic SELEX on the bacteriophage Ff gene 5 protein [64]. The authors hypothesized that constant nucleotides remaining in the library after an enzymatic digestion could also influence the course of selection. In their version of primer-free genomic SELEX, the Fok1 restriction site at the 5'-end was combined with a ribose linkage at the 3'-end of the library (Figure 3c). Enzymatic digestion followed by alkaline treatment provided a genomic insert free of any constant nucleotides. To regenerate the primer-binding sites for amplification at every SELEX round, the authors employed thermal cycles of hybridization-extension using the initial genomic library as a template.

Pan et al. [65–67] employed the possibility of using the second strand as a template for completely randomized libraries. The authors developed two similar approaches for primer-free SELEX, which allowed the use of DNA libraries with only two constant nucleotides or even without constant positions (Figure 3d). The first approach was based on the introduction of Nt.BbvCI and Nt.BstNBI restriction sites into the initial dsDNA library. These enzymes recognize dsDNA, but cleave only one strand. A subsequent digestion of the library resulted in the formation of 32-nt ssDNA (0 + 30 + 2), which was used for in vitro selection. The second DNA strand remained uncleaved and acted as a template for the ligation of primer-binding sites prior to amplification. The second protocol provides a completely primer-less DNA library. In this case, the authors supplied the initial DNA library with Nt.BstNBI and BspMI restriction sites. Digestion by both restrictases provided the 30-nt ssDNA library (0 + 30 + 0), while the treatment only by Nt.BstNBI gave an uncleaved second strand, which also acted as a ligation template.

The possibility of using primer-free SELEX for completely randomized RNA libraries was also shown in [68]. The authors developed a tailored SELEX approach, implying the use of primers/adapters added previously by ligation and removed within the amplification processes (Figure 3e). A randomized 40-nt region was flanked by two short constant sequences (4 and 6 nt) for annealing the adapter oligonucleotides, so the total length of the aptamers generated by this method was as low as 50 nt. Further development of the method led to the design of the dual RNA library [69]. An introduction of both T3 and T7 RNA promoters (Figure 3f) allowed the generation of two different RNA libraries. The transcription carried out by a T3 RNA polymerase provided a long "traditional" RNA library with 34-nt random regions and conventional primer binding sites. Alternatively, the use of T7 RNA polymerase obtained an RNA library for tailored SELEX, with the same N34 region flanked by two short fixed sequences forming a stem that excluded their involvement in active functional structures. The design of primer-binding sequences complementary to each other was also employed in [70]. It is noteworthy that such stem-forming flanking sequences could, in some cases, hamper the selection of aptamers [71].

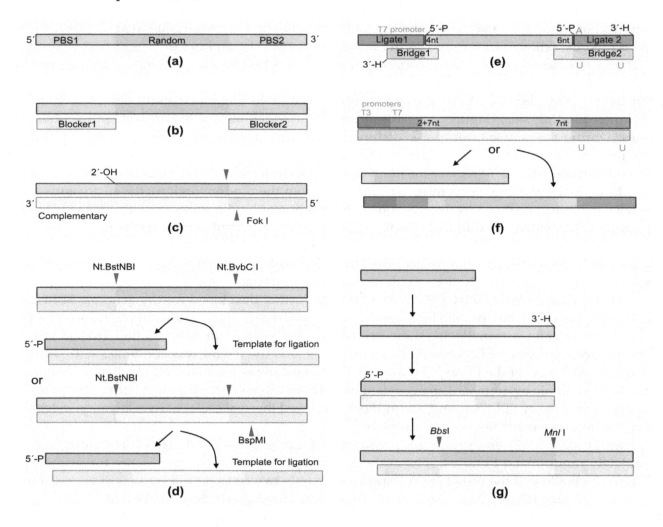

Figure 3. Different variants of design for NA libraries for a primer-free SELEX. (a) A conventional NA library; (b) blocked primer-binding sites for primer-annealing SELEX [62]; (c) the design of primer-binding sites for primer-free genomic SELEX [64]; (d) the design of a DNA library for primer-free SELEX from a completely randomized library [65]; (e) the RNA library for a tailored SELEX in a complex with auxiliary oligonucleotides [68]; (f) the DNA template for a dual-RNA library suitable for both conventional and tailored SELEX [69]; (g) DNA libraries lacking any constant nucleotides for the primer-free SELEX protocol of Lai et al. [72]. PBS: primer binding site, 2′-OH-ribonucleotide, 3′-H-dideoxynucleotide.

Another protocol for primer-free SELEX was developed by Lai et al. [72,73] for a totally randomized 30-nt DNA library aimed at selecting aptamers for HIV RT (Figure 3g). To amplify the library after target binding, the authors proposed the use of a non-template ligation of the 3′-primer-binding fragment containing the MnlI site by the thermostable RNA ligase at 60 °C.

These ligation conditions are supposed to lower the possibility of secondary structure formation and increase the efficiency of ligation as compared to the conventional T4 RNA ligase. The ligation of the 5'-primer-binding site as a duplex containing the BbsI site was performed by the T4 DNA ligase.

A drastic approach to avoid the use of primer-binding sites was recently proposed by Tsao et al. [74]. The Rotating Magnetic Field Magnetic-Assisted Rapid Aptamer Selection (RO-MARAS) method enables the one-step generation of high affinity aptamers, which relies on the sophisticated, but efficient, procedure of pool isolation. The protocol included an incubation of the starting library, free of any constant nucleotides, with a target protein immobilized on the surface of magnetic beads. This was followed by the employment of a rotating magnetic field to select the most tightly bound molecules. Notably, the amplification of the enriched library before sequencing required a very complex scheme to add primer-binding sites.

To summarize, a number of different initial libraries and selection schemes are now available to generate the aptamers lacking primer-binding sites. We would like to emphasize that the absence of fixed flanking sequences provides the important advantages of (1) decreasing the probability of SELEX artifacts; and (2) shortening the overall length of the aptamer sequence. At the same time, all primer-free SELEX protocols rely on the additional stages of ligation and restrictase digestion. Insufficient ligation, or deletion of restriction sites during PCR amplification could result in a loss of some potential binders, which can be considered as the pitfall of primer-free selection.

2.5. NA Libraries Containing Additional Constant Sequences

It should be mentioned that primer-binding sites are not the only possible constant regions of the library having an auxiliary role. NA libraries can also be supplied by additional constant sequences necessary for the immobilization within a capture SELEX approach. This approach, first proposed by Nutiu et al. [75,76] for the selection of structure-switching aptamers specific to ATP (adenosine triphosphate) or GTP (guanosine triphosphate), relies on the annealing of the so-called docking sequence within a library to the complementary capture oligonucleotide bound to a carrier through biotin–streptavidin interactions (Figure 4). In this way, prior to selection, the initial library is immobilized on a carrier, and target binding causes a structural rearrangement, which results in duplex dissociation and passing of the library to the solution. Therefore, the pool without target binding affinity remains immobilized and can be easily separated from the enriched one. Aptamers selected by this method gain the ability of structure-switching, which can be employed for engineering analytical systems (e.g., fluorescent beacons) for the detection of target molecules. A capture SELEX method turned out to be particularly suitable for selecting aptamers on small-molecule targets such as antibiotics, toxins, drugs or food contaminants (see the reviews in [32,77]). The problem of separating bound and unbound pools becomes crucial for these selections. After target binding, a change of the properties of NA molecules is not significant enough to isolate the complexes from unbound molecules in solution. Otherwise, the immobilization of small molecule targets masks potential binding sites and also increases the probability of selecting aptamers with an affinity to the target-carrier conjugate, but not to the target itself. Capture SELEX enables the selection of the target in its native state in solution, while employing the advantages of resin-based isolation. Some shortcomings of the method are connected with subsequent applications of structure-switching aptamers: during their binding with a target in solution, a rearrangement of the structure could be different from that of immobilized aptamers, which can influence binding affinity [32].

The design of a docking sequence for capture SELEX, namely the length and nucleotide composition, should provide both strong immobilization before target binding and sufficient dissociation afterwards [57]. As a rule, it is a heterosequence of 12–18 deoxynucleotides (see, e.g., Table 1) placed within the random region (as in [57,78,79]), or extending one of the primer-binding sites (as described in [80–83]). Currently, the capture SELEX strategy is generally employed for DNA selection, but also suits RNA libraries. For example, Morse et al. [84] isolated RNA beacon aptamers specific to tobramycin; interestingly, in this case, only a 6-nt capture deoxy oligomer was used for immobilization of the library.

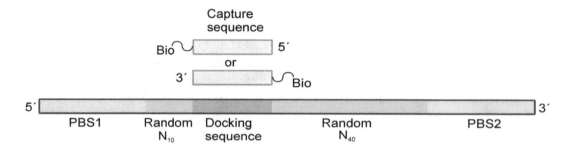

Figure 4. A general scheme of library design for a capture SELEX.

2.6. NA Libraries for a Genomic SELEX

Genomic SELEX is employed to screen sequences within a certain genome for aptamers or regulator sequences, which interact with proteins or other ligands [85], such as DNA sequences recognized by transcription factors [86], or RNA sites bound by splicing factors [87]. Initial libraries consist of genomic DNA fragments, and the motifs obtained by this method are called "genomic aptamers" [88].

Genomic SELEX libraries are derived from the genomic DNA of a given organism by means of random priming and transcription. This allows the representation of all possible genomic aptamers within a library. The first strand of a genomic DNA library is usually synthesized by the Klenow fragment in the presence of the random primer supplied by a fixed sequence at the 5'-end. After the reaction, the excess primer is thoroughly removed. The second strand is synthesized by the same method. As a result, a set of genomic sequences is obtained, flanked by constant regions. At this step, fragments of a certain length can be isolated, e.g., by electrophoretic separation. If RNA transcription is required, the T7 promotor sequence is introduced by means of PCR with the corresponding primers.

The benefits of the genomic SELEX approach over the conventional one include the use of much more restricted sequence space and the increased probability of selecting a biologically-relevant aptamer. Since the initial library is obtained from genomic DNA, RNA selection can be performed regardless of the expression level, thus making it possible to isolate RNA motifs with a low expression level, or those expressible only at certain stages of a cell cycle. Unfortunately, non-expressible RNAs can also be obtained [85].

3. The Design of Initial NA Libraries for More Affine Aptamers

One of the most important issues in the design of nucleic acids libraries is the maximal selection efficiency, i.e., the highest probability of selecting tight-binding aptamers. In contrast to proteins, nucleic acids possess a very limited repertoire of functional groups. Consequently, high binding affinity is reached by combining the diversity of spatial structures with the available functional groups. Otherwise, a toolkit of functionalities can be artificially expanded by adding extra chemical modifications. Below, we discuss both of these possibilities.

3.1. Expanding the Chemical Repertoire of NA Libraries

A more obvious (but definitely not simpler) way to generate higher-affinity aptamers is to use additional functional groups, thereby making nucleic acid aptamers more similar to proteins.

Expanding the chemical repertoire of NA libraries enables a selection of either better binders or aptamers directed to target epitopes inaccessible for unmodified pools. Additional chemical functions are generally introduced into heterocyclic bases (thoroughly reviewed in [35]).

SomaLogic, one of the world's leading companies in the development of aptamers, has created so-called SOMAmers, or Slow Off-rate Modified Aptamers. SOMAmers are selected from base-modified nucleic acids libraries [56,89–93] (see Table 1 for example sequences of the library and primers). Heterocyclic base modifications introduce protein-like functionalities, which provide a unique aptamer-target complex stability and even make it possible to select aptamers for previously inaccessible targets. Novel hydrophobic base modifications for DNA libraries have also been recently proposed by Chudinov et al. [94].

Heterocyclic base modification can also expand the genetic alphabet of nucleic acid libraries. The use of an extra artificial base pair Ds:Px (Figure 5) in the starting library was proposed by Kimoto et al. [95] to select VEGF-165 (vascular endothelial growth factor) binding aptamers. The selected aptamers, which contained several artificial base pairs, possessed 100-fold higher binding affinity as compared to the non-modified analogs. Sefah et al. [96] supplemented four natural bases with non-natural nucleosides Z and P (Figure 5) to generate DNA aptamers binding to liver cancer cells with nanomolar affinities.

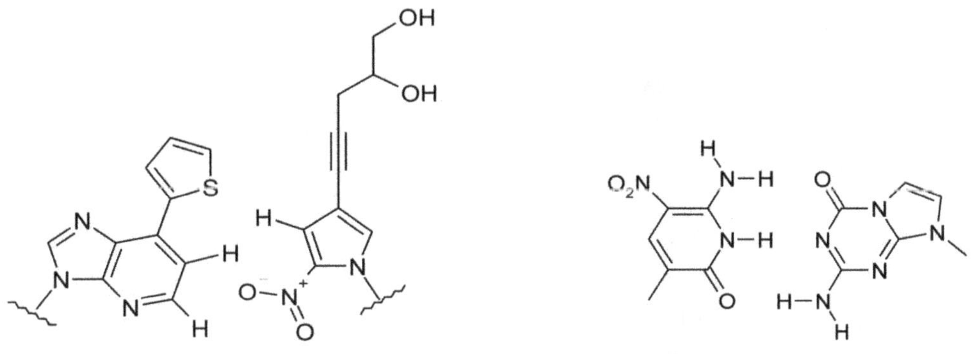

Figure 5. Chemical structures of artificial base pairs Ds:Px [95] and Z:P [96].

Click-SELEX represents a relatively new method for introducing chemical modifications into NA libraries. In this case, thymidine residues within a DNA library are replaced by C5-ethynyl-2'-deoxyuridine, followed by the Cu(I)-catalyzed cycloaddition of the azide component. The modified library is then employed in the modified SELMA (SELection with Modified Aptamers) protocol for different targets [97–102]. For example, this method was used to generate glycan-conjugated aptamers. Interestingly, in this case, the DNA aptamer served as a scaffold to provide an optimal tertiary structure and flexibility for the glycoclusters, which were then used as vaccine components.

Notably, expanding the chemical repertoire of NA libraries requires base-modified nucleotide monomers and mutant polymerases, as well as more complex SELEX protocols. That is probably why such a promising strategy has not yet become routine.

3.2. Structural Repertoire of Nucleic Acid Libraries

3.2.1. Uniformly Randomized Libraries

According to a widely-held point of view, all four nucleotides have to be uniformly represented in the random region of the library. An equal distribution is considered to provide the maximal sequence diversity, thus increasing the probability of selecting highly affine aptamers [103,104].

Currently, protocols for chemical synthesis have been developed to provide equal nucleotide distribution in the random region, which consider the different reactivities of corresponding phosphoramidites (see [53]). Methods of high-throughput sequencing and specially-developed program packages enable the estimation of the smoothness of the randomization in terms of nucleotides or short sequences, e.g., hexanucleotides [103,105]. In the latter case, a Gaussian profile is characteristic for the balanced library.

Unfortunately, today, only a few studies devoted to the impact of nucleotide composition on the structure of the library have been published. For example, the computer analysis of the structure distribution for random regions of RNA libraries revealed that for the 40-nt region, a shift to G and C (30% each) led to the predominant formation of structures with more stems when compared to the same A + U shift [106]. At the same time, for the 100-nt random region, such bias in nucleotide composition was not significant and did not markedly change the distribution of secondary structures.

On the other hand, several experiments on RNA SELEX from smoothly-randomized starting libraries have shown that the selection progress is accompanied by an accumulation of pyrimidine-rich sequences and the loss of adenosine [50,104], both for targeted and non-targeted selections. The loss of adenosine was observed for all adenosine-containing dinucleoside pairs. This corresponded to a decrease in the overall minimum free energy of the RNA library, which resulted in RNA sequences with higher predicted structural stability [50]. Therefore, a slight bias in the initial library, especially a pyrimidine bias, can be considered as acceptable, since over the course of selection, the nucleotide distribution will inevitably shift.

3.2.2. Doped and Segmented NA Libraries

When a starting library is designed to improve the properties of existing aptamers by determining their target binding sites or for a functional analysis of natural RNA, the task is not a total randomization, but a delicate varying of particular nucleotides within a certain sequence. To solve this problem, one should choose doped or segmented NA libraries.

In their pioneering work, Bartel et al. [107] generated a doped library on the basis of the viral RNA element of the Rev protein of human immunodeficiency virus 1 (HIV-1) to identify the binding site for the protein. The 66-nt fragment of Rev-responsive element (RRE) was generated in such a way that point mutations were introduced uniformly throughout the sequence at a rate of 30% with 5% deletions (which meant that every position contained 65% of a wild-type nucleotide, 10% of each other nucleotide and 5% deletions). An example of the use of doping strategy to explore the secondary structure of the aptamer and determine its conservative positions is given in [108]. The authors doped the sequence of the aptamer specific to the ricin A-chain (generated by the conventional SELEX) at a 15% mutation rate. The doping strategy also helps to improve the affinity of the aptamer. Burke et al. [106,109] employed it for a secondary SELEX of pseudoknot aptamers for an HIV reverse transcriptase: truncated aptamer motifs found by the primary SELEX were doped at a 30% mutation rate (70% of the wild-type base and 10% of each of the other bases).

Nevertheless, how can we choose the mutation rate suitable for a particular task and sequence? To answer this question, Knight et al. [110] performed a comprehensive theoretical analysis of doped selections and developed an algorithm to select the length of the doped sequence and mutation rate depending on a given task. To search for sequences close to the wild-type, the authors recommended a low mutation rate (about 5%). If the structure space had to be extended, the mutation rate increased up to 30–50%. The concrete values for the doping scheme could be calculated by the developed method.

Apart from the doping of certain positions, segmental randomization is employed to specify the sequence or optimize the structure of an aptamer. For this, certain parts of the sequence are replaced by randomized stretches of the appropriate length. In principle, the segmental randomization can be considered as a special case of a doped randomization with a mutation rate of 75%. Usually, segments represent rather short sequences placed within certain elements of the secondary structure or other wild-type context [53]. A contrary example is given in [60], where core RNA aptamer sequences were flanked by 40- and 45-nt random regions to improve the aptamer analogs of green fluorescent protein. Longer segments provide larger structural diversity, which increases the probability of generating a better binder.

3.2.3. Nonhomologous Recombination as an Alternative to the Doping Strategy

Bittker et al. [111] proposed an entirely different approach of varying the existing aptamer sequences to find conservative regions, identify binding sites or improve the affinity: a nonhomologous random recombination (NRR). This method enables variation of the length of the library, deletion of inactive fragments and alternation of the mutual location of different motifs. For this purpose, a sequential scheme of enzymatic synthesis of NRR libraries was developed (Figure 6), starting from the treatment of the dsDNA library by DNase I and T4 DNA polymerase, which gives a mixture of blunt-ended DNA fragments. During the recombination step, DNA fragments were treated with the T4 DNA ligase under conditions favoring intermolecular ligation. The presence of an additional 5′-phosphorylated hairpin DNA containing a restriction site enabled both introducing the fixed PBS to the ends of the library and regulating the length of the recombined molecules (by varying the stoichiometry of the hairpin). Digestion of the resulted circular DNAs gave a pool of dsDNA molecules with defined sequences at both ends.

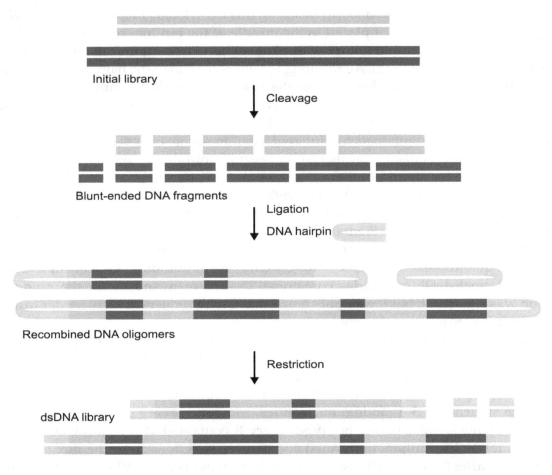

Figure 6. A scheme of the nonhomologous random recombination method [111].

When the NRR approach had been applied to a model partly-enriched aptamer library, the authors observed that NRR-derived aptamers accumulated several copies of the active motif. Therefore, the NRR strategy was considered as a more effective alternative for error-prone PCR or site-directed mutagenesis. This strategy might also be used instead of a synthesis of doped libraries. Although the NRR protocol seems to be more complex, the synthesis of the NRR library, otherwise, does not require a sophisticated doping scheme for chemical synthesis and enables almost unlimited exploration of the sequence space. We presume that the NRR strategy could also bring benefits when used as a basic SELEX protocol starting from an unselected random pool.

3.2.4. Nucleic Acid Libraries with Pre-Defined Secondary Structures

The design of starting libraries can also be performed in the framework of a paradigm that does not follow uniform randomization. An alternative concept arises from the facts that the number of productive structures providing the selection of effective binders is limited and the maximal accessible diversity of sequences folds in a restricted set of spatial structures (see [48] for a review). A computer analysis of uniformly-randomized libraries of different lengths (20–100 nt) [106] revealed that a limited set of secondary structures corresponded to every library. It was found that the complexity of the structures increased with the length of the library, and every length was characterized by three predominating structural motifs.

Thus, instead of a "smooth" randomization, it could be more beneficial to introduce secondary structure motifs into an initial library. A pioneering work in the field was published by Davis and Szostak [58]. Integrating structural data for aptamers that had been known at the time, the authors observed a common element for all structures: a stem-loop, which appeared to act as a structural anchor for recognition loops. Based on this knowledge, they designed an RNA library containing an 8-nt stem-loop motif placed in the middle of the random region (Figure 7a, Table 1). An equal mix of this pre-structured library with a conventionally-randomized one was employed in the SELEX of GTP-binding aptamers. All resulting aptamers contained the hairpin insert, thus proving the efficiency of the strategy. To further establish the proof-of-principle, the authors demonstrated that more complex structures provided more active RNAs (by examples of GTP-binding aptamers and ligase ribozymes) [112]. Notably, the hairpin motif derived in [58] was then successfully employed by other researchers to generate aptamers for different small-molecule targets [113,114].

Secondary structure elements can also be successfully introduced into DNA libraries. To form a hydrophobic pocket for steroid binding, Yang et al. [115] designed a DNA library containing a three-way junction structure with a total of eight randomized positions (Figure 7b). The same motif was also used in [116] to select structure-switchable aptamer beacons for the steroid hormone dehydroisoandrosterone 3-sulfate (Figure 7c).

Attempts were also made to design DNA libraries in a manner that provided a preferential formation of G-quadruplex structures. To generate hemin-binding G-quadruplex structures, Zhu et al. [117] created DNA libraries containing 25–45% of guanosine in the random region. The selection was successful, but the authors noted that G-rich sequences were harder to amplify by PCR, which may lead to a loss of the best binders.

Ruff et al. [118] developed a general approach for the design of pre-structured DNA libraries, also using a doping strategy. A structured DNA library with 60-nt random regions contained an RY pattern (alternating purines (R) and pyrimidines (Y)) that favors stem formation. To increase the frequency and diversity of loops and other non-stem structures within the patterned library, RY sequences alternated with stretches of 3–4 random nucleotides. Moreover, every position in the RY sites was slightly doped by nucleosides of another type: every R contained 45% A and G and 5% C and T, and vice versa for Y. The authors performed competitive selections from the mix of unpatterned and patterned libraries for three different target proteins (streptavidin, VEGF and IgE). The results proved that namely a combination of RY fragments and doping provided the selection of the highest affinity aptamers.

During the last decade, several approaches to in silico optimization of starting libraries have been developed to lower the fraction of poorly-structured (and thus low-affinity) sequences. Chushak et al. [119] developed a protocol for the computer optimization of RNA libraries prior to the selection of aptamers for small molecules. The algorithm included two main steps. First, the secondary structures of all possible sequences of a given length were analyzed. Based on secondary structure data for existing aptamers, the authors derived a set of criteria that allowed selecting an affective binder. At Step 2, 3D structures were built for all sequences meeting these criteria, followed by molecular docking with a given target molecule that resulted in a minimal free energy rating. Such high-throughput virtual screening enabled them to reduce a library of 2.5×10^8 sequences to 10^3–10^4 sequences suitable for the experimental screening and verification.

The concepts of doped and partly-structured RNA libraries complemented each other in the method developed by Kim et al. [120]. The approach included the use of a definite set of starting sequences and certain mutation rates in certain positions within a random region (mixing matrixes). To generate these two key sets of parameters, the authors employed graph theory and matrix analysis, respectively. Starting RNA pools obtained by the proposed algorithms ensured the selection of better binders when compared to the uniformly-randomized pools. The authors also developed the web server RAGPOOLS (RNA-As-Graph-Pools) for designing and analyzing structured pools for SELEX (http://rubin2.biomath.nyu.edu/home.html) [121,122]. It is worth noting that the synthesis of the initial pool according to the mixing matrix (i.e., with an individual mutation rate for every doped position) may be laborious and time-consuming.

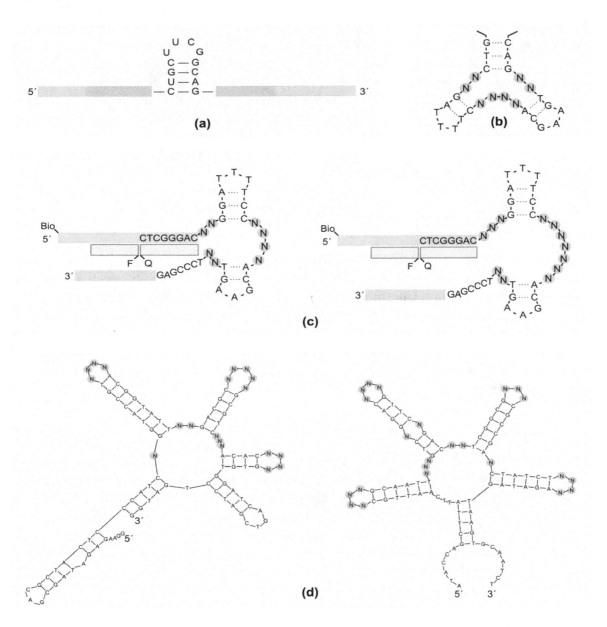

Figure 7. Partially-structured NA libraries. (**a**) The pre-structured RNA library with the stem-loop scaffold engineered in [58]; (**b**) the DNA library with three-way junction scaffolds for a steroid binding engineered in [115]; (**c**) DNA libraries with three-way junction scaffolds for a selection of steroid-binding DNA beacon aptamers [116]; and (**d**) highly structured RNA libraries engineered by the random filtering (left) and genetic filtering (right) approaches [123].

Luo et al. [123] developed two computational methods to generate starting DNA libraries with increased structural diversity: random filtering and genetic filtering. The random filtering approach is based on the secondary structure analysis of all sequences in the library and isolating those containing five-way junctions as the most structured elements. Then, for every such sequence, a set of mutant versions is generated with all four possible nucleotides at all positions not involved in base pairing. Random filtering thus pre-enriches the starting library with highly-structured motifs, hence increasing the probability of generating better binders. The genetic filtering approach aims to create a library with a desired distribution (either uniform or not) of all secondary structure elements (one-way, two-way, three-way, four-way and five-way junction). First, all secondary structures are analyzed for a library of a given length and the primer-binding sites. The authors recommended using 24 random positions for the pool design to provide complete sequence coverage. After secondary structure analysis, the pool is assigned a fitness score that indicates its proximity to the desired distribution of the structure elements. New generations of pool designs are obtained by selecting designs from previous generations with better (i.e., smaller) fitness scores and applying mutation, copy and crossover procedures. Typically, 500–3000 generations are needed for the best pool design. Examples of starting pools developed by random filtering and genetic filtering methods are given in Figure 7d. The pool with a uniform structure distribution was tested in a wet SELEX experiment aimed at finding ATP-binding DNA aptamers. Notably, the resulting aptamers possessed five-way junction structures, and their binding affinities were close to those for previously published aptamers from a conventionally-designed library. The authors concluded that although complexity alone could not guarantee better target binding, higher complexity structures possessed the potential to yield better aptamers. They also emphasized the importance of structural diversity, and not only structural complexity in the starting pool.

To sum up this section, the use of NA libraries with pre-designed secondary structures is a very promising strategy, which has been strongly underestimated until now. The inherent ability of nucleic acids to form complex spatial structures is used here to its full extent. A pre-structured library can be designed in silico considering the properties of a given molecular target. Once generated, the pre-structured initial library is further used in a routine SELEX protocol without any additional stages, modified nucleotides or unusual polymerases. However, it may be suggested that a combination of base-functionalized monomers with a pre-defined secondary structure would provide even more efficient starting libraries.

4. Conclusions

Nucleic acid aptamers generated by SELEX technology have proven themselves as highly selective and high-affinity, biospecific molecules for a number of applications. Aptamers are now considered as "chemical antibodies" with the advantages of chemical synthesis, long shelf-life and the ability to be built into almost any system of interest. In principle, aptamers can be selected for nearly any molecular or supramolecular target. However, to generate an efficient aptamer for a certain target, one should choose the most suitable SELEX protocol, and the most important issue in this case is the proper choice of an initial library. The design of a library is governed by the different parameters of a particular system such as the need for nuclease resistance, hydrophobicity, the molecular weight of a target molecule, etc. A classic design of a starting library, which still remains the most popular, uses a uniformly-randomized region flanked by two fixed primer-binding sequences. These universal "traditional" libraries are suitable for any SELEX target, from small molecules to proteins. Nevertheless, a number of alternative strategies has recently been developed. Primer-binding sites can be deleted to exclude their impact on the course of selection and to shorten the resulting aptamers. Different primer-free selection strategies have proven successful for protein targets. On the contrary, to generate an aptamer for a small-molecule target, it could be better to use the capture-SELEX technique, where the library is resin-immobilized through an additional docking sequence and the target retains its native structure. The smart design of random region enables the enrichment of a library with complex spatial structures favorable for the selection of tightly-binding motifs. The shape of the random region can be adjusted to fit the structure of the given molecule (or a class of molecules), so the "smart randomization" strategy might be recommended for any target. A chemical repertoire of initial NA libraries can also be expanded to generate better binders and to obtain aptamers for previously "SELEX-inaccessible" targets.

To summarize, a large variety of different approaches for library design is now available. A conscious choice from this diversity and the development of novel approaches to design the initial NA libraries would guarantee the generation of high-affinity aptamers for any desired ligand.

Acknowledgments: The work was supported by the Russian Science Foundation (Grant No. 16-14-10296).

Author Contributions: Mariya A. Vorobyeva and Anna S. Davydova wrote the manuscript, and Pavel E. Vorobjev generated the figures and wrote the manuscript. Dmitrii V. Pyshnyi conceived of the review topic and performed general revision of the manuscript. Alya G. Venyaminova performed general revision and editing of the manuscript. All authors read and approved the final manuscript.

Abbreviations

SELEX	Selective Evolution of Ligands by Exponential enrichment
NA	Nucleic Acid(s)
PBS	Primer-Binding Site(s)
PCR	Polymerase Chain Reaction
Pu, or R	Purine nucleotide
Py, or Y	Pyrimidine nucleotide

References

1. Ellington, A.D.; Szostak, J.W. In vitro selection of RNA molecules that bind specific ligands. *Nature* **1990**, *346*, 818–822. [CrossRef] [PubMed]
2. Ono, A.; Togashi, H. Highly selective oligonucleotide-based sensor for mercury (II) in aqueous solutions. *Angew. Chem. Int. Ed.* **2004**, *43*, 4300–4302. [CrossRef] [PubMed]
3. Ye, B.-F.; Zhao, Y.-J.; Cheng, Y.; Li, T.-T.; Xie, Z.-Y.; Zhao, X.-W.; Gu, Z.-Z. Colorimetric photonic hydrogel aptasensor for the screening of heavy metal ions. *Nanoscale* **2012**, *4*, 5998–6003. [CrossRef] [PubMed]
4. Jenison, R.D.; Gill, S.C.; Pardi, A.; Polisky, B. High-resolution molecular discrimination by RNA. *Science* **1994**, *263*, 1425–1429. [CrossRef] [PubMed]
5. Stojanovic, M.N.; de Prada, P.; Landry, D.W. Aptamer-based folding fluorescent sensor for cocaine. *J. Am. Chem. Soc.* **2001**, *123*, 4928–4931. [CrossRef] [PubMed]
6. Holeman, L.A.; Robinson, S.L.; Szostak, J.W.; Wilson, C. Isolation and characterization of fluorophore-binding RNA aptamers. *Fold. Des.* **1998**, *3*, 423–431. [CrossRef]
7. Grate, D.; Wilson, C. Laser-mediated, site-specific inactivation of RNA transcripts. *Proc. Natl. Acad. Sci. USA* **1999**, *96*, 6131–6136. [CrossRef] [PubMed]
8. Li, Y.; Geyer, R.; Sen, D. Recognition of anionic porphyrins by DNA aptamers. *Biochemistry* **1996**, *35*, 6911–6922. [CrossRef] [PubMed]
9. Leva, S.; Lichte, A.; Burmeister, J.; Muhn, P.; Jahnke, B.; Fesser, D.; Erfurth, J.; Burgstaller, P.; Klussmann, S. GnRH Binding RNA and DNA spiegelmers: A novel approach toward GnRH antagonism. *Chem. Biol.* **2002**, *9*, 351–359. [CrossRef]
10. Yoshida, W.; Mochizuki, E.; Takase, M.; Hasegawa, H.; Morita, Y.; Yamazaki, H.; Sode, K.; Ikebukuro, K. Selection of DNA aptamers against insulin and construction of an aptameric enzyme subunit for insulin sensing. *Biosens. Bioelectron.* **2009**, *24*, 1116–1120. [CrossRef] [PubMed]
11. Dupont, D.M.; Andersen, L.M.; Botkjaer, K.A.; Andreasen, P.A. Nucleic acid aptamers against proteases. *Curr. Med. Chem.* **2011**, *18*, 4139–4151. [CrossRef] [PubMed]
12. Cerchia, L.; De Franciscis, V. Nucleic acid aptamers against protein kinases. *Curr. Med. Chem.* **2011**, *18*, 4152–4158. [CrossRef] [PubMed]

13. Missailidis, S. Targeting of antibodies using aptamers. *Methods Mol. Biol.* **2004**, *248*, 547–555. [CrossRef] [PubMed]

14. Chen, M.; Yu, Y.; Jiang, F.; Zhou, J.; Li, Y.; Liang, C.; Dang, L.; Lu, A.; Zhang, G. Development of cell-SELEX technology and its application in cancer diagnosis and therapy. *Int. J. Mol. Sci.* **2016**, *17*, 2079. [CrossRef] [PubMed]

15. Khvorova, A.; Kwak, Y.G.; Tamkun, M.; Majerfeld, I.; Yarus, M. RNAs that bind and change the permeability of phospholipid membranes. *Proc. Natl. Acad. Sci. USA* **1999**, *96*, 10649–10654. [CrossRef] [PubMed]

16. Darmostuk, M.; Rimpelová, S.; Gbelcová, H.; Ruml, T. Current approaches in SELEX: An update to aptamer selection technology. *Biotechnol. Adv.* **2015**, *33*, 1141–1161. [CrossRef] [PubMed]

17. Tuerk, C.; Gold, L. Systematic evolution of ligands by exponential enrichment: RNA ligands to bacteriophage T4 DNA polymerase. *Science* **1990**, *249*, 505–510. [CrossRef] [PubMed]

18. Tremblay, R.; Mulhbacher, J.; Blouin, S.; Penedo, J.C.; Lafontaine, D.A. Natural functional nucleic acids: Ribozymes and riboswitches. In *Functional Nucleic Acids for Analytical Applications*; Yingfu, L., Yi, L., Eds.; Springer: New York, NY, USA, 2009; pp. 11–46. ISBN 978-0-387-73711-9.

19. Silverman, S.K. Artificial functional nucleic acids: Aptamers, ribozymes, and deoxyribozymes identified by in vitro selection. In *Functional Nucleic Acids for Analytical Applications*; Yingfu, L., Yi, L., Eds.; Springer: New York, NY, USA, 2009; pp. 47–108. ISBN 978-0-387-73711-9.

20. Blackwell, T.K.; Weintraub, H. Differences and similarities in DNA-binding preferences of MyoD and E2A protein complexes revealed by binding site selection. *Science* **1990**, *250*, 1104–1110. [CrossRef] [PubMed]

21. Jacoby, K.; Lambert, A.R.; Scharenberg, A.M. Characterization of homing endonuclease binding and cleavage specificities using yeast surface display SELEX (YSD-SELEX). *Nucleic Acids Res.* **2017**, *45*, e11. [CrossRef] [PubMed]

22. Babendure, J.R.; Adams, S.R.; Tsien, R.Y. Aptamers switch on fluorescence of triphenylmethane dyes. *J. Am. Chem. Soc.* **2003**, *125*, 14716–14717. [CrossRef] [PubMed]

23. Nakamura, Y. Aptamers as therapeutic middle molecules. *Biochimie* **2017**. [CrossRef] [PubMed]

24. Parashar, A. Aptamers in therapeutics. *J. Clin. Diagn. Res.* **2016**, *10*, BE01–BE06. [CrossRef] [PubMed]

25. Poolsup, S.; Kim, C.Y. Therapeutic applications of synthetic nucleic acid aptamers. *Curr. Opin. Biotechnol.* **2017**, *48*, 180–186. [CrossRef] [PubMed]

26. Catuogno, S.; Esposito, C.L.; de Franciscis, V. Aptamer-mediated targeted delivery of therapeutics: An update. *Pharmaceuticals* **2016**, *9*, 69. [CrossRef] [PubMed]

27. Ni, S.; Yao, H.; Wang, L.; Lu, J.; Jiang, F.; Lu, A.; Zhang, G. Chemical modifications of nucleic acid aptamers for therapeutic purposes. *Int. J. Mol. Sci.* **2017**, *18*, 1683. [CrossRef] [PubMed]

28. Ilgu, M.; Nilsen-Hamilton, M. Aptamers in analytics. *Analyst* **2016**, *141*, 1551–1568. [CrossRef] [PubMed]

29. Seo, H.B.; Gu, M.B. Aptamer-based sandwich-type biosensors. *J. Biol. Eng.* **2017**, *11*, 11. [CrossRef] [PubMed]

30. Vorobyeva, M.; Vorobjev, P.; Venyaminova, A. Multivalent aptamers: Versatile tools for diagnostic and therapeutic applications. *Molecules* **2016**, *21*, 1613. [CrossRef] [PubMed]

31. Zhang, H.; Zhou, L.; Zhu, Z.; Yang, C. Recent progress in aptamer-based functional probes for bioanalysis and biomedicine. *Chem. A Eur. J.* **2016**, *22*, 9886–9900. [CrossRef] [PubMed]

32. Ruscito, A.; DeRosa, M.C. Small-molecule binding aptamers: Selection strategies, characterization, and applications. *Front. Chem.* **2016**, *4*, 1–14. [CrossRef] [PubMed]

33. Volk, D.E.; Lokesh, G.L.R. Development of Phosphorothioate DNA and DNA thioaptamers. *Biomedicines* **2017**, *5*, 41. [CrossRef] [PubMed]

34. Chen, T.; Hongdilokkul, N.; Liu, Z.; Thirunavukarasu, D.; Romesberg, F.E. The expanding world of DNA and RNA. *Curr. Opin. Chem. Biol.* **2016**, *34*, 80–87. [CrossRef] [PubMed]

35. Lipi, F.; Chen, S.; Chakravarthy, M.; Rakesh, S.; Veedu, R.N. In vitro evolution of chemically-modified nucleic acid aptamers: Pros and cons, and comprehensive selection strategies. *RNA Biol.* **2016**, *13*, 1232–1245. [CrossRef] [PubMed]

36. McKeague, M.; McConnell, E.M.; Cruz-Toledo, J.; Bernard, E.D.; Pach, A.; Mastronardi, E.; Zhang, X.; Beking, M.; Francis, T.; Giamberardino, A.; et al. Analysis of In Vitro Aptamer Selection Parameters. *J. Mol. Evol.* **2015**, *81*, 150–161. [CrossRef] [PubMed]

37. Dunn, M.R.; Jimenez, R.M.; Chaput, J.C. Analysis of aptamer discovery and technology. *Nat. Rev. Chem.* **2017**, *1*, 76. [CrossRef]

38. Gilbert, W. The RNA world. *Nature* **1986**, *319*, 618. [CrossRef]

39. Ellington, A.; Szostak, J. Selection in vitro of single-stranded DNA molecules that fold into specific ligand-binding structures. *Nature* **1992**, *355*, 850–852. [CrossRef] [PubMed]

40. McKeague, M.; Derosa, M.C. Challenges and opportunities for small molecule aptamer development. *J. Nucleic Acids* **2012**, *2012*, 748913. [CrossRef] [PubMed]

41. Lapa, S.A.; Chudinov, A.V.; Timofeev, E.N. The Toolbox for Modified Aptamers. *Mol. Biotechnol.* **2016**, *58*, 79–92. [CrossRef] [PubMed]

42. Meek, K.N.; Rangel, A.E.; Heemstra, J.M. Enhancing aptamer function and stability via in vitro selection using modified nucleic acids. *Methods* **2016**, *106*, 29–36. [CrossRef] [PubMed]

43. Diafa, S.; Hollenstein, M. Generation of aptamers with an expanded chemical repertoire. *Molecules* **2015**, *20*, 16643–16671. [CrossRef] [PubMed]

44. Lin, Y.; Qiu, Q.; Gill, S.C.; Jayasena, S.D. Modified RNA sequence pools for in vitro selection. *Nucleic Acids Res.* **1994**, *22*, 5229–5234. [CrossRef] [PubMed]

45. Fitzwater, T.; Polisky, B. A SELEX primer. *Meth. Enzym.* **1996**, *267*, 275–301. [PubMed]

46. Lauridsen, L.H.; Rothnagel, J.A.; Veedu, R.N. Enzymatic recognition of 2'-modified ribonucleoside 5'-triphosphates: Towards the evolution of versatile aptamers. *ChemBioChem* **2012**, *13*, 19–25. [CrossRef] [PubMed]

47. Stovall, G.M.; BedenBaugh, R.; Singh, S.; Meyer, A.; Hatala, P.; Ellington, A.D.; Hall, B. In vitro selection using modified or unnatural nucleotides. *Curr. Protoc. Nucleic Acid Chem.* **2014**, *56*, 9.6.1–9.6.33. [PubMed]

48. Pobanz, K.; Luptak, A. Improving the odds: Influence of starting pools on in vitro selection outcomes. *Methods* **2016**, *106*, 14–20. [CrossRef] [PubMed]

49. Cowperthwaite, M.C.; Ellington, A.D. Bioinformatic analysis of the contribution of primer sequences to aptamer structures. *J. Mol. Evol.* **2008**, *67*, 95–102. [CrossRef] [PubMed]

50. Thiel, W.H.; Bair, T.; Wyatt Thiel, K.; Dassie, J.P.; Rockey, W.M.; Howell, C.A.; Liu, X.Y.; Dupuy, A.J.; Huang, L.; Owczarzy, R.; et al. Nucleotide Bias Observed with a Short SELEX RNA Aptamer Library. *Nucleic Acid Ther.* **2011**, *21*, 253–263. [CrossRef] [PubMed]

51. Famulok, M.; Szostak, J.W. Stereospecific Recognition of Tryptophan Agarose by in Vitro Selected RNA. *J. Am. Chem. Soc.* **1992**, *114*, 3990–3991. [CrossRef]

52. Sassanfar, M.; Szostak, J.W. An RNA motif that binds ATP. *Nature* **1993**, *364*, 550–553. [CrossRef] [PubMed]

53. Hall, B.; Micheletti, J.M.; Satya, P.; Ogle, K.; Pollard, J.; Ellington, A.D. Design, synthesis, and amplification of DNA pools for in vitro selection. *Curr. Protoc. Mol. Biol.* **2009**, *88*. [CrossRef]

54. Burmeister, P.E.; Lewis, S.D.; Silva, R.F.; Preiss, J.R.; Horwitz, L.R.; Pendergrast, P.S.; McCauley, T.G.; Kurz, J.C.; Epstein, D.M.; Wilson, C.; et al. Direct In Vitro Selection of a 2'-O-Methyl Aptamer to VEGF. *Chem. Biol.* **2005**, *12*, 25–33. [CrossRef] [PubMed]

55. Burmeister, P.E.; Wang, C.; Killough, J.R.; Lewis, S.D.; Horwitz, L.R.; Ferguson, A.; Thompson, K.M.; Pendergrast, P.S.; McCauley, T.G.; Kurz, M.; et al. 2'-Deoxy purine, 2'-O-methyl pyrimidine (dRmY) aptamers as candidate therapeutics. *Oligonucleotides* **2006**, *16*, 337–351. [CrossRef] [PubMed]

56. Gupta, S.; Hirota, M.; Waugh, S.M.; Murakami, I.; Suzuki, T.; Muraguchi, M.; Shibamori, M.; Ishikawa, Y.; Jarvis, T.C.; Carter, J.D.; et al. Chemically modified DNA aptamers bind interleukin-6 with high affinity and inhibit signaling by blocking its interaction with interleukin-6 receptor. *J. Biol. Chem.* **2014**, *289*, 8706–8719. [CrossRef] [PubMed]

57. Stoltenburg, R.; Nikolaus, N.; Strehlitz, B. Capture-SELEX: Selection of DNA aptamers for aminoglycoside antibiotics. *J. Anal. Methods Chem.* **2012**, *2012*, 415697. [CrossRef] [PubMed]

58. Davis, J.H.; Szostak, J.W. Isolation of high-affinity GTP aptamers from partially structured RNA libraries. *Proc. Natl. Acad. Sci. USA* **2002**, *99*, 11616–11621. [CrossRef] [PubMed]

59. Wiegand, T.W.; Williams, P.B.; Dreskin, S.C.; Jouvin, M.H.; Kinet, J.P.; Tasset, D. High-affinity oligonucleotide ligands to human IgE inhibit binding to Fc epsilon receptor I. *J. Immunol.* **1996**, *157*, 221–230. [PubMed]

60. Shui, B.; Ozer, A.; Zipfel, W.; Sahu, N.; Singh, A.; Lis, J.T.; Shi, H.; Kotlikoff, M.I. RNA aptamers that functionally interact with green fluorescent protein and its derivatives. *Nucleic Acids Res.* **2012**, *40*. [CrossRef] [PubMed]

61. Shtatland, T.; Gill, S.C.; Javornik, B.E.; Johansson, H.E.; Singer, B.S.; Uhlenbeck, O.C.; Zichi, D. A; Gold, L. Interactions of Escherichia coli RNA with bacteriophage MS2 coat protein: Genomic SELEX. *Nucleic Acids Res.* **2000**, *28*, E93. [CrossRef] [PubMed]

62. Ouellet, E.; Foley, J.H.; Conway, E.M.; Haynes, C. Hi-Fi SELEX: A high-fidelity digital-PCR based therapeutic aptamer discovery platform. *Biotechnol. Bioeng.* **2015**, *112*, 1506–1522. [CrossRef] [PubMed]

63. Ouellet, E.; Lagally, E.T.; Cheung, K.C.; Haynes, C.A. A simple method for eliminating fixed-region interference of aptamer binding during SELEX. *Biotechnol. Bioeng.* **2014**, *111*, 2265–2279. [CrossRef] [PubMed]

64. Wen, J.-D.; Gray, D.M. Selection of genomic sequences that bind tightly to Ff gene 5 protein: Primer-free genomic SELEX. *Nucleic Acids Res.* **2004**, *32*, e182. [CrossRef] [PubMed]

65. Pan, W.; Xin, P.; Patrick, S.; Dean, S.; Keating, C.; Clawson, G. Primer-Free Aptamer Selection Using A Random DNA Library. *J. Vis. Exp.* **2010**, *629*, 369–385. [CrossRef]

66. Pan, W.; Xin, P.; Clawson, G.A. Minimal primer and primer-free SELEX protocols for selection of aptamers from random DNA libraries. *Biotechniques* **2008**, *44*, 351–360. [CrossRef] [PubMed]

67. Pan, W.; Clawson, G.A. Primer-free aptamer selection using a random DNA library. *Meth. Mol. Biol.* **2010**, *629*, 367–383. [CrossRef]

68. Vater, A.; Jarosch, F.; Buchner, K.; Klussmann, S. Short bioactive Spiegelmers to migraine-associated calcitonin gene-related peptide rapidly identified by a novel approach: Tailored-SELEX. *Nucleic Acids Res.* **2003**, *31*, e130. [CrossRef] [PubMed]

69. Jarosch, F.; Buchner, K.; Klussmann, S. In vitro selection using a dual RNA library that allows primerless selection. *Nucleic Acids Res.* **2006**, *34*, e86. [CrossRef] [PubMed]

70. Skrypina, N.A.; Savochkina, L.P.; Beabealashvilli, R.S. In vitro selection of single-stranded DNA aptamers that bind human pro-urokinase. *Nucleosides Nucleotides Nucleic Acids* **2004**, *23*, 891. [CrossRef] [PubMed]

71. Legiewicz, M.; Lozupone, C.; Knight, R.; Yarus, M. Size, constant sequences, and optimal selection. *RNA* **2005**, *11*, 1701–1709. [CrossRef] [PubMed]

72. Lai, Y.T.; DeStefano, J.J. A primer-free method that selects high-affinity single-stranded DNA aptamers using thermostable RNA ligase. *Anal. Biochem.* **2011**, *414*, 246–253. [CrossRef] [PubMed]

73. Lai, Y.-T.; DeStefano, J.J. DNA aptamers to human immunodeficiency virus reverse transcriptase selected by a primer-free SELEX method: Characterization and comparison with other aptamers. *Nucleic Acid Ther.* **2012**, *22*, 162–176. [CrossRef] [PubMed]

74. Tsao, S.-M.; Lai, J.-C.; Horng, H.-E.; Liu, T.-C.; Hong, C.-Y. Generation of aptamers from a primer-free randomized ssDNA library using magnetic-assisted rapid aptamer selection. *Sci. Rep.* **2017**, *7*, 45478. [CrossRef] [PubMed]

75. Nutiu, R.; Li, Y. In vitro selection of structure-switching signaling aptamers. *Angew. Chem. Int. Ed.* **2005**, *44*, 1061–1065. [CrossRef] [PubMed]

76. Nutiu, R.; Li, Y. Structure-switching signaling aptamers. *J. Am. Chem. Soc.* **2003**, *125*, 4771–4778. [CrossRef] [PubMed]

77. Pfeiffer, F.; Mayer, G. Selection and biosensor application of aptamers for small molecules. *Front. Chem.* **2016**, *4*, 25. [CrossRef] [PubMed]

78. Nikolaus, N.; Strehlitz, B. DNA-aptamers binding aminoglycoside antibiotics. *Sensors* **2014**, *14*, 3737–3755. [CrossRef] [PubMed]

79. Paniel, N.; Istambouli, G.; Triki, A.; Lozano, C.; Barthelmebs, L.; Noguer, T. Selection of DNA aptamers against penicillin G using Capture-SELEX for the development of an impedimetric sensor. *Talanta* **2017**, *162*, 232–240. [CrossRef] [PubMed]

80. Zhang, A.; Chang, D.; Zhang, Z.; Li, F.; Li, W.; Wang, X.; Li, Y.; Hua, Q. In vitro selection of DNA aptamers that binds geniposide. *Molecules* **2017**, *22*, 383. [CrossRef] [PubMed]

81. Martin, J.A.; Smith, J.E.; Warren, M.; Chávez, J.L.; Hagen, J.A.; Kelley-Loughnane, N. A method for selecting structure-switching aptamers applied to a colorimetric gold nanoparticle assay. *J. Vis. Exp.* **2015**, e52545. [CrossRef] [PubMed]

82. Spiga, F.M.; Maietta, P.; Guiducci, C. More DNA-aptamers for small drugs: A capture-SELEX coupled with surface plasmon resonance and high-throughput sequencing. *ACS Comb. Sci.* **2015**, *17*, 326–333. [CrossRef] [PubMed]

83. Rajendran, M.; Ellington, A.D. Selection of fluorescent aptamer beacons that light up in the presence of zinc. *Anal. Bioanal. Chem.* **2008**, *390*, 1067–1075. [CrossRef] [PubMed]

84. Morse, D.P. Direct selection of RNA beacon aptamers. *Biochem. Biophys. Res. Commun.* **2007**, *359*, 94–101. [CrossRef] [PubMed]

85. Boots, J.L.; Matylla-Kulinska, K.; Zywicki, M.; Zimmermann, B.; Schroeder, R. Genomic SELEX. In *Handbook of RNA Biochemistry: Second, Completely Revised and Enlarged Edition*; Hartmann, R. K., Bindereif, A., Schön, A., Westhof, E., Eds.; Wiley-VCH Verlag GmbH & Co: Weinheim, Germany, 2014; pp. 1185–1206. ISBN 9783527327645.

86. Ogasawara, H.; Hasegawa, A.; Kanda, E.; Miki, T.; Yamamoto, K.; Ishihama, A. Genomic SELEX search for target promoters under the control of the PhoQP-RstBA signal relay cascade. *J. Bacteriol.* **2007**, *189*, 4791–4799. [CrossRef] [PubMed]

87. Kim, S.; Shi, H.; Lee, D.K.; Lis, J.T. Specific SR protein-dependent splicing substrates identified through genomic SELEX. *Nucleic Acids Res.* **2003**, *31*, 1955–1961. [CrossRef] [PubMed]

88. Zimmermann, B.; Bilusic, I.; Lorenz, C.; Schroeder, R. Genomic SELEX: A discovery tool for genomic aptamers. *Methods* **2010**, *52*, 125–132. [CrossRef] [PubMed]

89. Davies, D.R.; Gelinas, A.D.; Zhang, C.; Rohloff, J.C.; Carter, J.D.; O'Connell, D.; Waugh, S.M.; Wolk, S.K.; Mayfield, W.S.; Burgin, A.B.; et al. Unique motifs and hydrophobic interactions shape the binding of modified DNA ligands to protein targets. *Proc. Natl. Acad. Sci. USA* **2012**, *109*, 19971–19976. [CrossRef] [PubMed]

90. Gawande, B.N.; Rohloff, J.C.; Carter, J.D.; von Carlowitz, I.; Zhang, C.; Schneider, D.J.; Janjic, N. Selection of DNA aptamers with two modified bases. *Proc. Natl. Acad. Sci. USA* **2017**, *114*, 2898–2903. [CrossRef] [PubMed]

91. Gold, L.; Ayers, D.; Bertino, J.; Bock, C.; Bock, A.; Brody, E.N.; Carter, J.; Dalby, A.B.; Eaton, B.E.; Fitzwater, T.; Flather, D.; et al. Aptamer-based multiplexed proteomic technology for biomarker discovery. *PLoS ONE* **2010**, *5*, e15004. [CrossRef] [PubMed]

92. Naduvile Veedu, R.; AlShamaileh, H. Next generation nucleic acid aptamers with two base modified nucleotides improve the binding affinity and potency. *ChemBioChem* **2017**, *9*, 9–12. [CrossRef]

93. Ochsner, U.A.; Katilius, E.; Janjic, N. Detection of Clostridium difficile toxins A, B and binary toxin with slow off-rate modified aptamers. *Diagn. Microbiol. Infect. Dis.* **2013**, *76*, 278–285. [CrossRef] [PubMed]

94. Chudinov, A.V.; Kiseleva, Y.Y.; Kuznetsov, V.E.; Shershov, V.E.; Spitsyn, M.A.; Guseinov, T.O.; Lapa, S.A.; Timofeev, E.N.; Archakov, A.I.; Lisitsa, A.V.; et al. Structural and functional analysis of biopolymers and their complexes: Enzymatic synthesis of high-modified DNA. *Mol. Biol.* **2017**, *51*, 474–482. [CrossRef]

95. Kimoto, M.; Yamashige, R.; Matsunaga, K.; Yokoyama, S.; Hirao, I. Generation of high-affinity DNA aptamers using an expanded genetic alphabet. *Nat. Biotechnol.* **2013**, *31*, 453–457. [CrossRef] [PubMed]

96. Sefah, K.; Yang, Z.; Bradley, K.M.; Hoshika, S.; Jimenez, E.; Zhang, L.; Zhu, G.; Shanker, S.; Yu, F.; Turek, D.; et al. In vitro selection with artificial expanded genetic information systems. *Proc. Natl. Acad. Sci. USA* **2014**, *111*, 1449–1454. [CrossRef] [PubMed]

97. Horiya, S.; Macpherson, I.S.; Krauss, I.J. Recent Strategies Targeting HIV Glycans in Vaccine Design Satoru. *Nat. Chem. Biol.* **2015**, *10*, 990–999. [CrossRef] [PubMed]

98. Temme, J.S.; Krauss, I.J. SELMA: Selection with modified aptamers. *Curr. Protoc. Chem. Biol.* **2015**, *7*, 73–92. [CrossRef] [PubMed]

99. Temme, J.S.; MacPherson, I.S.; Decourcey, J.F.; Krauss, I.J. High temperature SELMA: Evolution of DNA-supported oligomannose clusters which are tightly recognized by HIV bnAb 2G12. *J. Am. Chem. Soc.* **2014**, *136*, 1726–1729. [CrossRef] [PubMed]

100. Tolle, F.; Brändle, G.M.; Matzner, D.; Mayer, G. A Versatile Approach Towards Nucleobase-Modified Aptamers. *Angew. Chem. Int. Ed.* **2015**, *54*, 10971–10974. [CrossRef] [PubMed]

101. Warner, W.A.; Sanchez, R.; Dawoodian, A.; Li, E.; Momand, J. Multivalent glycocluster design through directed evolution. *Angew. Chem. Int. Ed.* **2013**, *80*, 631–637. [CrossRef]

102. Warner, W.A.; Sanchez, R.; Dawoodian, A.; Li, E.; Momand, J. Directed Evolution of 2G12-Targeted Nonamannose Glycoclusters by SELMA. *Chemistry* **2013**, *19*, 17291–17295. [CrossRef]

103. Blind, M.; Blank, M. Aptamer Selection Technology and Recent Advances. *Mol. Ther. Acids* **2015**, *4*, e223. [CrossRef] [PubMed]

104. Takahashi, M.; Wu, X.; Ho, M.; Chomchan, P.; Rossi, J.J.; Burnett, J.C.; Zhou, J. High throughput sequencing analysis of RNA libraries reveals the influences of initial library and PCR methods on SELEX efficiency. *Sci. Rep.* **2016**, *6*, 33697. [CrossRef] [PubMed]

105. Blank, M. Next-Generation Analysis of Deep Sequencing Data: Bringing Light into the Black Box of SELEX Experiments. *Methods* **2016**, *1380*, 85–95. [CrossRef]

106. Gevertz, J.; Gan, H.H.; Schlick, T. In vitro RNA random pools are not structurally diverse: A computational analysis. *RNA* **2005**, *11*, 853–863. [CrossRef] [PubMed]

107. Bartel, D.P.; Zapp, M.L.; Green, M.R.; Szostak, J.W. HIV-1 Rev regulation involves recognition of non-Watson-Crick base pairs in viral RNA. *Cell* **1991**, *67*, 529–536. [CrossRef]

108. Hesselberth, J.R.; Miller, D.; Robertus, J.; Ellington, A.D. In vitro selection of RNA molecules that inhibit the activity of ricin A-chain. *J. Biol. Chem.* **2000**, *275*, 4937–4942. [CrossRef] [PubMed]

109. Burke, D.H.; Scates, L.; Andrews, K.; Gold, L. Bent pseudoknots and novel RNA inhibitors of type 1 human immunodeficiency virus (HIV-1) reverse transcriptase. *J. Mol. Biol.* **1996**, *264*, 650–666. [CrossRef] [PubMed]

110. Knight, R.; Yarus, M. Analyzing partially randomized nucleic acid pools: Straight dope on doping. *Nucleic Acids Res.* **2003**, *31*, e30. [CrossRef] [PubMed]

111. Bittker, J.A.; Le, B.V.; Liu, D.R. Nucleic acid evolution and minimization by nonhomologous random recombination. *Nat. Biotechnol.* **2002**, *20*, 1024–1029. [CrossRef] [PubMed]

112. Carothers, J.M.; Oestreich, S.C.; Davis, J.H.; Szostak, J.W. Informational complexity and functional activity of RNA structures. *Science* **2004**, *126*, 5130–5137. [CrossRef] [PubMed]

113. Paige, G.F.; Wu, K.; Jaffrey, S.R. RNA mimics of green fluorescent protein. *Science* **2011**, *333*, 642–646. [CrossRef] [PubMed]

114. Xu, J.; Carrocci, T.J.; Hoskins, A.A. Evolution and characterization of a benzylguanine-binding RNA aptamer. *Chem. Commun.* **2016**, 549–552. [CrossRef] [PubMed]

115. Yang, K.A.; Pei, R.; Stefanovic, D.; Stojanovic, M.N. Optimizing cross-reactivity with evolutionary search for sensors. *J. Am. Chem. Soc.* **2012**, *134*, 1642–1647. [CrossRef] [PubMed]

116. Trevino, S.G.; Levy, M. High-throughput bead-based identification of structure-switching aptamer beacons. *ChemBioChem* **2014**, *15*, 1877–1881. [CrossRef] [PubMed]

117. Zhu, L.; Li, C.; Zhu, Z.; Liu, D.; Zou, Y.; Wang, C.; Fu, H.; Yang, C.J. In vitro selection of highly efficient G-quadruplex-based DNAzymes. *Anal. Chem.* **2012**, *84*, 8383–8390. [CrossRef] [PubMed]

118. Ruff, K.M.; Snyder, T.M.; Liu, D.R. Enhanced functional potential of nucleic acid aptamer libraries patterned to increase secondary structure. *J. Am. Chem. Soc.* **2010**, *132*, 9453–9464. [CrossRef] [PubMed]

119. Chushak, Y.; Stone, M.O. In silico selection of RNA aptamers. *Nucleic Acids Res.* **2009**, *37*, e87. [CrossRef] [PubMed]

120. Kim, N.; Gan, H.H.; Schlick, T. A computational proposal for designing structured RNA pools for in vitro selection of RNAs. *RNA* **2007**, *13*, 478–492. [CrossRef] [PubMed]

121. Kim, N.; Shin, J.S.; Elmetwaly, S.; Gan, H.H.; Schlick, T. RagPools: RNA-As-Graph-Pools-a web server for assisting the design of structured RNA pools for in vitro selection. *Bioinformatics* **2007**, *23*, 2959–2960. [CrossRef] [PubMed]

122. RAGPOOLS (RNA-As-Graph-Pools)—A Web Server. Available online: http://rubin2.biomath.nyu.edu/home.html (accessed on 27 October 2017).

123. Luo, X.; McKeague, M.; Pitre, S.; Dumontier, M.; Green, J.; Golshani, A.; Derosa, M.C.; Dehne, F. Computational approaches toward the design of pools for the in vitro selection of complex aptamers. *RNA* **2010**, *16*, 2252–2262. [CrossRef] [PubMed]

8

Nucleic Acid Aptamers: Emerging Applications in Medical Imaging, Nanotechnology, Neurosciences, and Drug Delivery

Pascal Röthlisberger [1], Cécile Gasse [2],* and Marcel Hollenstein [1],*

[1]   Institut Pasteur, Department of Structural Biology and Chemistry, Laboratory for Bioorganic Chemistry of Nucleic Acids, CNRS UMR3523, 28, rue du Docteur Roux, 75724 Paris CEDEX 15, France; pascal.rothlisberger@pasteur.fr

[2]   Institute of Systems & Synthetic Biology, Xenome Team, 5 rue Henri Desbruères Genopole Campus 1, University of Evry, F-91030 Evry, France

*   Correspondence: cecile.gasse@univ-evry.fr (C.G.); marcel.hollenstein@pasteur.fr (M.H.)

Abstract: Recent progresses in organic chemistry and molecular biology have allowed the emergence of numerous new applications of nucleic acids that markedly deviate from their natural functions. Particularly, DNA and RNA molecules—coined aptamers—can be brought to bind to specific targets with high affinity and selectivity. While aptamers are mainly applied as biosensors, diagnostic agents, tools in proteomics and biotechnology, and as targeted therapeutics, these chemical antibodies slowly begin to be used in other fields. Herein, we review recent progress on the use of aptamers in the construction of smart DNA origami objects and MRI and PET imaging agents. We also describe advances in the use of aptamers in the field of neurosciences (with a particular emphasis on the treatment of neurodegenerative diseases) and as drug delivery systems. Lastly, the use of chemical modifications, modified nucleoside triphosphate particularly, to enhance the binding and stability of aptamers is highlighted.

Keywords: aptamers; systematic evolution of ligands by exponential enrichment (SELEX); modified triphosphates; medical imaging; drug delivery; gene regulation; DNA origami; neurodegenerative diseases

---

1. Introduction

Over the last decades, the repository of genetic information in living organisms—DNA—and the vector for gene expression—RNA—have seen an impressive expansion in applications that substantially deviate from their natural functions. Indeed, nucleic acids play a key role in the development of gene silencing therapeutic agents [1,2], the construction of novel nanomaterials [3], and the crafting of biocatalysts [4–6]. In this context, aptamers are rapidly developing nucleic acid tools that consist of single stranded DNA or RNA molecules comprising 20–100 nucleotides [7], and that are capable of selective binding to a broad array of targets with remarkable affinity [8,9]. Hence, these functional nucleic acids are often considered as the nucleic acid equivalent of protein antibodies. However, unlike their proteinaceous counterparts, aptamers are not plagued by physical or chemical instability or by potential immunogenicity and can be produced on a relatively large scale by standard chemical synthesis with little batch-to-batch variation [10]. Even though natural aptamers interacting with RNA polymerases have recently been identified [11], these functional nucleic acids are generally isolated in vitro by a combinatorial method coined SELEX (Systematic Evolution of Ligands by Exponential Enrichment) [9,12,13]. During SELEX, large libraries of oligonucleotides (typically 10^{14}–10^{15} individual molecules) are challenged to bind to the intended target and iterative rounds of selection-amplification cycles are utilized to enrich the populations with high binding species. Since the invention of the SELEX protocol in the early 1990s, thousands of aptamers have been selected for targets ranging from small molecules [14] to larger entities such as proteins [15,16] or cells [17–19], and databases have been created to canalize this exponential growth of aptameric sequences and wealth of information [20–22]. The binding capacity of an aptamer is best described by its dissociation constant K_d, which in turn is given by the ratio of the dissociation and association rate constants (k_{off}/k_{on}) [23]. Typically, values in the low nM or even pM range are observed for potent aptamers. These impressive properties are reflected by the numerous clinical trials involving aptamers and the first FDA-approved oligonucleotide-based drug (Macugen®) [24,25]. Moreover, the versatility of the selection protocol and the high binding affinities have propelled aptamers in the forefront of numerous applications including for instance biosensing [26,27], proteomics [28], purification and biotechnology [29–31], therapeutics [25,32], and diagnostics [33,34]. Herein, we have chosen to give an overview and a brief description of the emerging but rapidly growing applications of aptamers. Particularly, we will discuss recent implications of aptamers as radiopharmaceutical tools for medical imaging purposes (MRI and PET imaging) and in neurosciences for the treatment and detection of Alzheimer's and Parkinson's diseases. We also discuss the combined use of aptamers with DNA origamis to develop novel nanomaterials and biosensing platforms. Since the development of all these new therapeutic, imaging, and sensing agents require means of targeted delivery, we also cover the use of aptamers as drug delivery systems and as gene silencing agents. The last facet of this review will involve a discussion on the possibility of using chemical modifications to enhance the general properties of aptamers and we will focus particularly on the direct use of modified nucleoside triphosphates (dN*TPs) in SELEX experiments.

2. Medical Imaging (MRI and PET)

The ease of chemical modification at both 3′- and 5′-ends [35] combined with the high target affinity and selectivity dramatically increases the potential of aptamers to serve as molecular imaging agents, particularly for magnetic resonance imaging (MRI) and positron emission tomography (PET) [36].

2.1. Aptamers and MRI

MRI is a highly efficient technique that provides non-invasive three-dimensional images of living systems and of biological events with sub-millimeter spatial resolution [37]. In MRI, exogenous contrast agents—mainly small molecules based on Gd^{3+}-complexes—are used to enhance the image contrast by increasing the longitudinal (T_1) or transverse (T_2) relaxation times [38,39]. An important research avenue in the field of MRI consists in the development of smart or responsive contrast agents which consist either of systems that induce a change in magnetic relaxation in the presence of a biochemical stimuli (Figure 1A) or conjugates that vector MRI probes to their intended targets and sites (Figure 1B) [40]. Smart contrast agents based on aptamers have been devised by the application of both strategies. Indeed, in a proof-of-principle article, Yigit et al. developed a method for the detection and bisosensing of adenosine in vitro [41]. The contrast agent chosen in this system relied on biocompatible superparamagnetic iron oxide nanoparticles (SPIONs) due to their excellent capacity at changing the nuclear spin relaxation of neighbouring water protons [42]. The SPIONs were coated with cross-linked dextran which in turn could be functionalized with 3′- or 5′-thiol-modified DNA sequences that were designed so as to partially hybridize to a potent anti-adenosine aptamer [43]. In the presence of the adenosine analyte, the aptameric section refolded into its three-dimensional binding pocket concomitantly disrupting the hybridization to the SPION carrying oligonucleotides. The disruption of the SPION clusters led to the dispersion of single nanoparticles which display larger T_2 values compared to the initial bioconjugate [41]. The observed brightening of the MR images was ascribed to the resulting increase in T_2 values. In a related system, anti-thrombin aptamers [15,44] were immobilized on cross-linked dextran coated SPIONs and upon binding to the thrombin target, the nanoparticles assembled into larger aggregates which led to a decrease in T_2 values (and thus a reduction of the brightness of MR images); a strategy that is often preferred in T_2-weighted MR imaging [45]. More recently, a similar strategy was applied however by replacing the SPIONs with a Gd^{3+}-based T_1-weighted contrast agent [46]. Indeed, a Gd-DOTA complex was connected (by standard amide bond formation chemistry) to the 3′-amino modified end of a DNA oligonucleotide designed to be partially complementary to the adenosine aptamer. The aptameric part was connected to streptavidin and released the Gd-DOTA-modified oligonucleotide from the large streptavidin complex upon binding to the target adenosine. The release of the contrast agent bearing oligonucleotide in turn led to an increase in T_1 value (and thus of the brightness of the MRI signal). In a conceptually related strategy, a catalytic DNA molecule (DNAzyme) [4,5] was used to release Gd-DOTA from a bulky complex [47]. Indeed, the RNA substrate was equipped with the Gd-DOTA complex while the UO_2^{2+}-dependent DNAzyme was connected to the protein streptavidin via a biotin moiety anchored at its 3′-end. In the presence of the analyte (UO_2^{2+}), the DNAzyme adopted its catalytically active structure and hydrolyzed the single embedded rA unit, releasing the contrast agent.

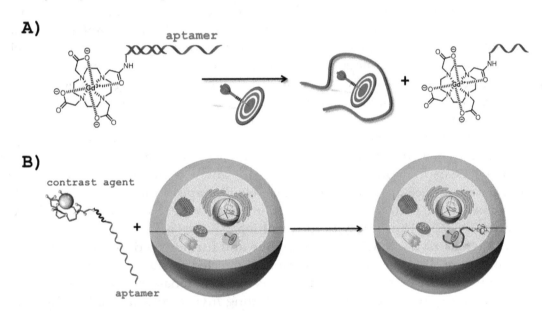

Figure 1. Strategies for the construction of aptamers acting as smart contrast agents: (**A**) Response to a biochemical stimuli: An oligonucleotide is equipped with a Gd^{3+}-DOTA complex. This oligonucleotide is complementary to part of the aptamer and upon binding to the target, the structural reorganization causes the Gd^{3+}-DOTA-labeled strand to dissociate from the duplex, which in turn increases the relaxation time and thus the brightness of the MRI signal [46]; (**B**) Vectoring to intended target: An aptamer is equipped with a contrast agent and will vector the probe directly to the intended target.

Monoclonal antibodies have been employed as tumor-specific ligands for the delivery of contrast agents [48,49]. By analogy to their proteinaceous counterparts, aptamers can play the role of vectors to transport contrast agents to specific regions of interest for MR imaging in vivo (Figure 1B) [50]. For instance, in a proof-of-principle study, the anti-thrombin aptamer was coupled to a Gd-DPTA (DPTA = diethylenetriaminepentaacetic acid) complex [51]. When the aptamer-Gd-DPTA bioconjugate was incubated with thrombin, significant relaxivity enhancements could be observed due to target interaction which increases the size of the contrast agent and concomitantly the rotational tumbling time [52]. In an ingenious system, Wang et al. bioconjugated a 2′-fluoropyrimidine-modified RNA aptamer specific for prostate cancer cells [53] on thermally cross-linked SPION [54]. The resulting construct not only allowed the transport of a contrast agent to the intended target and the concomitant MRI detection of prostate cancer cells in vitro but also served as a convenient scaffold for the selective delivery of the anticancer agent doxorubicin (DOX) [55]. The same RNA aptamer-SPION construct was then later used for the in vivo MRI detection of prostate tumors in a mouse model [56]. Related to this approach, a G-rich 26-nucleotide long aptamer coined AS1411 [57] was first conjugated to silver nanoclusters (Ag NCs) and then coupled to ultra-small gadolinium oxide (Gd_2O_3) nanoparticles [58]. The resulting Gd_2O_3-aptamer-Ag NCs system was successfully employed for the detection of MCF-7 tumor cells by MR and fluorescence imaging in vitro. Similarly, a variant of cell-SELEX was recently used to isolate aptamers that specifically bound inflamed human aortic endothelial (HAE) cells [59]. The resulting aptamer tightly bound the desired target (K_d = 82 and 460 nM for fixed and free HAE cells, respectively) and was conjugated to magnetic iron oxide particles for the efficient and selective in vitro detection of activated HAE cells.

2.2. Aptamers and PET Imaging

Positron emission tomography (PET) is another highly accurate biomedical imaging modality that is used worldwide in clinical diagnostic applications due to its capacity at providing tomographic resolution at any tissue depth [60,61]. Several radioisotopes (e.g., ^{18}F, ^{64}Cu, ^{11}C, ^{13}N, ^{124}I, and ^{68}Ga) display suitable properties for PET, namely a decay by emission of a positively charged particle (the positron (β^+)). Of these potential positron emitting radionuclides, ^{18}F is often preferred due to its rather convenient half-life ($t_{1/2}$ = 110 min), facile production, and favorable physical properties (clean decay and low emission energy) [62]. Besides the development of ^{18}F-based synthons and radiolabeling strategies, an important challenge in the field of PET imaging is the crafting of target-specific imaging agents [61]. The potential of aptamers at delivering radionuclide probes was realized early on by Lange et al. who photoconjugated an ^{18}F-labeled precursor on the 5′-amino-modified DNA thrombin aptamer [63]. More recently, an aptamer (sgc8) selective for the protein tyrosine kinase 7 (PTK7) [19] was ^{18}F-radiolabeled and the resulting bioconjugate was used for the detection and the quantification by PET imaging of the expression of PTK7 both in vitro and in different tumor mouse models [64]. Similarly, the very same sgc8 aptamer was radiolabeled by application of the copper (I)-catalyzed alkyne-azide cycloaddition (CuAAC or click reaction) using a metabolically stable ^{18}F-areene-arene derivative (Figure 2A) [65]. The affinity of the resulting ^{18}F-sgc8 aptamer for the PTK7 target could be determined (K_d = 1.1 nM) by PET imaging in vivo and further used for the mapping of tumoral PTK7 expression (Figure 2B). ^{18}F-arene tags were also connected by standard amide bond formation to an aptamer selective for the extracellular matrix glycoprotein tenascin-C which has been identified as a potential biomarker for various diseases, including myocarditis as well as different forms of cancer [66,67]. The anti-tenascin-C aptamer was also radiolabeled with a ^{64}Cu-NOTA complex and both the ^{64}Cu and ^{18}F-labeled aptamers were used for the in vitro and in vivo PET imaging analysis of the stability of the aptameric construct and for tumor localization in a mouse model [67]. These first examples of in vivo PET imaging guided by aptamer ligands were followed by a recent article by Zhu et al. where DNA aptamers were screened both in vitro and in vivo against the cell membrane HER2 which is overexpressed in various types of cancer [68]. In a first step, a traditional in vitro selection experiment was carried out using a His-tagged extracellular domain of HER2 to isolate aptamers against this biomarker. Following eight rounds of the protein-based selection, seven rounds of cell-SELEX were applied with live SKOV3 ovarian cancer cells as targets to ensure proper binding of the aptamer candidates under in vivo-like conditions. This dual selection strategy allowed for the isolation of different high affinity (K_D values in the low nM range) aptamers against SKOV3 cells, which were subsequently ^{18}F-radiolabeled by application of a click reaction protocol. The ^{18}F-labeled aptamers were injected intravenously into an SKOV3 xenograft tumor and their tumor uptake efficiency was evaluated by PET imaging analysis. The most efficient radiolabeled aptamer was then successfully used for the PET imaging detection of HER2 in an ovarian cancer mouse model [68].

In an alternative methodology, a hybridization reaction between a radiolabeled sequence that is partially complementary to an aptamer can be used to circumvent the tedious and material consuming purification step involved in the direct labeling of an aptamer. In this context, using click chemistry,

Park et al. ^{18}F-radiolabeled an oligonucleotide that recognizes the anti-nucleolin aptamer AS1411 [57] and used the resulting duplex for the in vitro and in vivo PET imaging detection and targeting of C6 tumors in a mouse model [69]. As clearly shown in this section, the potential of aptamers to serve as PET imaging agents only begins to be explored and additional and alternative ^{18}F-radiolabeling strategies [70–73] will certainly facilitate the application of aptamers in this imaging modality.

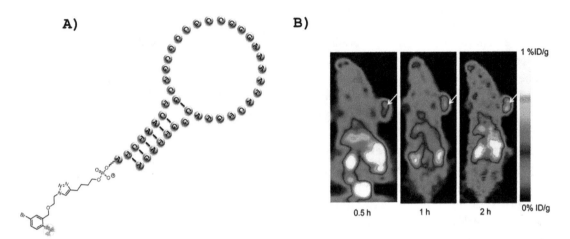

Figure 2. (A) Hypothetical secondary structure of the sgc8 aptamer and the ^{18}F-label; and (B) positron emission tomography (PET) images of a mouse model with HCT116 tumors, white arrows represent the HCT116 xenograft [65].

Aptamers have also been used in the related imaging technique SPECT (single photon emission computed tomography) where radionuclides (e.g., ^{99m}Tc or ^{111}In) decay by emitting a single γ-ray [74]. In a recent example, a 2'-fluoro-modified RNA aptamer (F3B) was raised against the human Matrix MetalloProtease-9 (hMMP-9) which is implicated in angiogenesis and believed to favor tumor cell formation [75]. The unmodified purine ribonucleotides were converted to 2'-O-methyl-modified units after SELEX and the resulting aptamer (F3Bomf) displayed a very high specificity and binding affinity for its intended hMMP-9 target (K_d = 20 nM). The fully modified aptamer F3Bomf could be connected to a ^{99m}Tc complex and successfully used for the detection of the tumor biomarker hMMP-9 in human glioblastoma sections [75]. The same aptamer F3Bomf was subsequently radiolabelled with ^{99m}Tc and ^{111}In complexes which revealed to be excellent candidates for the in vivo detection of hMMP-9 in mice bearing human melanoma tumors [76].

3. Aptamers for the Treatment and Diagnostics of Neurological Diseases

Aptamers can prevent protein-protein interactions, protein aggregation, and inhibit enzymes and thus represent alluring biomolecules for the modulation and mechanistic investigation of biological events related to neurodegenerative diseases. Surprisingly, the use of aptamers in the field of neurosciences is rather modest but is steadily increasing since new perspectives of traversing the blood brain barrier are rising for those molecules [77–79].

3.1. Aptamers and Neurotransmitters

The transmission of signals between two neurons is relayed by the exocytotic release of a battery of distinct chemical entities called neurotransmitters (see Figure 3A). Neurotransmitters consist mainly of single amino acids and their metabolites (e.g., glutamate and GABA, respectively) [80], biogenic monoamines (e.g., dopamine (DA), norepinephrine (NE), acetylcholine (Ach), and serotonin (5-HT)) [81], soluble gases (mainly NO, CO, and H_2S), and neuropeptides (e.g., neurokinin A and B, substance P or neuropeptide Y) [82]. In addition to their critical roles in numerous physiological functions, abnormal levels of neurotransmitters are indicators of various diseases including tumors [83].

tauopathies [84], and psychological and mood disorders such as schizophrenia [82,85], However, due to the presence of only low amounts, complex and delicate matrix composition, and the inherent chemical nature of neurotransmitters, detection of variation of their local concentrations is a rather difficult undertaking, even on samples obtained by ex vivo preparation [81], Aptamers have already demonstrated their capacity at recognizing and sensing various neurotransmitters [86]. Indeed, in an early report, Mannironi et al. have isolated an RNA aptamer that specifically recognized dopamine (K_d = 1.6 µM for the free molecule in solution) [87]. This RNA aptamer was fundamental in the development of a potent dopamine biosensing system. This approach exploited the three-dimensional folding produced by the binding event which in turn favored gold nanoparticle aggregation leading to a colorimetric change [88]. Similarly, the RNA aptamer was used for the selective (despite the presence of competitive catecholamines) and sensitive (100 nM to 5 µM concentration range) electrochemical detection of DA [89]. Surprisingly, when the sequence of this anti-DA RNA aptamer was converted into its DNA counterpart, the affinity of the aptamer was increased and the specificity retained [90]. This DNA version of the DA aptamer was recently used in an in vivo study assessing its capacity at reversing cognitive deficits caused by the non-competitive NMDA-receptor antagonist, MK-801 in a rat model [91]. However, the specificity and binding capacity of the DNA homolog was seriously questioned recently, and the authors even suggested that it was not acting as a true aptamer [92].

Additionally, aptamers were also raised against the biogenic monoamines norepinephrine [93], acetylcholine [94], and serotonin (developed by Base Pair Biotechnologies, Inc., Pearland, TX, USA) [95].

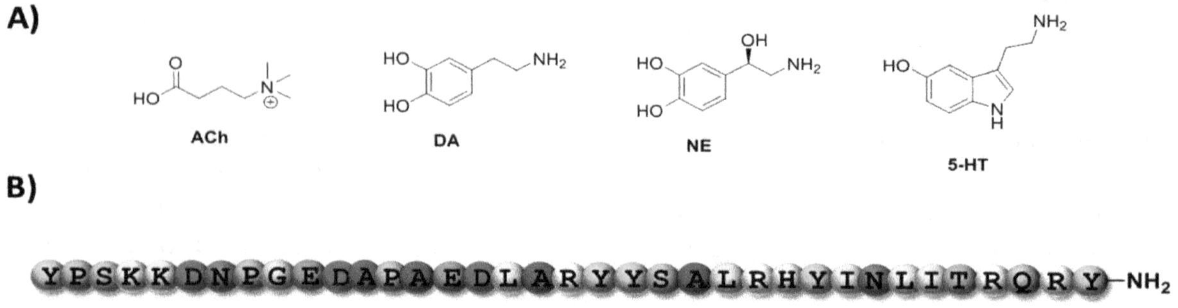

Figure 3. (A) Chemical structure of the main biogenic monoamine neurotransmitters; and (B) amino acid sequence of neuropeptide Y [96].

Neuropeptides represent the largest family of neuromessengers and can modulate both gene expression and synaptic communication [82,97]. Due to their larger size and broader chemical diversity, neuropeptides bind to their targets with higher affinities than biogenic monoamines and are consequently present in even lower quantities. Neuropeptides thus represent attractive targets for aptamer selection to devise potent sensing and quantification systems. In this context, first selection campaigns aimed at raising aptamers against neuropeptide Y (Figure 3B), which is negatively charged (pI = 5.52) at pH 7.0, and thus represents a challenging target [98,99]. First, an RNA aptamer was isolated and shown to bind tightly to the C terminus of neuropeptide Y (K_d = 370 nM) and displayed no cross-reactivity with the closely related (~50% sequence homology) human pancreatic polypeptide (hPP) [99]. More recently, DNA aptamers were also raised against neuropeptide Y and displayed similar affinities (K_d values in the 0.3–1 µM range) and selectivities to the RNA counterpart [98]. One DNA aptamer was integrated in a graphene-gold nanocomposite-based sensing platform for the fast, selective, and precise in vitro detection of neuropeptide Y [100]. This sensing platform displays a detection limit of 10 pM as well as high selectivity and fast response. Similarly, Banerjee et al. developed an aptasensor based on carbon fiber amperometry to detect neuropeptide Y in pheochromocytoma 12 cells [101].

The undecapeptide substance P is a member of the tachykinin family and is an essential excitatory transmitter involved in numerous important biological activities and functions. In light of its high biological and neurological significance, an RNA aptamer was isolated against substance P [102]. Indeed, an automated SELEX procedure with the D-peptide of substance P as target was applied to isolate an L-RNA aptamer which could be converted to its corresponding Spiegelmer (D-RNA) [103] which bound to the naturally occurring L-substance P with high affinity (K_d = 40 nM). This Spiegelmer was also efficiently used to inhibit the substance P-mediated calcium release in human AR42J cells (IC_{50} = 45 nM). A similar strategy was applied in the isolation of a Spieglemer (D-RNA) aptamer against the neuropeptide nociceptin/orphanin FQ (N/OFQ), involved in numerous capital biological and neurological responses such as anxiety, pain, and stress [104]. The most potent Spiegelmer, NOX 2149, recognized N/OFQ with high affinity (K_d = 0.2 μM) and concomitantly prevented N/OFQ from binding to its receptor (IC_{50} = 110 nM). Aptamers have also been raised against the neuropeptides somatostatin [105], ghrelin [106], Glucagon [107], angiotensin II [108], and calcitonin gene-related peptide 1 (α-CGRP) [109], as well as against certain receptors such as neurotensin receptors [110,111], and the cholecystokinin B receptor [112].

3.2. Aptamers and Tauopathies

Tauopathies are progressive neurodegenerative disorders including Alzheimer's disease (AD), Parkinson's disease (PD), Huntington's and prion diseases, and are characterized by the presence of aggregates of the microtubule-associated protein tau in the brain [113]. Even though the exact origins and the molecular mechanisms are vastly unknown, it is believed that misfolded and abnormal forms (often hyperphosphorylated) of the wild-type proteins are involved in the physiopathology of these diseases by acting as seeds for the aggregation of these proteins. Aptamers could thus contribute to this field as tools for the investigation of the origin of tauopathies and for the detection, the prevention, and the treatment of these disorders [77,78], as highlighted in this section for AD and PD.

3.2.1. Alzheimer's Disease

In AD, the combined accumulation and deposition of abnormal forms of tau protein and amyloid β (Aβ) peptides in the human brain is followed by a progressive functional disruption of neuronal networks [114]. Consequently, both tau and Aβ proteins represent valid targets for aptamer selection experiments.

Tau proteins play an important role in the stabilization and assembly of microtubules and display little propensity at aggregating and oligomerizing when found in their native folds and states. In AD, aggregates of hyperphosphorylated tau are thought to be transmitted in a prion-like manner that proceeds along connected neurons throughout the brain (the so-called tau-hypothesis) [115]. The understanding of how and why tau protein aggregates are capable of propagating in the brain is an important issue in neurosciences. In order to develop new tools to investigate and prevent this aggregation, Kim et al. have recently reported the isolation of an RNA aptamer against the longest isoform of human tau (tau40, 2N4R) [116]. The resulting aptamer efficiently prevented the oligomerization of tau monomers in vitro without affecting its degradation but on the other hand was not capable of disentangling pre-existing tau oligomers. Interestingly, the RNA aptamer could also delay tau oligomerization in DOX-inducible tau HEK293 cells and significantly reduced interneuronal tau propagation in primary rat neuronal cells, underscoring the inhibitory potential of aptamers for the regulation of tau oligomerization. A ssDNA sequence capable of binding to the human tau isoforms 381 and 410 (K_d = 0.19 and 0.35 μM, respectively) was identified not by SELEX but by kinetic capillary electrophoresis [117]. This DNA oligonucleotide was subsequently used in an aptamer-antibody sandwich assay for the detection of tau 381 in human plasma (limit of detection of 10 fM) [118].

In addition to abnormal tau protein accumulation, the deposition of rather short (~4 kDa) Aβ peptides is believed to be a key step in the progression of AD—known as the Aβ-hypothesis [119]. Aβ peptides stem from the sequential cleavage of the larger transmembrane glycoprotein amyloid precursor protein (APP) mainly mediated by the combined action of a β-secretase (also known as β-site APP cleaving enzyme-1; BACE1) and a γ-secretase complex (Figure 1). Both aspartyl proteases hydrolyze APP into several Aβ isoforms (mainly Aβ38, Aβ40, and Aβ42) which first assemble into synaptotoxic oligomers and then into amyloid fibrils, often considered as one of the major toxic agents in AD [119]. Lastly, Aβ oligomers also seem capable of binding to normal prion protein (PrP^C) with high affinity which might be at the origin of the toxicity of these oligomers [120]; this hypothesis, however, remains somewhat controversial [121–123]. Consequently, PrP^C, Aβ monomers and oligomers, BACE1, and the γ-secretase complex all represent relevant targets for aptamer selections to modulate, inhibit, or understand their functions [78,79].

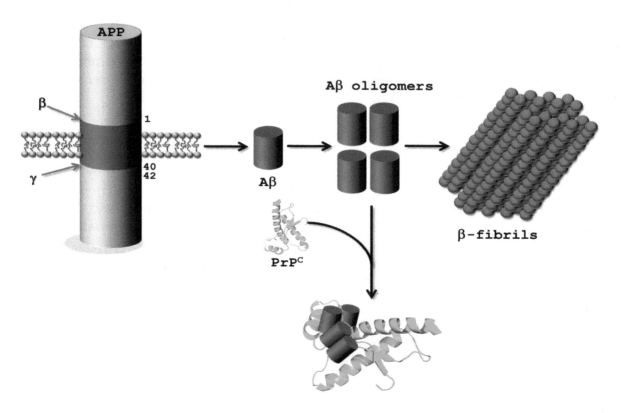

Figure 4. Schematic depiction of the formation of Aβ fibrils: BACE1 cleaves APP (Aβ sequence numbers are shown only) in the extracellular domain and the resulting fragment remains membrane-bound where it is cleaved by the γ-secretase complex into Aβ peptides (only the main Aβ40 and Aβ42 isoforms are shown). Monomeric Aβ peptides then aggregate to form oligomers and eventually β-fibrils. The cellular prion protein PrP^C can also bind to Aβ oligomers [78,119].

Inhibition of the catalytic activity of BACE1 or the γ-secretase complex could directly hinder the formation of the toxic Aβ oligomers and β-fibrils. Consequently, both a natural and a modified DNA aptamer have been raised against the transmembrane protease BACE1 [124]. The natural DNA aptamer binds BACE1 with high affinity (K_d = 69 nM) and was shown by a FRET assay to inhibit its activity both in vitro (IC$_{50}$ = 242 nM) and in an AD cell model [124]. The modified aptamers, selected with the triphosphate 5-chloro-dUTP and 7-deaza-dATP [125], displayed similar properties to the unmodified sequence, albeit with a slightly higher binding affinity and an interesting agonist/antagonist behavior [126]. Similarly, an RNA aptamer (called S10) was selected against the cytoplasmic tail B1-CT of BACE1 and showed a remarkable affinity for both phosphorylated and nonphosphorylated BACE1 (K_d = 330 and 360 nM, respectively) [127]. Aptamer S10 was also capable of binding to cellular BACE1 and did not prevent binding of the protein factor GGA1 to the cytoplasmic tail or the casein kinase-mediated phosphorylation of the single serine S498 located in the B1-CT tail. Thus far, no aptamers have been raised against the membrane bound γ-secretase nor any of its constitutive proteins (i.e., presenilins, nicastrin, APH-1, and PEN-2) [119].

The Aβ40 and Aβ42 isoforms readily associate to form soluble but toxic Aβ oligomers. In this context, an RNA aptamer coined β55 was identified by SELEX and shown to recognize its intended target Aβ40 with high affinity (K_d = 29 nM), albeit not in its monomeric or oligomeric forms but as fibrillar assemblies [128]. Furthermore, Farrar et al. also demonstrated that β55 was capable of binding to amyloid plaques in ex vivo human AD brain tissue slices. Remarkably, this binding event was also confirmed in vivo using a fluorescently labeled β55 in a transgenic mouse model [129]. Having realized that β55 only bound to Aβ40 polymers despite using a monomeric species in SELEX, Rahimi et al. set up an in vitro selection experiment that involved covalently-stabilized oligomers of Aβ40 [130]. Surprisingly, the resulting RNA aptamers did not bind to Aβ40 oligomers but only to Aβ40 fibrils, showing the high and natural propensity of nucleic acids to recognize fibrillar motifs in protein assemblies. This propensity was further confirmed by the substantial cross-reactivity of the isolated aptamers with Aβ42 and other amyloid fibrils. Another set of RNA aptamers was obtained through a selection experiment that used monomeric Aβ40 conjugated to gold nanoparticles; this target was hypothesized to act as a model of Aβ oligomerization and to allow both binding to the Aβ40 peptide and facile separation from unbound material [131]. Two aptamers, obtained by different elution protocols from the Aβ40-gold nanoparticles, indeed recognized monomeric Aβ40 (K_d values of 22 and 11 μM) and were capable of inhibiting Aβ fibrilization.

Lastly, the prion protein (native PrP^C or infectious isoform PrP^{Sc}) is an interesting target for aptamer selection due to its implication in AD (Figure 4) and prion diseases such as spongiform encephalopathies. Consequently, various DNA and RNA aptamers have been isolated against bovine [132], mouse [133], and human [134] PrP^C as well as PrP^{Sc} [135]. However, we refer the interested reader to a recent and concise review article thoroughly covering this topic [77].

3.2.2. Parkinson's Disease

As for AD, the origin of the PD pathology is believed to be connected to the formation of misfolded protein aggregates. However, in PD, the presynaptic neuronal, 140 amino acid-long protein α-synuclein and not Aβ has been identified as a possibly responsible agent for the pathogenesis [136]. The aggregation of α-synuclein proteins into oligomers eventually leads to the formation of fibrils which then accumulate in cytosolic filamentous inclusions called Lewy bodies, which are the hallmark of PD [137]. Consequently, the first example of a DNA aptamer against α-synuclein—coined M5-15—was obtained by SELEX using the monomeric protein as target in the selection protocol [138]. However, M5-15 preferentially bound to α-synuclein oligomers and did not recognize the monomeric form. This inherent conformation specificity prompted the authors to isolate other DNA aptamers using a competitive screening method based on aptamer blotting and α-synuclein oligomers as target [139,140]. The resulting aptamers presented K_d values in the low nM range that selectively recognized the oligomeric form of α-synuclein over monomers and fibrils [140]. Interestingly, the isolated aptamers also bound to Aβ40 oligomers (with slightly lower K_d values), hinting at the possibility of a selective recognition of aggregates with β-sheet-rich proteins. One of the isolated aptamers (T-SO517) was recently integrated in a potent label-free aptasensor system for the selective detection of α-synuclein oligomers (the limit of detection was in the low nM or pM depending on the analysis method) [141]. Aptamer T-SO517 was also integrated in a fluorescent sensing platform that enabled the detection of Aβ40 oligomers down to 12.5 nM [142]. Similarly, a nanocomposite of aptamer T-SO508, gold nanoparticles, and thionine was used as probe for the selective and sensitive detection of Aβ40 oligomers (100 pM limit of detection) [143]. Lastly, variants of T-SO508 were either grafted on magnetic nanoparticles to detect Aβ oligomers (detection limit of 36 pM) [144] or combined with abasic site-containing molecular beacons to monitor Aβ aggregation [145].

Lastly, low levels of dopamine are also frequently found in patients suffering from PD and thus, all the anti-DA aptamers described in Section 3.1 could be of use for the detection, monitoring, and treatment of PD. Other potential targets are the other two members of the synuclein family, namely β-synuclein and γ-synuclein which might be involved in neurodegenerative diseases.

4. Aptamers as Key Components in the Fabrication of Smart DNA Origami Objects

The predictable nature and the high degree of fidelity of DNA base pairing are the origin of the development of a plethora of DNA-based nanomaterials [3,146]. Of particular interest are DNA origamis which are created by combining large single-stranded DNA frameworks (e.g., M13 bacteriophage genomic DNA (7249 nucleotides) [147]) with hundreds of shorter (20–60 nucleotides) oligonucleotides (called staple strands) partially complementary to particular sequences on the genomic DNA scaffold so as to form pre-designed two- and/or three-dimensional folds [3,148]. Since both aptamers and DNA origami are made out of the very same biopolymer and because of the inherent binding capacity of aptamers, a combination of both nucleic acid-based materials has been used for different applications including: (1) defined spatial positioning of proteins (and other ligands of interest) on DNA arrays for medical diagnostic, biosensing, or tissue and material engineering purposes [149,150]; (2) development of nanorobots for the delivery of drugs and other payloads [151,152]; and (3) the creation of multimodal sensing platforms [153].

Chhabra et al. were the first to report on the immobilization of proteins at specific locations of two-dimensional DNA nanoarrays [149]: two tile double-crossed (DX) DNA molecules [154] were equipped either with the thrombin aptamer [44] or an aptamer raised against the Platelet-derived growth factor (PDGF) [155] and combined with another set of DX tiles to form a two-dimensional DNA network with interspaced alternating lines of both aptamers. An atomic force microscopy (AFM) analysis showed that the corresponding target proteins successfully bound to their respective aptamers following their sequential addition on the network. This approach was then extended to a DNA origami constructed with 200 staple strands which resulted in the formation of rectangular two-dimensional DNA nanoarrays. The inclusion of the two aforementioned aptamers controlled the spatial positioning of the respective targets (i.e., thrombin and PDGF proteins) and therefore the generation of programmable high-density protein-DNA nanoarrays [149]. In a conceptually related construct, two aptamers binding to different epitopes of thrombin [15,44] were appended at different locations of rigid four- or five-helix bundle DNA tiles to evaluate the optimal inter-aptamer distance that ensures the highest binding to thrombin after formation of the hetero-aptamer system [150]. The optimal inter-aptamer distance appears to be at around 5.3 nM (i.e., slightly over the size of thrombin) for high affinity bivalent binding (K_d ~10 nM). The same strategy was then applied for the construction of a rectangular-shaped two-dimensional DNA origami object [156] two lines of each aptamer separated by 5.3 nM were included in the scaffold of the origami tile resulting in efficient bivalent binding to thrombin by dual-aptamer lines. These initial proof-of-principle studies paved the way for the development of smart DNA origami objects. For instance, in a landmark work, Douglas and co-workers designed a DNA nanorobot in the form of a hexagonal barrel (Figure 5A) consisting of 196 staple oligonucleotides and the 7308 nucleotide-long genomic DNA of an M13-like phage [151]. The two domains that constitute the barrel are connected by single-stranded oligonucleotidic hinges at the rear and closed in the front by an aptamer sequence hybridized to a partially complementary sequence. In the presence of protein tyrosine kinase 7 (PTK7), the sgc8c aptamers (specific for PTK7) [19] dissociate from the lock-duplexes and the DNA nanorobot opens as a direct consequence of the formation of the aptamer-ligand complex. The different payloads (i.e., either 5-nm gold nanoparticles or antibody fragments against the human leukocyte antigen–A/B/C) are affixed on the inner walls of the barrel (either through 5′-thiol or 5′-amino modified linkers, respectively) and can be released upon opening of the DNA nanorobot. This ingenious system was incubated with different cell lines expressing the HLA–A/B/C antigen with a combination of molecular inputs. An increase in fluorescence caused by the binding of the antibody fragments on the cell surface could only be detected in the presence of the correct "key", namely PTK7. Lastly, the DNA origami system could be extended to yield a more discriminatory system by incorporating combinations of aptamers recognizing different targets.

Godonoga et al. recently reported a DNA origami-aptamer construct for the specific recognition of a malaria biomarker [152]. In this system, a rectangular DNA origami [156] was fabricated by including the aptamer 2008s that specifically recognizes the malaria biomarker *Plasmodium falciparum* lactate dehydrogenase (*Pf*LDH) with high affinity (K_d = 42 nM) through the formation of a 2:1 aptamer:ligand complex [157]. An AFM analysis clearly revealed that only *Pf*LDH and not the related human homolog (hLDH) bound to the surface of the DNA origami decorated with the aptamer 2008s (Figure 5B), thus clearly demonstrating that large supramolecular DNA constructs could be used as diagnostic tools for diseases.

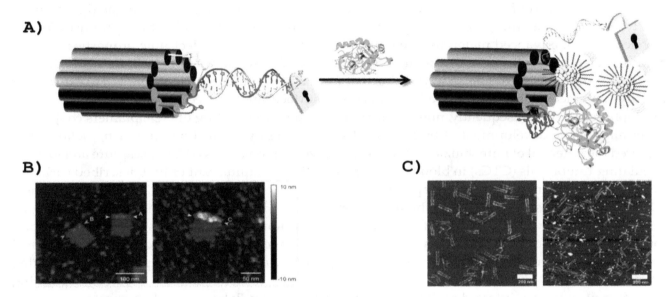

Figure 5. (**A**) Schematic representation of the aptamer-gated DNA nanorobot [151]: the aptamer (magenta) is locked in a double-stranded form by a partially complementary sequence (yellow) and both are grafted on the nanorobot. The payloads (either gold nanoparticles (shown) or antibody fragments) are constrained to remain within the DNA construct and only the recognition of intended target by the aptamer (green) will unlock the DNA nanorobot and enable the delivery of the payload; (**B**) AFM images of a DNA origami-aptamer construct in the presence of a non-target protein (hLDH; left-hand side) and presence of the target protein (*Pf*LDH; right-hand side) [152]; (**C**) AFM analysis of a DNA origami equipped with a split aptamer system in the closed (left-hand side) and open (right-hand side) forms [153].

Lastly, DNA origami-aptamers systems can also be used as smart biosensing platforms. In a recent contribution, Walter et al. blended the pinching capacity of a nanomechanical DNA origami forceps [158] with the biosensing capacity of split aptamer systems [159]. Indeed, the constituting sequences of the ATP-specific split aptamer were equipped with green- and red-emitting photostable cyanine dyes that act as energy donor and acceptor, respectively [160] and were subsequently appended on one arm of the DNA origami forceps. In the absence of the analyte (i.e., ATP) the constructs remained in an open form (Figure 5C) and emitted green light (absence of energy transfer). On the other hand, in the presence of ATP, the split aptamer sequences bound together which resulted in the closure of the DNA origami forceps which concomitantly led to red light emission due to an efficient energy transfer (Figure 5C). The sensing event can be observed by both AFM (Figure 5C) or by the change in fluorescence from green to red, hence following the concept of DNA traffic lights [161].

5. Aptamers as Drug Carriers and Gene Regulating Agents

Repair of the biological damage or chemical imbalance caused by a disease often requires an intervention in the form of a drug at the precise location of the inflicted disorder. Depending on the disease, the nature and the location of the damaged biomaterial can vary substantially ranging from simpler constructs such as proteins or nucleic acids to larger systems such as cells or entire tissues. In order to improve the therapeutic index of drugs and reduce offside effects and some inherent toxicity, numerous systems have been devised to carry drugs in a stable form to the intended sites [162,163]. In this context, aptamers represent ideal candidates as delivery systems for conventional or encapsulated drugs, but also for therapeutic oligonucleotides and peptides due to their properties (vide supra) [25].

5.1. Nanomaterials Conjugated Aptamers

Due to the ease of functionalization of oligonucleotides, aptamers can be conjugated to virtually any type of nanomaterial [164]. Amongst these, graphene oxide (GO) holds high promises for the development of advanced materials due to its interesting optical and electronic properties but also because of the possibility of constructing barrier films and new classes of membranes. In addition, graphene oxide consists of interspaced sheets that form 2D structures which are known to exhibit low toxicity, high mechanical flexibility, a large accessible surface, and an excellent quenching capacity of fluorophores [165]. Consequently, numerous reports exist on the crafting of GO-aptamers conjugates, especially for the development of biosensors [166]. In a highly interesting approach, Nellore et al. explored the potential of immobilizing aptamers selective for biomarkers on GO to capture and identify circulating tumor cells (CTCs) in blood—an approach that is reminiscent of that described earlier in this review for the immobilization of aptamers on DNA origamis (see Section 4) [167]. Thus, aptamers selective for the prostate-specific membrane antigen (PSMA), HER2, and the carcinoembryonic antigen (CEA) biomarkers were covalently affixed on two-dimensional GO sheets by standard amide bond formation with the carboxylic acid residues present on GO and then converted to a three-dimensional architecture using polyethyleneglycol (PEG) as a cross-linking agent. The high capacity of the resulting aptamer-coated 3D GO foam-based membrane at capturing tumor cells was proven by separating the different cells from infected rabbit blood. In a related study, Bahreyni et al. connected aptamers selective against the membrane protein mucin MUC1 which is overexpressed in various epithelial carcinomas, through π-π stacking interactions with the GO surface [168]. GO was charged either with the anti-MUC1 aptamer or a fluorescently-labeled aptamer selective for MUC1 cytochrome C. Upon successful internalization of the labeled aptamer nano complex into target MDA-MB-231 and MCF-7 cells, a strong fluorescent signal was observed indicating binding and release of the fluorescein labeled aptamers to cytochrome C, while no effects were observed with non-targeted cell lines (HepG2). This theranostic system appeared to be non-invasive and selective for the targeted cells inducing apoptosis.

An elegant "on and off" strategy reported by Tang et al. required a non-covalent assembly of the Cy5.5-labeled AS1411 aptamer, targeting nucleolin, on the surface of a GO-wrapped, DOX-loaded mesoporous silica nanoparticle [169]. This system allowed a light induced administration of DOX that could be monitored by fluorometric measurements: a first fluorescent signal of the Cy5.5-labeled aptamer indicated real time endocytosis of the aptamers into the target cell, while a second fluorescent signal was induced by laser irradiation because absorption and transduction of the near infrared light by the GO structures lead to local heat and expansion of the GO sheets and thus the release of the DOX molecules in a light dependent manner.

Besides GO, gold-based nanomaterials and derivatives have attracted considerable attention for the crafting of aptamer-based materials and devices due to their stability and their advantageous optical and electronic properties [170–172]. In particular, gold nanoparticles have served as drug delivery systems for aptamers since they generally increase the stability of ssDNA towards nuclease degradation, the cellular uptake capacity of oligonucleotides, as well as their biocompatibility and usually induce a limited immune response [173,174]. For instance, Huang and coworkers used gold nanoparticles to attach thrombin-binding aptamers on their surface [175]. The complexed thrombin on the gold nanoparticles (d = 13 nM) efficiently inhibited the thrombin activity against fibrinogen upon activation with green laser light. The anticoagulant activity of these complexes was found to be 30 times more potent than recent commercially available drugs (heparin, argatroban, hirudin, or warfarin) exhibiting a good biocompatibility, low toxicity and showed excellent half-life stability in serum ($t_{1/2}$ >14 d) [175].

Besides the use of gold nanoparticles as drug delivery systems and diagnostic tools, Niu et al. demonstrated that conjugation of the sgc8c aptamer [19] with a gold N-heterocyclic gold (I) complex (NHC-Au^I-aptamer), known to induce cell apoptosis, resulted in internalization of the bioconjugate specifically into CCRF-CEM leukemia cells and exhibited excellent cytotoxicity [171]. The sgc8c-7aptamer employed in this strategy targeted the receptor protein tyrosine kinase 7 (PTK-7) that is overexpressed in CCRF-CEM leukemia cells and allowed for a 30-fold increase in cytotoxicity compared to the unmodified NHC-Au^I complex. No toxicity to off-target cells (in this case, K526 cells) was observed with the bioconjugate, on the other hand, a dose-dependent toxicity was observed with CCRF-CEM cells, underscoring the usefulness of this approach.

Quantum dots (QDs) represent another very attractive class of inorganic nanoparticles for the formation of aptamer bioconjugates [176–178]. QDs have unique photophysical properties, including color tunability and bright and extremely photostable fluorescence, that is keeping them in the forefront of numerous sensing applications [178]. In addition, QDs have also been used in dual imaging-drug delivery systems based on aptamers [179]. For instance, Su and co-workers covalently attached a capture sequence on a near infrared CuInS$_2$-QD via amide coupling to connect a MUC1-aptamer to this nanoparticle through the formation of Watson–Crick base pairs [180]. Multiple CG-motifs present in the resulting duplex served as intercalation sites for daunorubicin (DNR), which is a drug for the treatment of acute myeloid leukemia. This system allowed a specific delivery of the DNR to prostate cancer cells in vitro and a concomitant sensing of the presence of DNR in the construct due to a marked drop in fluorescence intensity upon binding of the drug. A high cytotoxicity was found for MUC1 positive PC-3M cells but not for MUC1 negative HepG2 cells when treated with the DNR-loaded bioconjugate [180].

In addition to inorganic frameworks, aptamers can be coupled to various organic nanomaterials including DNA constructs [164], DNA micelles [181,182], aptamer-based hydrogels [183], lipids [184], or even vitamins [185,186]. Recently, Dai et al. developed a DNA tetrahedron (Td) labeled with an aptamer targeting MUC1. In this drug delivery system, the aptamer serves for the targeted delivery to MUC1-positive breast cancer cells while the DNA tetrahedron is instrumental for the intercalation of DOX [187]. A fluorescence-based drug loading experiment showed that each aptamer-Td construct with an average size of 12.4 nM could carry up to 25 DOX molecules. A high red fluorescent signal of free DOX molecules could be observed upon binding of the negatively charged complex to the MUC1 positive breast cancer cells but not with MUC1 negative cells. Accordingly, a very high cytotoxicity was observed when MUC1 positive cancer cells were treated with the aptamer-Td conjugate [187].

A last category of nanoparticles that will be considered consist of linear block copolymers and dendritic polymers which have been extensively used as encapsulating devices in drug delivery systems [188,189]. Aptamers have also been used to decorate the surface of these polymeric entities and to facilitate targeted delivery. This concept was developed by the Langer laboratory who reported a strategy for a drug encapsulation by a triblock copolymer comprising a controlled release polymer (poly (lactic-co-glycolic-acid); PLGA), a hydrophilic polymer (PEG), and the 2′-fluoro-modified RNA A10 aptamer (see Sections 5.5 and 6) that selectively recognizes the prostate cancer specific membrane antigene (PSMA) [55,190]. By systematically changing the composition of the complex, a narrow ratio between PEG and aptamer could be determined that enabled maximal specific binding to the target and cellular uptake, efficient drug (in this case the anticancer drug Docetaxel) release, high antibiofouling properties, and minimized self-aggregation and thus undesired accumulation in the spleen [190]. A similar approach was also used to selectively deliver a Pt(IV)-prodrug [191], a cisplatin prodrug [192], and other anticancer drugs [193–195], while dendritic polymers-aptamer systems have also successfully been used for in vivo [196] and in vitro [197] tumor imaging and drug delivery [198].

5.2. Micelles and Liposomes Conjugated Aptamers

Liposomes are small spherical and artificial vesicles with one or multiple lipid bilayers and have already been suggested as potentially highly efficient drug deliver systems at an early stage [199]. Indeed, Huwyler and co-workers already described in the early 1990s the use of liposomes decorated with monoclonal antibodies to delivery encapsulated tritium-radiolabeled daunomycin in a directed manner to target cells [200]. This strategy has many benefits including the modulation of the affinity of the bionconjugate complex by simply changing the ratio of antibodies present on the lipid bilayer surface along with the impressive quantity of drug molecules (>10,000) that can be delivered in a selective manner [200,201]. In a more recent study, Ara et al. used liposomes composed of lipid bilayers that are labeled with an aptamer instead of a monoclonal antibody [202] to ensure selective cellular uptake [203]. The aptamers were covalently linked on PEG 2000 disteraoyl phosphotheanolamine and targeted the primary cultured mouse tumor endothelial cells (mTEC). By application of a standard lipid hydration method, the corresponding aptamer-labeled PEG2000 liposomes and PEG200 liposomes (negative control) were formed. Fluorescent measurements and confocal laser scanning microscopy showed an increased uptake in mTEX cells of the aptamer-PEG liposomes compared to the unlabeled liposomes. In vitro experiments revealed that approximately 39% of the aptamer-PEG-liposomes could escape the endosomes in a receptor mediated way followed by clathrin-mediated endocytosis. Lastly, in vivo experiments with the aptamer-conjugated liposomes were performed with human renal cell carcinoma (OS-RC-2 cells) inoculating mice using confocal laser scanning microscopy. These experiments clearly indicated that the aptamer modified liposomes strongly accumulated (co-localization ratio of 25%) on tumor vasculature compared to non-labeled liposomes, where a co-localization ratio of only 3% accumulation was observed [203]. In a similar approach, a 2′-fluoro-modified RNA aptamer against the cancer stem cells (CSC) surface markers CD44 was affixed on the surface of a PEG-functionalized liposome by application of the thiol-maleimide click reaction [204]. The resulting construct was shown to act as a very efficient potential drug delivery system due to its high selectivity and specificity for CD44 positive cells.

In a recent study by Plourde et al., the high binding capacity of aptamers served as a driving force for the incorporation of DOX into cationic liposomes (Figure 6A) [205]. Indeed, different versions of an anti-DOX DNA aptamer [206] were designed by either adding an additional base-pair on the binding motif or by including two binding motifs into one construct. Upon intercalation of DOX into the aptamer its intrinsic fluorescence was quenched and was used to determine both the loading efficiency and the binding affinity. The drug aptamer complexes were incorporated into cationic liposomes (d < 200 nM) via electrostatic interactions, where encapsulation of DOX into liposomes was greatly enhanced (ten times higher) when using an aptamer compared to negative controls without any specific DOX binding sites. This aptameric-encapsulation was compared to the Doxil-like formulation, which is a commercialized version of liposomes of DOX that allows the loading of up to 10,000 molecules through strong entrapment but concomitantly reduces the therapeutic efficiency due to a slow release [207]. The aptamer-based loading strategy offered a number of advantages over the Doxil-like formulation including similar loading capacity, faster release, and a higher therapeutic efficiency. Cytotoxicity assays revealed that liposomes with aptamers having an intermediate affinity (K_d = 334 nM) for DOX exhibited the higher therapeutic effects. The same approach was then extended to the amphiphilic drug tobramycin used for the treatment of lung infections. It is known that encapsulation of tobramycin in liposomes increases the in vivo efficacy compared to the free drug. The low encapsulation efficiency of tobramycin in liposomes is a limiting factor in this approach. The active loading with an aptamer allowed the encapsulation of up to six times more drug compared to passive encapsulation. Therefore, the authors reasoned that this strategy could find a broad application to load a large variety of drugs into liposomes.

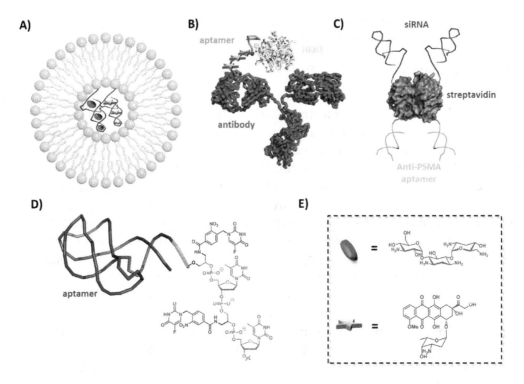

Figure 6. Illustrative examples of aptameric-based systems used as drug delivery systems: (**A**) encapsulation of aptamer-drug complexes in liposome [205]; (**B**) antibody-aptamer pincers (AAPs) for selective targeting of HER2 and delivery of DOX [208]; (**C**) streptavidin serves as the core for the connection of two anti-PSMA aptamers and two siRNA molecules [209]; (**D**) delivery of the 5-fluorouracil (5-FU; shown in red) drug connected to an aptamer-oligonucleotide scaffold via a photo-cleavable linker (shown in blue) [210]; and (**E**) chemical structures of tobramycin (red) and DOX (blue).

5.3. Aptamer-Drug Conjugates

Aptamers have also been considered to guide drugs directly and selectively to their intended targets by direct conjugation to the therapeutic agent or through a small linker arm [211]. An early example of a covalent attachment of a drug on an aptamer is the appendage of DOX on the sgc8c-aptamer via a short linker. Indeed, the linker was connected to the aptamer by means of the thiol-ene click reaction and to DOX via a hydrazone moiety. The hydrazone unit was chosen because of the compatibility with the ketone group found on DOX and a possible hydrolysis in the acidic environment (pH 4.5–5.5) of endosomes [212]. In vitro tests showed that the administration of unconjugated DOX exhibited a higher toxicity towards non-targeted cells compared to the aptamer-DOX conjugate.

In an attempt to overcome low aptamers-drug ratios, Wang et al. synthesized an aptamer containing multiple copies of 5-fluorouracil (5-FU), a drug against colorectal and pancreatic cancer, via a photo-cleavable linker in order to spatially and temporally control the drug release (Figure 6D) [210]. Besides covalent attachment of drugs on aptamers, a few examples of non-covalently attached drug-aptamer conjugates have also been reported, mainly by intercalation. The benefit of this strategy stems from the steady increase in loading capacity of the intercalating drug on aptamers with aptamer vs. drug ratios that improved from 1/1.2 in 2006 [55] to nearly 1/50 in 2013 by the use of self-assembling dsDNA drug intercalation sites directly appended to the desired aptamers [213].

Most drug delivery systems rely on the presence of a single type of aptamer which results in an efficient escorting of drugs to the intended target. However, since aptamers are usually obtained by in vitro selection experiments, small conformational changes of their targets caused by their in vivo environments might interfere or even impede their binding activities [68,214,215]. A multimeric aptamer-drug delivery strategy was developed in order to face this potential drop in binding affinity.

The system consists of two distinct aptamers specific for different cancer subtypes and DOX intercalated in the double-stranded portions created by the self-assembly of the two aptameric species [216].

5.4. Aptamer-Antibody Conjugate

The conjugation of aptamers with their proteinaceous counterpart antibodies is used to increase the affinity of the resulting construct for a single target [7]. The rationale behind this strategy is based on the observation that antibody dimerization causes a decrease in K_d value (caused by lower k_{off} and higher k_{on} rates) compared to the two individual antibodies [208,217]. This working hypothesis was validated by an inhibition study of the VEGF-A and PDGF-B signaling pathways [218] as well as for the fluorescence detection of human CD4 [217]. More recently, Kang and Hah reported a drug delivery strategy based on the formation of an antibody-aptamer hybrid complex in order to improve the specificity of the construct for thrombin or the anti-human epidermal growth factor 2 (HER2) as model systems (Figure 6B). The resulting so-called antibody-aptamer pincers (AAPs) were found to increase the affinity for thrombin of the conjugate compared to the individual aptamers or antibody by 35-fold or 100-fold, respectively (K_d value of 567 pM). The DOX loaded AAP constructed with the anti-HER2 aptamer and antibody was also found to exhibit a 3–6 fold higher cytotoxicity than the individual antibody DOX conjugate or unconjugated DOX [208].

The conjugation of an aptamer with an antibody (a so-called oligobody) can also be helpful to overcome both the poor pharmacokinetics for systemic administration of the small aptamers and the limited tissue penetration of the rather large antibodies (~150 kDa) [219]. The combination of an anti-cotinine antibody (cot-body) with a cotinine labeled vascular endothelial growth factor (VEGF) targeting aptamer (cot-pega) led to the formation of an oligobody that exhibited no loss in affinity to cancer cells compared to the aptamer only, penetrated deep into tumor tissue of an A549-xenograft mouse model, displayed extended half-life times in serum ($t_{1/2}$ = 8.3 h) and reduced tumor growth. All these studies clearly demonstrate the potential of aptamer-antibody conjugates in anticancer therapeutics.

5.5. Aptamers and Gene Regulating Agents

Aptamers, being of nucleic acid nature, can easily be conjugated to relevant RNA or DNA sequences (for instance to sgRNA in the gene editing system CRISPR/Cas9 [220,221]) and particularly to therapeutic oligonucleotides (siRNA, miRNA, or antisense agents) [222–224].

The first siRNA-aptamer conjugate was reported by McNamara et al. who tried to improve the therapeutic efficacy of siRNAs by constructing a chimera with aptamers as vectors to achieve targeted delivery [225]. The aptamer-siRNA chimera was constructed to target the cell surface receptor prostate specific membrane antigen (PSMA) overexpressed in prostate cancer cells via the A10 RNA aptamer [53], while the delivered siRNA targeted the polo-like kinase 1 (PLK1) and B-cell lymphoma 2 (BCL2) genes, which are overexpressed in numerous human tumors [225]. Upon internalization and processing of the RNAs by Dicer, the siRNA is directed to the RNAi pathway and silence their cognate mRNAs which in turn leads to the depletion of the PLK1 and BCL2 survival genes and ultimately cell death. The benefit of that system compared to chimeras of siRNA with antibodies is the low immunogenicity of the RNA aptamers and also an increased in vivo tissue penetration due to the small size. Additionally, it was demonstrated that the genes were only regulated in cells expressing PSMAs on their surface. Dassie et al. further improved this system by optimizing both parts of the chimeric species in order to allow systemic administration, which should simplify clinical applications [226]. The fabrication of the second generation of optimized PSMA-Plk1 chimeras involved a reduction of the length of the aptamers (from 71 down to 39 nucleotides) to facilitate chemical synthesis while introduction of 2′ fluoropyrimidines residues in the longer RNA strand increased serum stability. In addition, the gene silencing activity was enhanced by fine-tuning the siRNA sequence for Dicer recognition, RISC complex formation, and mimicking endogenous miRNA precursors. This optimization included the inclusion of a UU 3′-overhang, engineering of a wobble base pair, swapping of the passenger and guide strands, and introduction of a short stem loop chimera. The second-generation chimeras were found to be active at concentration 50-fold lower than the first generation chimeras and produced a target-specific apoptotic activity [226].

A similar approach was followed in a recent study by Liu et al. where an RNA based aptamer siRNA chimera was engineered to target PSMAs of prostate cancer (PCa) [227]. The bivalent aptamer-dual siRNA chimeric system that was used consisted of an anti-PSMA aptamer [228] connected to an siRNA targeted against the survivin oncogene while the second moiety was composed of the same aptamer but linked to an siRNA for the EGFR gene. Upon binding to prostate cancer cells and internalization, the bivalent chimera was divided into the aptamer and the siRNA parts by digestion of the stem-loops by the Dicer activating RNAi machinery. The siRNAs then inhibited both EGFR and survivin simultaneously by selective mRNA cleavage and ultimately induced apoptosis. Both in vitro and in vivo studies revealed that the combination therapy where two oncogenic pathways are targeted simultaneously is a highly efficient strategy for the eradication of tumor growth and angiogenesis. In another similar approach followed by Jeong et al., DOX was intercalated into a multivalent aptamer-siRNA chimera in order to target multidrug resistant mucin1-overexpressing breast cancer cells (MCF7) [229]. A multimeric antisense siRNA construct was first built by covalent attachment of about 18 single 3′- and 5′-end dithiolated BCL2-specific siRNA sequences through a dithio-bis-maleimidoethane linker and by application of the thiol-ene click reaction. The resulting multivalent template was designed to bind to a chimera comprising a complementary siRNA sequence (the sense strand) and an anti-MUC1 aptamer for the selective delivery of the siRNA to the intended target. DOX was loaded on the construct in the double-stranded regions and the subsequent in vitro tests showed that the DOX-aptamer-siRNA efficiently reduced the viability of the cancer cells after one day by activating the apoptotic caspase-3/7 and releasing DOX as a therapeutic agent. Compared to monovalent DOX-aptamer-siRNA or free DOX, administration of the multivalent complex was the only one that led to a low recovery rate of the cancer cells. The construct thus fulfilled its intended purposes: selective delivery of DOX and efficient antisense activity.

Wilner et al. targeted the transferrin receptor CD71 (TfR) which is overexpressed in malignant cells with a liposome labeled with aptamers and loaded with an anti-enhanced green fluorescent protein siRNA [230]. To achieve specific delivery, they developed a nuclease-resistant aptamer that was obtained through application of a modified SELEX protocol. Indeed, in the first step of the hybrid in vitro selection protocol, a traditional SELEX was carried out using 2′-fluoro-pyrimidine NTPs (in lieu of the natural UTP and CTP) and recombinant hTfR immobilized on a solid support. After the 5th round of selection, a second, "internalization selection", step was included which consisted of incubation of the enriched RNA libraries stemming from the different generations with HeLa cells and extraction and amplification of the RNA molecules internalized by the cells. This ingenious selection protocol allowed for the isolation of a highly potent RNA aptamer (K_d = 17 nM) that could efficiently penetrate Jurkat cells and retain its high binding affinity. The selected anti-TfR aptamer could be engineered into a truncated version (K_d = 102 nM) and was then used to functionalize stable nucleic acid lipid particles (SNALP) containing the siRNA by connection with a thiol maleimide linker. Upon internalization of the functionalized SNALPs, an efficient gene knockdown activity was observed in HeLa cells in vitro, thus providing evidence for the successful activation of the RNAi pathway of the siRNA.

Paralleling these efforts, an elegant approach for aptamer-mediated siRNA delivery was developed by Chu et al. in 2006 in which streptavidin served as a tetravalent core to connect two identical biotinylated anti-PSMA aptamers and two identical biotinylated siRNAs directed against lamin A/C (Figure 6C). In vitro studies with LNCaP cells overexpressing PSMA showed that the chimeric construct was internalized within 30 min and that the gene expression was significantly reduced only when both the siRNA and the anti PSMA aptamers were present on the construct [209]. Control experiments with PSMA-negative PC3 cells demonstrated no cytotoxic effects and no reduction of gene expression, highlighting the potential of such a set up as a gene therapeutic agent despite a potential immunogenicity of the streptavidin adducts which could limit the delivery of such constructs [231]. The same anti-PSMA aptamer was recently involved in the construction of a pRNA-3WJ core based system for the specific delivery of a miRNA LNA to LNCaP prostate cancer cells and to knock down the oncogenes miR17 and miR21 [232]. In addition to the miRNA and the aptamer, a Cy5 dye was introduced in the framework to follow the in vitro internalization into cells. The ultra-stable and serum resistant constructs bound to PSMA in an excellent manner at RNA concentrations as low as 100 nM and were shown to deliver the miRNA specifically to the LNCap cells but not to PC-3 cells. Systemic in vivo administration to xenograft tumors in nude mice showed that the aptamer-pRNA-siRNA construct specifically bound to tumor cells with little or no accumulation in healthy cells. A reduced tumor growth could be found even days post administration without any indication of toxicity as a result of the specific delivery.

Lastly, aptamers have also been linked to catalytic DNA (DNAzymes [4]) and RNA (ribozymes [233]) molecules. For instance, a two-step selection protocol was developed to generate RNA molecules capable of both recognizing and binding to the inteRNAl ribosome entry site (IRES) of hepatitis C virus (HCV) and cleavage of the genomic viral RNA at a specific location [234]. This selection experiment led to the identification of seven distinct groups of aptamer-ribozyme chimeras that selectively bound to the intended target (K_d values of ~5–200 nM) and cleaved the viral RNA with appreciable rate constants ($k_{obs} \cong 0.01$–0.04 min^{-1}) [234,235]. Recently, a partial randomization of the sequence of an isolated aptamer-ribozyme followed by reselection allowed improving the inhibitory properties of such RNA molecules [236].

In the context of catalytic DNA molecules, a hemin/G-quadruplex (hGQ) horseradish peroxidase-mimicking DNAzyme [237] was linked to various aptamers specific for certain small molecules—yielding constructs coined nucleoapzymes—to expand the catalytic repertoire of these biocatalysts [238]. Particularly, when the DNAzyme was conjugated with DA- and arginine-binding aptamers, increased yields of oxidation (with multiple turnover kinetics) could be observed with the respective substrates (i.e., DA and N-hydroxy-L-arginine) compared to the individual functional nucleic acids.

6. Recent Chemical Modifications of Aptamers

Aptamers are often referred to as nucleic acids antibodies, however the chemical arsenal of DNA and RNA is rather limited when compared to that of proteins. In addition, unmodified, natural nucleic acids are highly prone to hydrolytic degradation by nucleases. The possibility of using chemical modifications in SELEX might help to alleviate these shortcomings [239]. Over the last quarter century, it was shown by several groups that incorporation of nucleoside triphosphates modified at the level of the α-phosphate, the sugar scaffold (mainly at the 2′ position), or at the level of the nucleobase (5 position of pyridines or 7 position of purines) can enhance the target affinity as well as the serum stability compared to aptamers that are restricted to natural nucleotides. Since the principle of using modified nucleotides in SELEX is a known strategy that has been reviewed extensively [240–244], this section will only highlight recent advances in this field.

6.1. Base Modified Aptamers

Gawande et al. recently investigated the influence on the outcome of a selection process when using a library comprising two amino acid-like 5′-modified pyrimidine bases (dC^X 1, dU^X; Figure 7) or a similar population of oligonucleotides but prepared with only one base modification [245]. A systematic study with all the possible pairwise combinations (i.e., 18 different libraries) of dCTPs equipped with two different side-chains (Nap and Pp) and five different modifications on dUTP (Nap, Tyr, Moe, Thr, and Pp) to find high affinity ligands for proprotein convertase substilisin/kexin type 9 (PCSK9) showed that after six rounds of selection, the libraries made with a single dN*TP were enriched with the modification while libraries prepared with two modifications displayed an increased content of the modified dC^X only. Ligands that showed a high affinity for PCSK9 with two modifications were in general more frequent than ligands with only one modification, with the selection experiments exploiting the combination of Tyr-dU and Pp-dC or Nap-dC performing best. A synergistic behavior of the two modifications with respect to affinity could be observed in some cases when compared to aptamers containing only one of the modifications separately. Interestingly, a useful property of double modified aptamers is the high abundance of the modified nucleotides within the sequence. It could be demonstrated that this allows truncating the aptamers with a lower loss of functionality compared to aptamers bearing only one modification. In addition, the serum stability was higher for SOMAmers with two distinct modification compared to natural or ligands with only a single modification. Lastly, another favorable property of using multiple modifications is a higher epitope coverage compared to natural aptamers or ligands containing only a single modification.

With the rationale of using nucleobases equipped with an amino acid-like modification to mimic the hypervariable domains of certain antibodies where tyrosine is overrepresented, Perrin et al. used a phenol modified 5'-deoxyuridine triphosphate (d^yUTP) [246] to raise aptamers against *E. coli* DH5α cells [247]. Despite the lower incorporation efficiency by the Vent (exo^-) DNA polymerase of d^yUTP compared to natural dTTP, a relative high abundance of the modified nucleotide (much larger than if introduced randomly) was found in the ligands obtained and sequenced after 12 rounds of a whole cell-SELEX. Unexpectedly, sequencing of the enriched library revealed a vast diversity of sequences that could not be classified into families, thus showing a low abundance of sequence identity. It was hypothesized that this lack of sequence identity was caused by the vast diversity of targets expressed on the cell surface of such a bacteria. Analysis of the binding specificity of the four most abundant aptamers compared with an unmodified aptamer as control for the gram positive cells showed that: (1) the modification is required for high affinity binding; and (2) that cross-reactivity with other cell strains was minimal. The apparent dissociation constant of the most promising aptamer was determined by saturation experiments to be 27.4 nM, which is about 10-fold lower than unmodified RNA aptamers recently reported for DH5α cells [248].

Minagawa et al. used a base-appended base (BAB) in a 75 nucleotide long modified library with a 30 mer randomized region to select aptamers against salivary α-amylase (sAA) [249]. Indeed, replacement of the thymidine nucleotide with a triphosphate displaying an adenine attached to the C5 position of the nucleobase (dU^{ad}TP [250], Figure 7) in the selection protocol led to the isolation of seven aptamers against sAA with an abundance of over 5% of the enriched pool after eight rounds of selection. SPR measurements revealed that the most potent aptamer bound to the target with an affinity of 559 pM, which is sufficient for potential applications as a biosensor of the stress biomarker sAA [251]. When the selection experiment was carried out in the absence of dU^{ad}TP (only natural dNTPs), no enrichment of the library was observed, further highlighting the usefulness and the potential of modified nucleoside triphosphates in SELEX. An optimization study of the initial 75 mer aptamer showed that the full-length sequence could be truncated down to a 36 mer species without inducing a substantial loss of binding affinity. Additionally, imino-proton NMR spectra were recorded at different temperatures to elucidate the structural properties of the sAA binding aptamer. This NMR analysis suggested that the imino-protons of the adenine part of dU^{ad} were engaged in additional hydrogen-bonding interactions, thus leading to the assumption that the BAB modification induced a well-defined and compact secondary structure. Finally, the potential of the selected aptamer was highlighted with the detection of human sAA in human saliva using capillary electrophoresis, pull down, and lateral flow assays.

Examples of base-modified RNA aptamers are not as common as for DNA, which reflects both the lower tolerance of natural RNA polymerases for base-modified NTPs and the restricted choice of engineered RNA polymerases. Despite these limitations, allyl-amino-UTP (U^{aa}TP) was used in an in vitro selection experiment aimed at raising an anti-ATP aptamer [252]. More recently, Kabza and Sczepanski used the same U^{aa}TP to isolate an aptamer against the oncogenic precursor microRNA 19a (pre-miR-19a) which was converted to its Spiegelmer via solid-phase synthesis using the L-U^{aa}-phosphoramidite [253]. The resulting L-RNA aptamer bound its target with a slightly lower affinity than its D-counterpart (K_d values of 2.2 and 0.72 nM, respectively) but efficiently inhibited the Dicer-mediated processing of the miR target ($IC_{50} \cong 4$ nM) in vitro.

Lastly, the Hili laboratory is currently exploring an interesting approach—coined LOOPER (Ligase-catalyzed OligOnucleotide PolymERization)—for the increase of chemical diversity without the involvement of dN*TPs and relying on the ligation of base-modified pentanucleotides [254,255]. This strategy was recently expanded to the Darwinian evolution of aptamers against human α-thrombin [256]. Indeed, a library containing 16 different pentanucleotidic codons (and as many functional groups ranging from hydrophobic to Brønsted acids and bases) was subjected to six rounds of SELEX. The most highly represented sequence displayed a remarkable affinity ($K_d = 1.6$ nM) and selectivity for its target (no cross-reactivity with BSA) and strictly required the presence of the modifications for binding. In addition to ablating the need for dN*TPs (and to an extent engineered polymerases), this strategy allows for the introduction of a broad variety of functional groups and will certainly be used for the evolution of other aptamers and potentially DNAzymes.

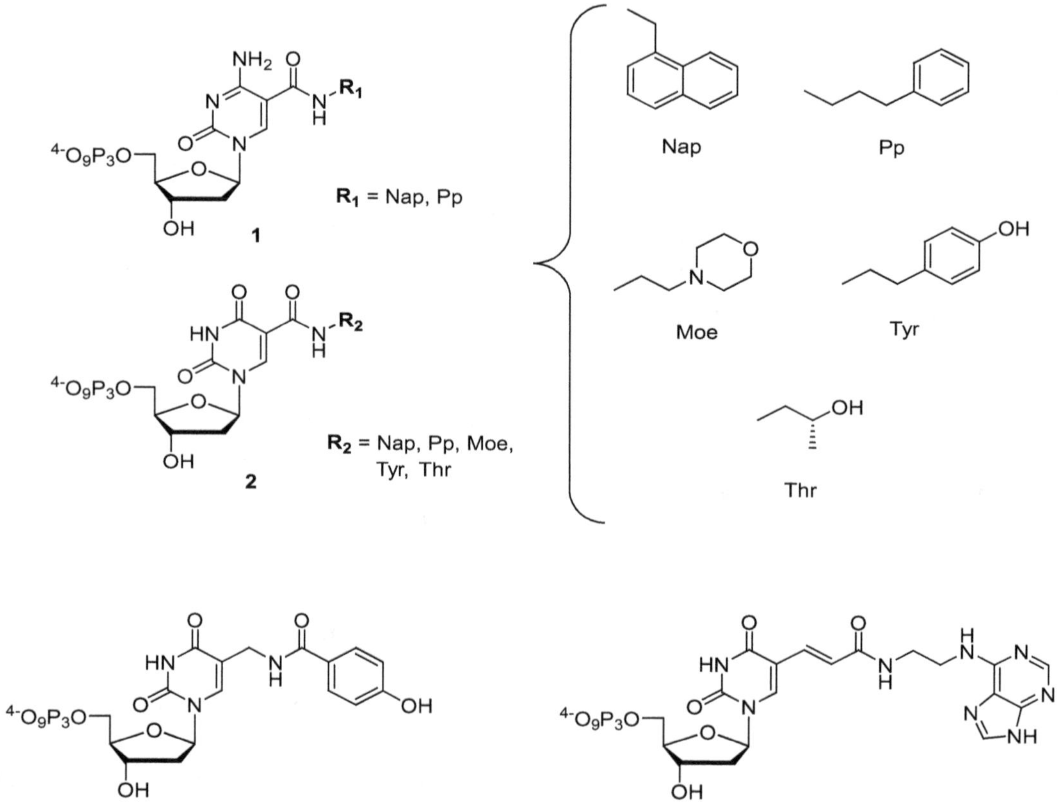

Figure 7. Chemical structures of base-modified nucleoside triphosphates used in selection experiments of aptamers with an expanded chemical repertoire.

6.2. Aptamers with an Extended Genetic Alphabet

Expanding the genetic code from a two- to a three- or even a four-base-pair system is a long standing goal in synthetic biology since this would enable the creation of functional nucleic acids with improved properties and ultimately to semi-synthetic organisms with proteins that potentially display novel structures and/or functions [257]. In this context, Hirao and colleagues have developed the ex-SELEX (genetic alphabet expansion for systematic evolution of ligands by exponential enrichment) method where an additional artificial Ds-Px base pair (Figure 8) complements the two natural Watson–Crick base pairs to isolate aptamers against the vascular endothelial growth factor (VEGF_{165}) [258,259]. In the ex-SELEX strategy, the hydrophobic base Ds is introduced into DNA by solid-phase synthesis at predetermined locations and specifically binds to its unnatural partner Px in PCR for the amplification of the generations during SELEX. The localization of the Ds modifications in each sublibrary—a key step in the ex-SELEX protocol—is possible through the combination of a unique barcode system and replacement PCR. In a new approach aimed at exploring a larger chemical space, a fully randomized library containing the unnatural base was used to isolate aptamers targeting the Willebrand factor A1 domain (vWF) [260]. Although the selection with the randomized additional base in the sequences led to higher affinity ligands compared to the selection with natural base pairs or the ex-SELEX selection procedure with limited sublibraries, it also increased the complexity of the system and led to scaling problems. The new complexity and the possibility of misincorporation of the Ds base during PCR experiments required the introduction of a novel sequencing method to elucidate the exact location of the Ds modification. The authors took advantage of the fact that only dATP misincorporated opposite Px by the Taq polymerase but none of the dye-carrying 2′,3′-dideoxynucleoside-5′-triphosphates were incorporated at these positions, which led to the formation of a gap in the sequencing peak pattern at the position of the unnatural base. By comparing the sequencing pattern with that of a clone after a replacement PCR, the exact positions of the modifications could be inferred. Lastly, both the serum stability and the thermostability of the anti-vWF aptamers could be enhanced by introducing additional mini-hairpin motifs containing GNA loops (N = A, G, C or T) [261] without a loss of affinity in both the natural (K_d = 182 pM) and the modified aptamer (K_d = 61 pM), which thus represents a convenient strategy for the improvement and construction of functional aptamers [259,262,263].

In the context of an expansion of the genetic alphabet, Benner and co-workers developed a third base pair dZ-dP (Figure 5) that can be used in SELEX and the related strategy was coined "laboratory in vitro evolution based on an artificial expanded genetic information system" (LIVE-AEGIS) [264,265]. Initial selection experiments using a 20 nucleotide long randomized library with a nucleotide composition of T/G/A/C/Z/P $\cong$ 1.5/1.2/1.0/1.0/1.0/0.5 by solid-phase synthesis and amplified in the selection rounds Hot Start Taq polymerase in the presence of natural and modified dNTPs led to the isolation of a high affinity aptamer (K_d = 30 nM) against MDA-MB 231 breast cancer cells [266]. In order to perform deep sequencing of the enriched populations at the end of the LIVE-AEGIS protocol, a conversion technique was used where the dZ nucleotides were converted into dC and dT and dP into dA and dG, allowing to assign the positions of the two modifications. Analysis of the sequence composition showed that only one dZ was present in the most potent aptamer which additionally lost its binding affinity when the modification was replaced with a natural nucleotide. The depletion of dZ and dP nucleotides in the isolated aptamer (an average of three dZ and 1.5 dP would be expected for a 20 nucleotide long sequence) was ascribed to a slight disadvantage of the modified triphosphates during PCR compared to the natural nucleotides. A similar observation was made in another study by Zhang et al. using the AEGIS system for cells engineered to place glypican 3 (hGPC3) on their surface where a ligand with only one dZ was found that bound with an affinity of 6 nM [267]. Analogues without the dZ showed a significant loss in activity again demonstrating the importance of the modification. The most recent AEGIS-LIVE study by Biondi et al. targeting the protective antigen (PA63), a cleaved version of the precursor protein PA83 from *Bacillus anthracis*, brought forth an aptamer that binds to PA63 and thus blocks the toxin channel by displacing the lethal factor and finally inhibiting translocation of toxins into infected cell [268]. Analysis of the sequences stemming from the AEGIS-LIVE selections revealed that no survivor contained a dZ modification but different dP-containing sequences were isolated [268]. The most abundant sequence of the enriched pool comprised two dP nucleotides and this aptameric species displayed a very high affinity for PA63 (K_d ~35 nM). This underrepresentation of modifications in the surviving sequences was ascribed to the conformational constraint imposed by the modified nucleotides. An intensive analysis of the secondary structure of the aptamer revealed interesting features resembling the formation of a higher folded structure [269] upon addition of cations that was not described in the literature before and which was also found to increase the stability against nuclease degradation.

Other prominent base-modified nucleoside and nucleotide analogs that have not (yet) been used to generate aptamers with enhanced properties include the DNAM-5dSICS unnatural base pair (UBP) developed by the Romesberg laboratory and used in the development of a semi-synthetic organism [270,271], artificial metallo-base pairs [272,273], and nucleotides modified with other functional groups [125,274–278].

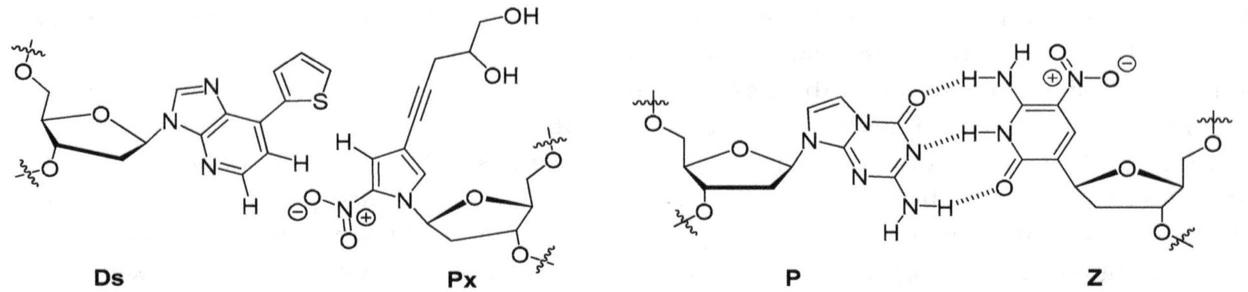

Figure 8. Chemical structures of the unnatural base pairs Ds Px and dP dZ used in the expansion of the genetic code.

6.3. Sugar Modified Aptamers

2′-fluoro-nucleotides (**3** in Figure 9) are popular modifications that are often introduced into aptameric scaffolds to increase their nuclease resistance, a strategy that culminated in the development of pegaptanib sodium (Macugen®) [25,240]. However, 2′-fluoro-modified nucleoside triphosphates are rather poor substrates for DNA polymerases which restricts their use in selection experiments and often require a time consuming reverse transcription in order to convert the 2′-modified sequences into natural oligonucleotides that can be amplified and then transcribed back into modified sequences [279]. In order to circumvent these shortcomings, the Romesberg laboratory evolved a thermostable polymerase that could PCR-amplify oligonucleotides containing 2′-fluoro- and 2′-OMe-nucleotides [280]. This polymerase, SFM4-3, was then recently used in a selection experiment with 2′-fluorine-modified purine nucleotides to raise aptamers that could bind human neutrophil elastase (NHE) [281]. With the engineered thermostable DNA polymerase SFM4-3, the need for a reverse transcription step was ablated which significantly shortened and simplified the selection process. After six rounds of selection, two aptamers were identified that displayed very high affinity (K_d = 20–170 nM) for the NHE target, albeit with slightly lower affinity than that of the sequences that did not contain the 2′-fluoro-modifications (K_d = 11–17 nM). In order to exclude nonspecific electrostatic interactions of the aptamers with the positively charged HNE, an experiment suppressing the charges with high salt concentrations (i.e., 1 M NaCl) was performed, where the natural variant of the aptamers lost its affinity to HNE but the ligand with the 2′-fluoro modification remained bound. Lastly, ^{19}F NMR experiments revealed that the secondary structure of the aptamer was influenced by the presence of the 2′-modifications which disfavours duplex formation and therefore allows the formation of a more soluble species that specifically recognized the target. A similar selection experiment was carried out to isolate fully-modified 2′-OMe-aptamers by combining an engineered polymerase and a reverse transcriptase. One particular aptamer, 2mHNE-4, bound its intended target, human neutrophil elastase, with good affinity (K_d = 45 nM) [282].

Other sugar modifications such as HNA (hexitol nucleic acid **8**) [283], LNA (locked nucleic acid **5**) [283], FANA (2′-fluoro-arabinose **4**) [284], and TNA (threose nucleic acid **6**) [285] have all been used for the generation of highly potent modified aptamers, while other nucleotides such as 7′,5′-bicyclo-DNA **9** [286], xylonucleic acids **7** [287], 2′-selenomethyl nucleotides [288], and 4′thio-DNA [289,290] have recently been suggested as potential candidates to explore new chemistries in selection experiments (Figure 9) [279,291].

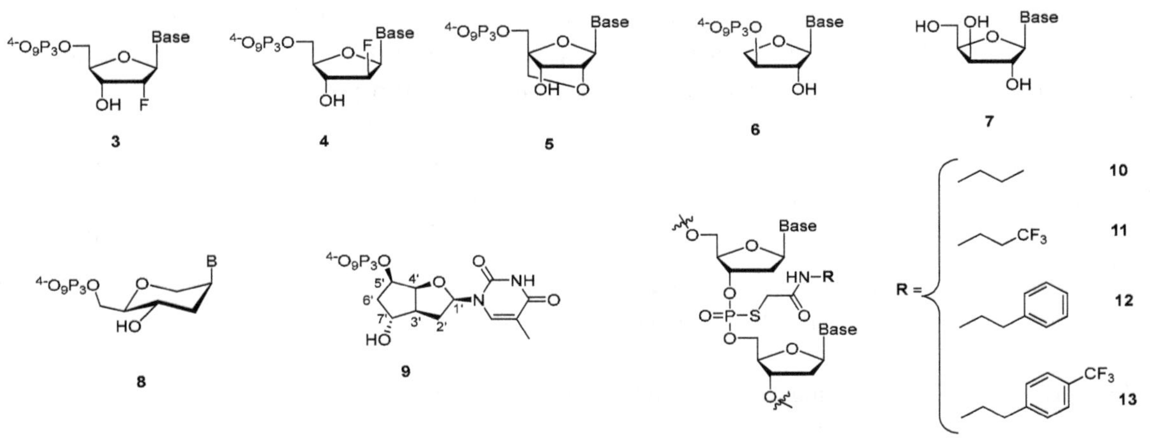

Figure 9. Chemical structures of sugar and phosphate modified nucleotides.

6.4. Phosphate Modified Aptamers

The possibilities for the modification at the level of the phosphate unit are more limited than in the case of the sugar and nucleobases moieties and most efforts have focused on the α-phosphate, particularly α-phosphorothioates [242,279,292–294]. In this context, Yang et al. have explored the alkylation of phosphorothioated thrombine-binding aptamers (TBA) with aim of improving the antitumor properties of the ligands by reducing the thrombin binding affinity [295]. Alkylation of the phosphorothiate moieties was achieved by a simple substitution reaction with four different brominated substrates (**10** to **13** in Figure 9). Previous X-ray [296] and NMR studies [297] of the TBA showed that the TT or TGT loop of the G-quadruplex structure was responsible for effective binding. Therefore, the modifications were systematically introduced at these positions. The presence of the phenyl moiety (**12** in Figure 9) was shown to reduce the flexibility of the loop regions through interaction with nucleobases which resulted in an inhibition of G-quadruplex formation which in turn is believed to cause a lowering of the binding affinity (K_d = 0.5 μM) for thrombin compared to the unmodified aptamer (K_d = 0.19 μM). The higher dissociation constants for the aptamers modified with the phenyl moiety also resulted in a reduction of the anticoagulation properties. The antiproliferation experiments with the natural aptamers as control revealed that the TBAs with the phenyl modification exhibited excellent inhibition of the proliferation of about 80% with the lung carcinoma cell line A549 but no activity against the human breast cancer cell line MCF-7.

Substitution of an oxygen atom on the α-phosphate with a BH_3 moiety instead of a sulfur atom was used for the selection of a potent anti-ATP aptamer but no further examples have been reported since [298].

7. Conclusions and Prospects

In view of their impressive functional properties, nucleic acid aptamers certainly deserve their description as chemical antibodies [25]. The high degree of structural flexibility associated with an impressive target specificity and selectivity has propelled aptamers into the forefront of numerous therapeutic and diagnostic applications. Moreover, aptamers are continuously employed in proof-of-concept studies to further expand the boundaries of the realm of aptamer-based technologies. Particularly, aptamers have recently started to infiltrate the fields of medical imaging and there is no doubt that aptamers will mature into valuable and smart imaging agents. In addition, aptamers commence to be used as tools in neuroscience for the detection of the variation of small concentrations of various neurotransmitters and abnormal protein folds. Hopes for the in vivo use of aptamers in neurosciences are spurred by a recent selection experiment of an aptamer for its capacity at penetrating the blood–brain barrier (BBB) [299] as well as a novel bioconjugation method to polymeric nanoparticles which has allowed the in vivo BBB passage [300]. Furthermore, due to their chemical malleability, aptamers can easily be conjugated to small molecules, other nucleic acid oligonucleotides, or larger constructs such as antibodies, liposomes, and DNA origamis that undoubtedly will improve their cellular uptake and help in the development of aptamer-mediated drug delivery systems as well as biosensing platforms. However, aptamers still suffer from temperature- and nuclease-mediated degradation and the restricted chemical arsenal available to natural nucleic acids precludes binding to difficult targets such as single enantiomers of small organic molecules or glycosylated proteins [240,301]. The use of modified nucleoside triphosphates, along with engineered polymerases [291,302] and new selection strategies [256,285,303–305], will certainly help in alleviating these predicaments.

All the emerging applications described in this Review along with technological and synthetic progress will certainly improve the limited commercial success of aptamers in the near future.

Acknowledgments: The authors gratefully acknowledge financial support from the Institut Pasteur start-up funds.

Author Contributions: Pascal Röthlisberger, Cécile Gasse, and Marcel Hollenstein contributed to the conception of the article, literature collection, and preparation of the manuscript. Marcel Hollenstein contributed to reviewing and revising the manuscript. All authors read and approved the final version of the manuscript.

Abbreviations

CTC	Circulating Cancer Cells
GO	Graphene oxide
AFM	Atomic Force Microscopy
SELEX	Systematic Evolution of Ligands by Exponential Enrichment
PET	Positron Emission Tomography
MRI	Magnetic Resonance Imaging
DOTA	1,4,7,10-tetraazacyclododecane-1,4,7,10-tetraacetic acid
K_D	Dissociation constant
XNA	Xeno nucleic acid
SPIONs	Superparamagnetic iron oxide nanoparticles
NOTA	1,4,7-triazacyclononane-triacetic acid
DNAzyme	DNA enzyme or catalytic DNA
SPECT	Single Photon Emission Computed Tomography
HER2	Human Epidermal Growth Factor Receptor 2
DOX	Doxorubicin
QD	Quantum Dot
PEG	Polyethyleneglycol
siRNA	Small interfering RNA
miRNA	Micro RNA
sgRNA	Single guide RNA
pRNA	Packaging RNA
3WJ	Three-way junction
GABA	γ-Aminobutyric acid
BSA	Bovine Serum Albumin
Aβ	Amyloid β

References

1. Goyenvalle, A.; Griffith, G.; Babbs, A.; El Andaloussi, S.; Ezzat, K.; Avril, A.; Dugovic, B.; Chaussenot, R.; Ferry, A.; Voit, T.; et al. Functional correction in mouse models of muscular dystrophy using exon-skipping tricyclo-DNA oligomers. *Nat. Med.* **2015**, *21*, 270–275. [CrossRef] [PubMed]
2. Khvorova, A.; Watts, J.K. The chemical evolution of oligonucleotide therapies of clinical utility. *Nat. Biotechnol.* **2017**, *35*, 238–248. [CrossRef] [PubMed]
3. Wang, P.; Meyer, T.A.; Pan, V.; Dutta, P.K.; Ke, Y. The beauty and utility of DNA origami. *Chem* **2017**, *2*, 359–382. [CrossRef]

4. Hollenstein, M. DNA catalysis: The chemical repertoire of DNAzymes. *Molecules* **2015**, *20*, 20777–20804. [CrossRef] [PubMed]

5. Silverman, S.K. Catalyic DNA: Scope, applications, and biochemistry of deoxyribozymes. *Trends Biochem. Sci.* **2016**, *41*, 595–609. [CrossRef] [PubMed]

6. Rioz-Martinez, A.; Roelfes, G. DNA-based hybrid catalysis. *Curr. Opin. Chem. Biol.* **2015**, *25*, 80–87. [CrossRef] [PubMed]

7. Yu, Y.Y.; Liang, C.; Lv, Q.X.; Li, D.F.; Xu, X.G.; Liu, B.Q.; Lu, A.P.; Zhang, G. Molecular selection, modification and development of therapeutic oligonucleotide aptamers. *Int. J. Mol. Sci.* **2016**, *17*, 358. [CrossRef] [PubMed]

8. Ellington, A.D.; Szostak, J.W. In vitro selection of RNA molecules that bind specific ligands. *Nature* **1990**, *346*, 818–822. [CrossRef] [PubMed]

9. Tuerk, C.; Gold, L. Systematic evolution of ligands by exponential enrichment: RNA ligands to bacteriophage T4 DNA polymerase. *Science* **1990**, *249*, 505–510. [CrossRef] [PubMed]

10. Zhang, L.Q.; Wan, S.; Jiang, Y.; Wang, Y.Y.; Fu, T.; Liu, Q.L.; Cao, Z.J.; Qiu, L.P.; Tan, W.H. Molecular elucidation of disease biomarkers at the interface of chemistry and biology. *J. Am. Chem. Soc.* **2017**, *139*, 2532–2540. [CrossRef] [PubMed]

11. Sedlyarova, N.; Rescheneder, P.; Magan, A.; Popitsch, N.; Rziha, N.; Bilusic, I.; Epshtein, V.; Zimmermann, B.; Lybecker, M.; Sedlyarov, V.; et al. Natural RNA polymerase aptamers regulate transcription in *E. coli*. *Mol. Cell* **2017**, *67*, 30–43. [CrossRef] [PubMed]

12. Joyce, G.F. Forty years of in vitro evolution. *Angew. Chem. Int. Ed.* **2007**, *46*, 6420–6436. [CrossRef] [PubMed]

13. Robertson, D.L.; Joyce, G.F. Selection in vitro of an RNA enzyme that specifically cleaves single-stranded DNA. *Nature* **1990**, *344*, 467–468. [CrossRef] [PubMed]

14. Pfeiffer, F.; Mayer, G. Selection and biosensor application of aptamers for small molecules. *Front. Chem.* **2016**, *4*. [CrossRef] [PubMed]

15. Bock, L.C.; Griffin, L.C.; Latham, J.A.; Vermaas, E.H.; Toole, J.J. Selection of single-stranded-DNA molecules that bind and inhibit human thrombin. *Nature* **1992**, *355*, 564–566. [CrossRef] [PubMed]

16. Nimjee, S.M.; White, R.R.; Becker, R.C.; Sullenger, B.A. Aptamers as therapeutics. *Annu. Rev. Pharmacol. Toxicol.* **2017**, *57*, 61–79. [CrossRef] [PubMed]

17. Chen, M.; Yu, Y.Y.; Jiang, F.; Zhou, J.W.; Li, Y.S.; Liang, C.; Dang, L.; Lu, A.P.; Zhang, G. Development of cell-SELEX technology and its application in cancer diagnosis and therapy. *Int. J. Mol. Sci.* **2016**, *17*, 2079. [CrossRef] [PubMed]

18. Sefah, K.; Shangguan, D.; Xiong, X.L.; O'Donoghue, M.B.; Tan, W.H. Development of DNA aptamers using cell-SELEX. *Nat. Protoc.* **2010**, *5*, 1169–1185. [CrossRef] [PubMed]

19. Shangguan, D.; Li, Y.; Tang, Z.W.; Cao, Z.H.C.; Chen, H.W.; Mallikaratchy, P.; Sefah, K.; Yang, C.Y.J.; Tan, W.H. Aptamers evolved from live cells as effective molecular probes for cancer study. *Proc. Natl. Acad. Sci. USA* **2006**, *103*, 11838–11843. [CrossRef] [PubMed]

20. Lee, J.F.; Hesselberth, J.R.; Meyers, L.A.; Ellington, A.D. Aptamer database. *Nucleic Acids Res.* **2004**, *32*, D95–D100. [CrossRef] [PubMed]

21. Cruz-Toledo, J.; McKeague, M.; Zhang, X.R.; Giamberardino, A.; McConnell, E.; Francis, T.; DeRosa, M.C.; Dumontier, M. Aptamer base: A collaborative knowledge base to describe aptamers and SELEX experiments. *Database* **2012**, *8*. [CrossRef] [PubMed]

22. Wu, Y.X.; Kwon, Y.J. Aptamers: The "evolution" of SELEX. *Methods* **2016**, *106*, 21–28. [CrossRef] [PubMed]

23. Gupta, S.; Hirota, M.; Waugh, S.M.; Murakami, I.; Suzuki, T.; Muraguchi, M.; Shibamori, M.; Ishikawa, Y.; Jarvis, T.C.; Carter, J.D.; et al. Chemically modified DNA aptamers bind interleukin-6 with high affinity and inhibit signaling by blocking its interaction with interleukin-6 receptor. *J. Biol. Chem.* **2014**, *289*, 8706–8719. [CrossRef] [PubMed]

24. Sundaram, P.; Kurniawan, H.; Byrne, M.E.; Wower, J. Therapeutic RNA aptamers in clinical trials. *Eur. J. Pharm. Sci.* **2013**, *48*, 259–271. [CrossRef] [PubMed]

25. Zhou, J.H.; Rossi, J. Aptamers as targeted therapeutics: Current potential and challenges. *Nat. Rev. Drug Discov.* **2017**, *16*, 181–202. [CrossRef] [PubMed]

26. Meng, H.M.; Liu, H.; Kuai, H.L.; Peng, R.Z.; Mo, L.T.; Zhang, X.B. Aptamer-integrated DNA nanostructures for biosensing, bioimaging and cancer therapy. *Chem. Soc. Rev.* **2016**, *45*, 2583–2602. [CrossRef] [PubMed]

27. Ku, T.H.; Zhang, T.T.; Luo, H.; Yen, T.M.; Chen, P.W.; Han, Y.Y.; Lo, Y.H. Nucleic acid aptamers: An emerging tool for biotechnology and biomedical sensing. *Sensors* **2015**, *15*, 16281–16313. [CrossRef] [PubMed]

28. Gold, L.; Ayers, D.; Bertino, J.; Bock, C.; Bock, A.; Brody, E.N.; Carter, J.; Dalby, A.B.; Eaton, B.E.; Fitzwater, T.; et al. Aptamer-based multiplexed proteomic technology for biomarker discovery. *PLoS ONE* **2010**, *5*, e15004. [CrossRef] [PubMed]

29. Bunka, D.H.J.; Stockley, P.G. Aptamers come of age—At last. *Nat. Rev. Microbiol.* **2006**, *4*, 588–596. [CrossRef] [PubMed]

30. Famulok, M.; Hartig, J.S.; Mayer, G. Functional aptamers and aptazymes in biotechnology, diagnostics, and therapy. *Chem. Rev.* **2007**, *107*, 3715–3743. [CrossRef] [PubMed]

31. Forier, C.; Boschetti, E.; Ouhammouch, M.; Cibiel, A.; Duconge, F.; Nogre, M.; Tellier, M.; Bataille, D.; Bihoreau, N.; Santambien, P.; et al. DNA aptamer affinity ligands for highly selective purification of human plasma-related proteins from multiple sources. *J. Chromatogr. A* **2017**, *1489*, 39–50. [CrossRef] [PubMed]

32. Keefe, A.D.; Pai, S.; Ellington, A. Aptamers as therapeutics. *Nat. Rev. Drug Discov.* **2010**, *9*, 537–550. [CrossRef] [PubMed]

33. Zhou, W.Z.; Huang, P.J.J.; Ding, J.S.; Liu, J. Aptamer-based biosensors for biomedical diagnostics. *Analyst* **2014**, *139*, 2627–2640. [CrossRef] [PubMed]

34. Jo, H.; Ban, C. Aptamer–nanoparticle complexes as powerful diagnostic and therapeutic tools. *Exp. Mol. Med.* **2016**, *48*, e230. [CrossRef] [PubMed]

35. Uzawa, T.; Tada, S.; Wang, W. Expansion of the aptamer library from a "natural soup" to an "unnatural soup". *Chem. Commun.* **2013**, *49*, 1786–1795. [CrossRef] [PubMed]

36. Wang, A.Z.; Farokhzad, O.C. Current progress of aptamer-based molecular imaging. *J. Nucl. Med.* **2014**, *55*, 353–356. [CrossRef] [PubMed]

37. Boros, E.; Gale, E.M.; Caravan, P. MR imaging probes: Design and applications. *Dalton Trans.* **2015**, *44*, 4804–4818. [CrossRef] [PubMed]

38. Hingorani, D.V.; Bernstein, A.S.; Pagel, M.D. A review of responsive MRI contrast agents: 2005–2014. *Contrast Media Mol. Imaging* **2015**, *10*, 245–265. [CrossRef] [PubMed]

39. Que, E.L.; Chang, C.J. Responsive magnetic resonance imaging contrast agents as chemical sensors for metals in biology and medicine. *Chem. Soc. Rev.* **2010**, *39*, 51–60. [CrossRef] [PubMed]

40. Gale, E.M.; Jones, C.M.; Ramsay, I.; Farrar, C.T.; Caravan, P. A janus chelator enables biochemically responsive MRI contrast with exceptional dynamic range. *J. Am. Chem. Soc.* **2016**, *138*, 15861–15864. [CrossRef] [PubMed]

41. Yigit, M.V.; Mazumdar, D.; Kim, H.K.; Lee, J.H.; Dintsov, B.; Lu, Y. Smart "turn-on" magnetic resonance contrast agents based on aptamer-functionalized superparamagnetic iron oxide nanoparticles. *ChemBioChem* **2007**, *8*, 1675–1678. [CrossRef] [PubMed]

42. Thomas, R.; Park, I.K.; Jeong, Y.Y. Magnetic iron oxide nanoparticles for multimodal imaging and therapy of cancer. *Int. J. Mol. Sci.* **2013**, *14*, 15910–15930. [CrossRef] [PubMed]

43. Huizenga, D.E.; Szostak, J.W. A DNA aptamer that binds adenosine and ATP. *Biochemistry* **1995**, *34*, 656–665. [CrossRef] [PubMed]

44. Tasset, D.M.; Kubik, M.F.; Steiner, W. Oligonucleotide inhibitors of human thrombin that bind distinct epitopes. *J. Mol. Biol.* **1997**, *272*, 688–698. [CrossRef] [PubMed]

45. Yigit, M.V.; Mazumdar, D.; Lu, Y. MRI detection of thrombin with aptamer functionalized superparamagnetic iron oxide nanoparticles. *Bioconjug. Chem.* **2008**, *19*, 412–417. [CrossRef] [PubMed]

46. Xu, W.C.; Lu, Y. A smart magnetic resonance imaging contrast agent responsive to adenosine based on a DNA aptamer-conjugated gadolinium complex. *Chem. Commun.* **2011**, *47*, 4998–5000. [CrossRef] [PubMed]

47. Xu, W.C.; Xing, H.; Lu, Y. A smart T-1-weighted MRI contrast agent for uranyl cations based on a DNAzyme-gadolinium conjugate. *Analyst* **2013**, *138*, 6266–6269. [CrossRef] [PubMed]

48. Artemov, D.; Mori, N.; Ravi, R.; Bhujwalla, Z.M. Magnetic resonance molecular imaging of the Her-2/neu receptor. *Cancer Res.* **2003**, *63*, 2723–2727. [PubMed]

49. Zhou, Z.X.; Lu, Z.R. Gadolinium-based contrast agents for magnetic resonance cancer imaging. *Wiley Interdiscip. Rev. Nanomed. Nanobiotechnol.* **2013**, *5*, 1–18. [CrossRef] [PubMed]

50. Hicke, B.J.; Stephens, A.W.; Gould, T.; Chang, Y.F.; Lynott, C.K.; Heil, J.; Borkowski, S.; Hilger, C.S.; Cook, G.; Warren, S.; et al. Tumor targeting by an aptamer. *J. Nucl. Med.* **2006**, *47*, 668–678. [PubMed]

51. Bernard, E.D.; Beking, M.A.; Rajamanickam, K.; Tsai, E.C.; DeRosa, M.C. Target binding improves relaxivity in aptamer-gadolinium conjugates. *J. Biol. Inorg. Chem.* **2012**, *17*, 1159–1175. [CrossRef] [PubMed]

52. Caravan, P. Protein-targeted gadolinium-based magnetic resonance imaging (MRI) contrast agents: Design and mechanism of action. *Acc. Chem. Res.* **2009**, *42*, 851–862. [CrossRef] [PubMed]

53. Lupold, S.E.; Hicke, B.J.; Lin, Y.; Coffey, D.S. Identification and characterization of nuclease-stabilized RNA molecules that bind human prostate cancer cells via the prostate-specific membrane antigen. *Cancer Res.* **2002**, *62*, 4029–4033. [PubMed]

54. Wang, A.Z.; Bagalkot, V.; Vasilliou, C.C.; Gu, F.; Alexis, F.; Zhang, L.; Shaikh, M.; Yuet, K.; Cima, M.J.; Langer, R.; et al. Superparamagnetic iron oxide nanoparticle-aptamer bioconjugates for combined prostate cancer imaging and therapy. *Chem. Med. Chem.* **2008**, *3*, 1311–1315. [CrossRef] [PubMed]

55. Farokhzad, O.C.; Cheng, J.J.; Teply, B.A.; Sherifi, I.; Jon, S.; Kantoff, P.W.; Richie, J.P.; Langer, R. Targeted nanoparticle-aptamer bioconjugates for cancer chemotherapy in vivo. *Proc. Natl. Acad. Sci. USA* **2006**, *103*, 6315–6320. [CrossRef] [PubMed]

56. Yu, M.K.; Kim, D.; Lee, I.H.; So, J.S.; Jeong, Y.Y.; Jon, S. Image-guided prostate cancer therapy using aptamer-functionalized thermally cross-linked superparamagnetic iron oxide nanoparticles. *Small* **2011**, *7*, 2241–2249. [CrossRef] [PubMed]

57. Bates, P.J.; Laber, D.A.; Miller, D.M.; Thomas, S.D.; Trent, J.O. Discovery and development of the G-rich oligonucleotide AS1411 as a novel treatment for cancer. *Exp. Mol. Pathol.* **2009**, *86*, 151–164. [CrossRef] [PubMed]

58. Li, J.J.; You, J.; Dai, Y.; Shi, M.L.; Han, C.P.; Xu, K. Gadolinium oxide nanoparticles and aptamer-functionalized silver nanoclusters-based multimodal molecular imaging nanoprobe for optical/magnetic resonance cancer cell imaging. *Anal. Chem.* **2014**, *86*, 11306–11311. [CrossRef] [PubMed]

59. Ji, K.L.; Lim, W.S.; Li, S.F.Y.; Bhakoo, K. A two-step stimulus-response cell-SELEX method to generate a DNA aptamer to recognize inflamed human aortic endothelial cells as a potential in vivo molecular probe for atherosclerosis plaque detection. *Anal. Bioanal. Chem.* **2013**, *405*, 6853–6861. [CrossRef] [PubMed]

60. Ametamey, S.M.; Honer, M.; Schubiger, P.A. Molecular imaging with PET. *Chem. Rev.* **2008**, *108*, 1501–1516. [CrossRef] [PubMed]

61. Perrin, D.M. [18F]-Organotrifluoroborates as radioprosthetic groups for PET imaging: From design principles to preclinical applications. *Acc. Chem. Res.* **2016**, *49*, 1333–1343. [CrossRef] [PubMed]

62. Bernard-Gauthier, V.; Bailey, J.J.; Liu, Z.B.; Wangler, B.; Wangler, C.; Jurkschat, K.; Perrin, D.M.; Schirrmacher, R. From unorthodox to established: The current status of F-18-trifluoroborate- and F-18-SiFA-based radiopharmaceuticals in PET nuclear imaging. *Bioconjug. Chem.* **2016**, *27*, 267–279. [CrossRef] [PubMed]

63. Lange, C.W.; VanBrocklin, H.F.; Taylor, S.E. Photoconjugation of 3-azido-5-nitrobenzyl-[18F] fluoride to an oligonucleotide aptamer. *J. Label. Compd. Radiopharm.* **2002**, *45*, 257–268. [CrossRef]

64. Jacobson, O.; Weiss, I.D.; Wang, L.; Wang, Z.; Yang, X.Y.; Dewhurst, A.; Ma, Y.; Zhu, G.Z.; Niu, G.; Kiesewetter, D.O.; et al. 18F-Labeled single-stranded DNA aptamer for PET imaging of protein tyrosine kinase-7 expression. *J. Nucl. Med.* **2015**, *56*, 1780–1785. [CrossRef] [PubMed]

65. Wang, L.; Jacobson, O.; Avdic, D.; Rotstein, B.H.; Weiss, I.D.; Collier, L.; Chen, X.Y.; Vasdev, N.; Liang, S.H. Ortho-stabilized 18F-Azido click agents and their application in PET imaging with single-stranded DNA aptamers. *Angew. Chem. Int. Ed.* **2015**, *54*, 12777–12781. [CrossRef] [PubMed]

66. Daniels, D.A.; Chen, H.; Hicke, B.J.; Swiderek, K.M.; Gold, L. A tenascin-c aptamer identified by tumor cell-SELEX: Systematic evolution of ligands by exponential enrichment. *Proc. Natl. Acad. Sci. USA* **2003**, *100*, 15416–15421. [CrossRef] [PubMed]

67. Jacobson, O.; Yan, X.F.; Niu, G.; Weiss, I.D.; Ma, Y.; Szajek, L.P.; Shen, B.Z.; Kiesewetter, D.O.; Chen, X.Y. PET imaging of tenascin-c with a radio labeled single-stranded DNA aptamer. *J. Nucl. Med.* **2015**, *56*, 616–621. [CrossRef] [PubMed]

68. Zhu, G.Z.; Zhang, H.M.; Jacobson, O.; Wang, Z.T.; Chen, H.J.; Yang, X.Y.; Niu, G.; Chen, X.Y. Combinatorial screening of DNA aptamers for molecular imaging of her2 in cancer. *Bioconjug. Chem.* **2017**, *28*, 1068–1075. [CrossRef] [PubMed]

69. Park, J.Y.; Lee, T.S.; Song, I.H.; Cho, Y.L.; Chae, J.R.; Yun, M.; Kang, H.; Lee, J.H.; Lim, J.H.; Cho, W.G.; et al. Hybridization-based aptamer labeling using complementary oligonucleotide platform for PET and optical imaging. *Biomaterials* **2016**, *100*, 143–151. [CrossRef] [PubMed]

70. Schulz, J.; Vimont, D.; Bordenave, T.; James, D.; Escudier, J.M.; Allard, M.; Szlosek-Pinaud, M.; Fouquet, E. Silicon-based chemistry: An original and efficient one-step approach to F-18 -nucleosides and F-18 -oligonucleotides for PET imaging. *Chem. Eur. J.* **2011**, *17*, 3096–3100. [CrossRef] [PubMed]

71. James, D.; Escudier, J.M.; Amigues, E.; Schulz, J.; Vitry, C.; Bordenave, T.; Szlosek-Pinaud, M.; Fouquet, E. A 'click chemistry' approach to the efficient synthesis of modified nucleosides and oligonucleotides for PET imaging. *Tetrahedron Lett.* **2010**, *51*, 1230–1232. [CrossRef]

72. Li, Y.; Schaffer, P.; Perrin, D.M. Dual isotope labeling: Conjugation of P-32-oligonucleotides with F-18-aryltrifluoroborate via copper(I) catalyzed cycloaddition. *Bioorg. Med. Chem. Lett.* **2013**, *23*, 6313–6316. [CrossRef] [PubMed]

73. Kuhnast, B.; de Bruin, A.; Hinnen, F.; Tavitian, B.; Dolle, F. Design and synthesis of a new F-18 fluoropyridine-based haloacetamide reagent for the labeling of oligonucleotides: 2-bromo-N-3-(2-F-18-fluoropyridin-3-yloxy)propyl acetamide. *Bioconjug. Chem.* **2004**, *15*, 617–627. [CrossRef] [PubMed]

74. James, M.L.; Gambhir, S.S. A molecular imaging primer: Modalities, imaging agents, and applications. *Physiol. Rev.* **2012**, *92*, 897–965. [CrossRef] [PubMed]

75. Gomes, S.D.; Miguel, J.; Azema, L.; Eimer, S.; Ries, C.; Dausse, E.; Loiseau, H.; Allard, M.; Toulme, J.J. Tc-99m-MAG3-aptamer for imaging human tumors associated with high level of matrix metalloprotease-9. *Bioconjug. Chem.* **2012**, *23*, 2192–2200. [CrossRef] [PubMed]

76. Kryza, D.; Debordeaux, F.; Azema, L.; Hassan, A.; Paurelle, O.; Schulz, J.; Savona-Baron, C.; Charignon, E.; Bonazza, P.; Taleb, J.; et al. Ex vivo and in vivo imaging and biodistribution of aptamers targeting the human matrix metalloprotease-9 in melanomas. *PLoS ONE* **2016**, *11*, e0149387. [CrossRef] [PubMed]

77. Macedo, B.; Cordeiro, Y. Unraveling prion protein interactions with aptamers and other PrP-binding nucleic acids. *Int. J. Mol. Sci.* **2017**, *18*, 1023. [CrossRef] [PubMed]

78. Qu, J.; Yu, S.Q.; Zheng, Y.; Zheng, Y.; Yang, H.; Zhang, J.L. Aptamer and its applications in neurodegenerative diseases. *Cell. Mol. Life Sci.* **2017**, *74*, 683–695. [CrossRef] [PubMed]

79. Wolter, O.; Mayer, G. Aptamers as valuable molecular tools in neurosciences. *J. Neurosci.* **2017**, *37*, 2517–2523. [CrossRef] [PubMed]

80. Nedergaard, M.; Takano, T.; Hansen, A.J. Beyond the role of glutamate as a neurotransmitter. *Nat. Rev. Neurosci.* **2002**, *3*, 748–755. [CrossRef] [PubMed]

81. Perry, M.; Li, Q.; Kennedy, R.T. Review of recent advances in analytical techniques for the determination of neurotransmitters. *Anal. Chim. Acta* **2009**, *653*, 1–22. [CrossRef] [PubMed]

82. Hokfelt, T.; Bartfai, T.; Bloom, F. Neuropeptides: Opportunities for drug discovery. *Lancet Neurol.* **2003**, *2*, 463–472. [CrossRef]

83. Li, B.R.; Hsieh, Y.J.; Chen, Y.X.; Chung, Y.T.; Pan, C.Y.; Chen, Y.T. An ultrasensitive nanowire-transistor biosensor for detecting dopamine release from living pc12 cells under hypoxic stimulation. *J. Am. Chem. Soc.* **2013**, *135*, 16034–16037. [CrossRef] [PubMed]

84. Kumar, A.; Singh, A.; Ekavali. A review on Alzheimer's disease pathophysiology and its management: An update. *Pharmacol. Rep.* **2015**, *67*, 195–203. [CrossRef] [PubMed]

85. Castren, E. Is mood chemistry? *Nat. Rev. Neurosci.* **2005**, *6*, 241–246. [CrossRef] [PubMed]

86. McConnell, E.M.; Holahan, M.R.; DeRosa, M.C. Aptamers as promising molecular recognition elements for diagnostics and therapeutics in the central nervous system. *Nucleic Acid Ther.* **2014**, *24*, 388–404. [CrossRef] [PubMed]

87. Mannironi, C.; DiNardo, A.; Fruscoloni, P.; TocchiniValentini, G.P. In vitro selection of dopamine RNA ligands. *Biochemistry* **1997**, *36*, 9726–9734. [CrossRef] [PubMed]

88. Zheng, Y.; Wang, Y.; Yang, X.R. Aptamer-based colorimetric biosensing of dopamine using unmodified gold nanoparticles. *Sens. Actuator B Chem.* **2011**, *156*, 95–99. [CrossRef]

89. Farjami, E.; Campos, R.; Nielsen, J.S.; Gothelf, K.V.; Kjems, J.; Ferapontova, E.E. RNA aptamer-based electrochemical biosensor for selective and label-free analysis of dopamine. *Anal. Chem.* **2013**, *85*, 121–128. [CrossRef] [PubMed]

90. Walsh, R.; DeRosa, M.C. Retention of function in the DNA homolog of the RNA dopamine aptamer. *Biochem. Biophys. Res. Commun.* **2009**, *388*, 732–735. [CrossRef] [PubMed]

91. Holahan, M.R.; Madularu, D.; McConnell, E.M.; Walsh, R.; DeRosa, M.C. Intra-accumbens injection of a dopamine aptamer abates MK-801-induced cognitive dysfunction in a model of schizophrenia. *PLoS ONE* **2011**, *6*, e22239. [CrossRef] [PubMed]

92. Alvarez-Martos, I.; Ferapontova, E.E. A DNA sequence obtained by replacement of the dopamine RNA aptamer bases is not an aptamer. *Biochem. Biophys. Res. Commun.* **2017**, *489*, 381–385. [CrossRef] [PubMed]

93. Kammer, M.N.; Olmsted, I.R.; Kussrow, A.K.; Morris, M.J.; Jackson, G.W.; Bornhop, D.J. Characterizing aptamer small molecule interactions with backscattering interferometry. *Analyst* **2014**, *139*, 5879–5884. [CrossRef] [PubMed]

94. Bruno, J.G.; Carrillo, M.P.; Phillips, T.; King, B. Development of DNA aptamers for cytochemical detection of acetylcholine. *In Vitro Cell. Dev. Biol. Anim.* **2008**, *44*, 63–72. [CrossRef] [PubMed]

95. Chavez, J.L.; Hagen, J.A.; Kelley-Loughnane, N. Fast and selective plasmonic serotonin detection with aptamer-gold nanoparticle conjugates. *Sensors* **2017**, *17*, 8. [CrossRef] [PubMed]

96. Tatemoto, K. Neuropeptide Y: Complete amino-acid-sequence of the brain peptide. *Proc. Natl. Acad. Sci. USA* **1982**, *79*, 5485–5489. [CrossRef] [PubMed]

97. Van den Pol, A.N. Neuropeptide transmission in brain circuits. *Neuron* **2012**, *76*, 98–115. [CrossRef] [PubMed]

98. Mendonsa, S.D.; Bowser, M.T. In vitro selection of aptamers with affinity for neuropeptide Y using capillary electrophoresis. *J. Am. Chem. Soc.* **2005**, *127*, 9382–9383. [CrossRef] [PubMed]

99. Proske, D.; Hofliger, M.; Soll, R.M.; Beck-Sickinger, A.G.; Famulok, M. A Y2 receptor mimetic aptamer directed against neuropeptide Y. *J. Biol. Chem.* **2002**, *277*, 11416–11422. [CrossRef] [PubMed]

100. Fernandez, R.E.; Sanghavi, B.J.; Farmehini, V.; Chavez, J.L.; Hagen, J.; Kelley-Loughnane, N.; Chou, C.F.; Swami, N.S. Aptamer-functionalized graphene-gold nanocomposites for label-free detection of dielectrophoretic-enriched neuropeptide Y. *Electrochem. Commun.* **2016**, *72*, 144–147. [CrossRef]

101. Banerjee, S.; Hsieh, Y.J.; Liu, C.R.; Yeh, N.H.; Hung, H.H.; Lai, Y.S.; Chou, A.C.; Chen, Y.T.; Pan, C.Y. Differential releases of dopamine and neuropeptide Y from histamine-stimulated pc12 cells detected by an aptamer-modified nanowire transistor. *Small* **2016**, *12*, 5524–5529. [CrossRef] [PubMed]

102. Eulberg, D.; Buchner, K.; Maasch, C.; Klussmann, S. Development of an automated in vitro selection protocol to obtain RNA-based aptamers: Identification of a biostable substance P antagonist. *Nucleic Acids Res.* **2005**, *33*, e45. [CrossRef] [PubMed]

103. Vater, A.; Klussmann, S. Turning mirror-image oligonucleotides into drugs: The evolution of spiegelmer therapeutics. *Drug Discov. Today* **2015**, *20*, 147–155. [CrossRef] [PubMed]

104. Faulhammer, D.; Eschgfaller, B.; Stark, S.; Burgstaller, P.; Englberger, W.; Erfurth, J.; Kleinjung, F.; Rupp, J.; Vulcu, S.D.; Schroder, W.; et al. Biostable aptamers with antagonistic properties to the neuropeptide nociceptin/orphanin FQ. *RNA* **2004**, *10*, 516–527. [CrossRef] [PubMed]

105. Takenaka, M.; Amino, T.; Miyachi, Y.; Ogino, C.; Kondo, A. Screening and evaluation of aptamers against somatostatin, and sandwich-like monitoring of somatostatin based on atomic force microscopy. *Sens. Actuator B Chem.* **2017**, *252*, 813–821. [CrossRef]

106. Kobelt, P.; Helmling, S.; Stengel, A.; Wlotzka, B.; Andresen, V.; Klapp, B.F.; Wiedenmann, B.; Klussmann, S.; Monnikes, H. Anti-ghrelin spiegelmer NOX-B11 inhibits neurostimulatory and orexigenic effects of peripheral ghrelin in rats. *Gut* **2006**, *55*, 788–792. [CrossRef] [PubMed]

107. Vater, A.; Sell, S.; Kaczmarek, P.; Maasch, C.; Buchner, K.; Pruszynska-Oszmalek, E.; Kolodziejski, P.; Purschke, W.G.; Nowak, K.W.; Strowski, M.Z.; et al. A mixed mirror-image DNA/RNA aptamer inhibits glucagon and acutely improves glucose tolerance in models of type 1 and type 2 diabetes. *J. Biol. Chem.* **2013**, *288*, 21136–21147. [CrossRef] [PubMed]

108. Heiat, M.; Ranjbar, R.; Latifi, A.M.; Rasaee, M.J. Selection of a high-affinity and in vivo bioactive ssDNA aptamer against angiotensin II peptide. *Peptides* **2016**, *82*, 101–108. [CrossRef] [PubMed]

109. Vater, A.; Jarosch, F.; Buchner, K.; Klussmann, S. Short bioactive spiegelmers to migraine-associated calcitonin gene-related peptide rapidly identified by a novel approach: Tailored-SELEX. *Nucleic Acids Res.* **2003**, *31*, e130. [CrossRef] [PubMed]

110. Kahsai, A.W.; Wisler, J.W.; Lee, J.; Ahn, S.; Cahill, T.J.; Dennison, S.M.; Staus, D.P.; Thomsen, A.R.B.; Anasti, K.M.; Pani, B.; et al. Conformationally selective RNA aptamers allosterically modulate the β(2)-adrenoceptor. *Nat. Chem. Biol.* **2016**, *12*, 709–716. [CrossRef] [PubMed]

111. Daniels, D.A.; Sohal, A.K.; Rees, S.; Grisshammer, R. Generation of RNA aptamers to the G-protein-coupled receptor for neurotensin, NTS-1. *Anal. Biochem.* **2002**, *305*, 214–226. [CrossRef] [PubMed]

112. Clawson, G.A.; Abraham, T.; Pan, W.H.; Tang, X.M.; Linton, S.S.; McGovern, C.O.; Loc, W.S.; Smith, J.P.; Butler, P.J.; Kester, M.; et al. A cholecystokinin B receptor-specific DNA aptamer for targeting pancreatic ductal adenocarcinoma. *Nucleic Acid Ther.* **2017**, *27*, 23–35. [CrossRef] [PubMed]

113. Costanzo, M.; Zurzolo, C. The cell biology of prion-like spread of protein aggregates: Mechanisms and implication in neurodegeneration. *Biochem. J.* **2013**, *452*, 1–17. [CrossRef] [PubMed]

114. Verwilst, P.; Kim, H.-R.; Seo, J.; Sohn, N.-W.; Cha, S.-Y.; Kim, Y.; Maeng, S.; Shin, J.-W.; Kwak, J.H.; Kang, C.; et al. Rational design of in vivo tau tangle-selective near-infrared fluorophores: Expanding the BODIPY universe. *J. Am. Chem. Soc.* **2017**, *139*, 13393–13403. [CrossRef] [PubMed]

115. Guo, J.L.; Lee, V.M.Y. Seeding of normal tau by pathological tau conformers drives pathogenesis of Alzheimer-like tangles. *J. Biol. Chem.* **2011**, *286*, 15317–15331. [CrossRef] [PubMed]

116. Kim, J.H.; Kim, E.; Choi, W.H.; Lee, J.; Lee, J.H.; Lee, H.; Kim, D.E.; Suh, Y.H.; Lee, M.J. Inhibitory RNA aptamers of tau oligomerization and their neuroprotective roles against proteotoxic stress. *Mol. Pharm.* **2016**, *13*, 2039–2048. [CrossRef] [PubMed]

117. Krylova, S.M.; Musheev, M.; Nutiu, R.; Li, Y.F.; Lee, G.; Krylov, S.N. Tau protein binds single-stranded DNA sequence specifically—The proof obtained in vitro with non-equilibrium capillary electrophoresis of equilibrium mixtures. *FEBS Lett.* **2005**, *579*, 1371–1375. [CrossRef] [PubMed]

118. Kim, S.; Wark, A.W.; Lee, H.J. Femtomolar detection of tau proteins in undiluted plasma using surface plasmon resonance. *Anal. Chem.* **2016**, *88*, 7793–7799. [CrossRef] [PubMed]

119. Hamley, I.W. The amyloid β peptide: A chemist's perspective. Role in Alzheimer's and fibrillization. *Chem. Rev.* **2012**, *112*, 5147–5192. [CrossRef] [PubMed]

120. Lauren, J.; Gimbel, D.A.; Nygaard, H.B.; Gilbert, J.W.; Strittmatter, S.M. Cellular prion protein mediates impairment of synaptic plasticity by amyloid-β oligomers. *Nature* **2009**, *457*, 1128–1132. [CrossRef] [PubMed]

121. Balducci, C.; Beeg, M.; Stravalaci, M.; Bastone, A.; Sclip, A.; Biasini, E.; Tapella, L.; Colombo, L.; Manzoni, C.; Borsello, T.; et al. Synthetic amyloid-β oligomers impair long-term memory independently of cellular prion protein. *Proc. Natl. Acad. Sci. USA* **2010**, *107*, 2295–2300. [CrossRef] [PubMed]

122. Kessels, H.W.; Nguyen, L.N.; Nabavi, S.; Malinow, R. The prion protein as a receptor for amyloid-β. *Nature* **2010**, *466*, E3–E4. [CrossRef] [PubMed]

123. Benilova, I.; De Strooper, B. Prion protein in alzheimer's pathogenesis: A hot and controversial issue. *EMBO Mol. Med.* **2010**, *2*, 289–290. [CrossRef] [PubMed]

124. Liang, H.Y.; Shi, Y.S.; Kou, Z.W.; Peng, Y.H.; Chen, W.J.; Li, X.W.; Li, S.J.; Wang, Y.; Wang, F.; Zhang, X.M. Inhibition of bace1 activity by a DNA aptamer in an Alzheimer's disease cell model. *PLoS ONE* **2015**, *10*, e0140733. [CrossRef] [PubMed]

125. Eremeeva, E.; Abramov, M.; Margamuljana, L.; Rozenski, J.; Pezo, V.; Marliere, P.; Herdewijn, P. Chemical morphing of DNA containing four noncanonical bases. *Angew. Chem. Int. Ed.* **2016**, *55*, 7515–7519. [CrossRef] [PubMed]

126. Gasse, C.; Zaarour, M.; Noppen, S.; Abramov, M.; Marliere, P.; Liekens, S.; De Strooper, B.; Herdewijn, P. Modulation of BACE1 activity by chemically modified aptamers. *Chembiochem* **2017**. submitted.

127. Rentmeister, A.; Bill, A.; Wahle, T.; Walter, J.; Famulok, M. RNA aptamers selectively modulate protein recruitment to the cytoplasmic domain of β-secretase BACE1 in vitro. *RNA* **2006**, *12*, 1650–1660. [CrossRef] [PubMed]

128. Ylera, F.; Lurz, R.; Erdmann, V.A.; Furste, J.P. Selection of RNA aptamers to the Alzheimer's disease amyloid peptide. *Biochem. Biophys. Res. Commun.* **2002**, *290*, 1583–1588. [CrossRef] [PubMed]

129. Farrar, C.T.; William, C.M.; Hudry, E.; Hashimoto, T.; Hyman, B.T. RNA aptamer probes as optical imaging agents for the detection of amyloid plaques. *PLoS ONE* **2014**, *9*, e89901. [CrossRef] [PubMed]

130. Rahimi, F.; Murakami, K.; Summers, J.L.; Chen, C.-H.B.; Bitan, G. RNA aptamers generated against oligomeric aβ40 recognize common amyloid aptatopes with low specificity but high sensitivity. *PLoS ONE* **2009**, *4*, e7694. [CrossRef] [PubMed]

131. Takahashi, T.; Tada, K.; Mihara, H. RNA aptamers selected against amyloid β-peptide (Aβ) inhibit the aggregation of ab. *Mol. Biosyst.* **2009**, *5*, 986–991. [CrossRef] [PubMed]

132. Murakami, K.; Nishikawa, F.; Noda, K.; Yokoyama, T.; Nishikawa, S. Anti-bovine prion protein RNA aptamer containing tandem gga repeat interacts both with recombinant bovine prion protein and its β isoform with high affinity. *Prion* **2008**, *2*, 73–80. [CrossRef] [PubMed]

133. Ogasawara, D.; Hasegawa, H.; Kaneko, K.; Sode, K.; Ikebukuro, K. Screening of DNA aptamer against mouse prion protein by competitive selection. *Prion* **2007**, *1*, 248–254. [CrossRef] [PubMed]

134. Proske, D.; Gilch, S.; Wopfner, F.; Schatzl, H.M.; Winnacker, E.L.; Famulok, M. Prion-protein-specific aptamer reduces PrPSc formation. *ChemBioChem* **2002**, *3*, 717–725. [CrossRef]

135. Wang, P.; Hatcher, K.L.; Bartz, J.C.; Chen, S.G.; Skinner, P.; Richt, J.; Liu, H.; Sreevatsan, S. Selection and characterization of DNA aptamers against PrPSc. *Exp. Biol. Med.* **2011**, *236*, 466–476. [CrossRef] [PubMed]

136. Goedert, M. Alpha-synuclein and neurodegenerative diseases. *Nat. Rev. Neurosci.* **2001**, *2*, 492–501. [CrossRef] [PubMed]

137. Spillantini, M.G.; Schmidt, M.L.; Lee, V.M.Y.; Trojanowski, J.Q.; Jakes, R.; Goedert, M. Alpha-synuclein in Lewy bodies. *Nature* **1997**, *388*, 839–840. [CrossRef] [PubMed]

138. Tsukakoshi, K.; Harada, R.; Sode, K.; Ikebukuro, K. Screening of DNA aptamer which binds to alpha-synuclein. *Biotechnol. Lett.* **2010**, *32*, 643–648. [CrossRef] [PubMed]

139. Hasegawa, H.; Sode, K.; Ikebukuro, K. Selection of DNA aptamers against VEGF(165) using a protein competitor and the aptamer blotting method. *Biotechnol. Lett.* **2008**, *30*, 829–834. [CrossRef] [PubMed]

140. Tsukakoshi, K.; Abe, K.; Sode, K.; Ikebukuro, K. Selection of DNA aptamers that recognize alpha-synuclein oligomers using a competitive screening method. *Anal. Chem.* **2012**, *84*, 5542–5547. [CrossRef] [PubMed]

141. Sun, K.; Xia, N.; Zhao, L.J.; Liu, K.; Hou, W.J.; Liu, L. Aptasensors for the selective detection of alpha-synuclein oligomer by colorimetry, surface plasmon resonance and electrochemical impedance spectroscopy. *Sens. Actuator B Chem.* **2017**, *245*, 87–94. [CrossRef]

142. Liu, L.; Chang, Y.; Yu, J.; Jiang, M.S.; Xia, N. Two-in-one polydopamine nanospheres for fluorescent determination of β-amyloid oligomers and inhibition of β-amyloid aggregation. *Sens. Actuator B Chem.* **2017**, *251*, 359–365. [CrossRef]

143. Zhou, Y.L.; Zhang, H.Q.; Liu, L.T.; Li, C.M.; Chang, Z.; Zhu, X.; Ye, B.X.; Xu, M.T. Fabrication of an antibody-aptamer sandwich assay for electrochemical evaluation of levels of β-amyloid oligomers. *Sci. Rep.* **2016**, *6*, 35186. [CrossRef] [PubMed]

144. Jiang, L.F.; Chen, B.C.; Chen, B.; Li, X.J.; Liao, H.L.; Huang, H.M.; Guo, Z.J.; Zhang, W.Y.; Wu, L. Detection of ab oligomers based on magnetic-field-assisted separation of aptamer-functionalized Fe_3O_4 magnetic nanoparticles and $bayf_5$:Yb,er nanoparticles as upconversion fluorescence labels. *Talanta* **2017**, *170*, 350–357. [CrossRef] [PubMed]

145. Zhu, L.L.; Zhang, J.Y.; Wang, F.Y.; Wang, Y.; Lu, L.L.; Feng, C.C.; Xu, Z.A.; Zhang, W. Selective amyloid β oligomer assay based on abasic site-containing molecular beacon and enzyme-free amplification. *Biosens. Bioelectron.* **2016**, *78*, 206–212. [CrossRef] [PubMed]

146. McLaughlin, C.K.; Hamblin, G.D.; Sleiman, H.F. Supramolecular DNA assembly. *Chem. Soc. Rev.* **2011**, *40*, 5647–5656. [CrossRef] [PubMed]

147. Rothemund, P.W.K. Folding DNA to create nanoscale shapes and patterns. *Nature* **2006**, *440*, 297–302. [CrossRef] [PubMed]

148. Endo, M.; Yang, Y.; Sugiyama, H. DNA origami technology for biomaterials applications. *Biomater. Sci.* **2013**, *1*, 347–360. [CrossRef]

149. Chhabra, R.; Sharma, J.; Ke, Y.G.; Liu, Y.; Rinker, S.; Lindsay, S.; Yan, H. Spatially addressable multiprotein nanoarrays templated by aptamer-tagged DNA nanoarchitectures. *J. Am. Chem. Soc.* **2007**, *129*, 10304–10305. [CrossRef] [PubMed]

150. Rinker, S.; Ke, Y.G.; Liu, Y.; Chhabra, R.; Yan, H. Self-assembled DNA nanostructures for distance-dependent multivalent ligand-protein binding. *Nat. Nanotechnol.* **2008**, *3*, 418–422. [CrossRef] [PubMed]

151. Douglas, S.M.; Bachelet, I.; Church, G.M. A logic-gated nanorobot for targeted transport of molecular payloads. *Science* **2012**, *335*, 831–834. [CrossRef] [PubMed]

152. Godonoga, M.; Lin, T.Y.; Oshima, A.; Sumitomo, K.; Tang, M.S.L.; Cheung, Y.W.; Kinghorn, A.B.; Dirkzwager, R.M.; Zhou, C.S.; Kuzuya, A.; et al. A DNA aptamer recognising a malaria protein biomarker can function as part of a DNA origami assembly. *Sci. Rep.* **2016**, *6*, 21266. [CrossRef] [PubMed]

153. Walter, H.K.; Bauer, J.; Steinmeyer, J.; Kuzuya, A.; Niemeyer, C.M.; Wagenknecht, H.A. "DNA origami traffic lights" with a split aptamer sensor for a bicolor fluorescence readout. *Nano Lett.* **2017**, *17*, 2467–2472. [CrossRef] [PubMed]

154. Liu, F.R.; Sha, R.J.; Seeman, N.C. Modifying the surface features of two-dimensional DNA crystals. *J. Am. Chem. Soc.* **1999**, *121*, 917–922. [CrossRef]

155. Green, L.S.; Jellinek, D.; Jenison, R.; Ostman, A.; Heldin, C.H.; Janjic, N. Inhibitory DNA ligands to platelet-derived growth factor B-Chain. *Biochemistry* **1996**, *35*, 14413–14424. [CrossRef] [PubMed]

156. Ke, Y.G.; Lindsay, S.; Chang, Y.; Liu, Y.; Yan, H. Self-assembled water-soluble nucleic acid probe tiles for label-free RNA hybridization assays. *Science* **2008**, *319*, 180–183. [CrossRef] [PubMed]

157. Cheung, Y.W.; Kwok, J.; Law, A.W.L.; Watt, R.M.; Kotaka, M.; Tanner, J.A. Structural basis for discriminatory recognition of plasmodium lactate dehydrogenase by a DNA aptamer. *Proc. Natl. Acad. Sci. USA* **2013**, *110*, 15967–15972. [CrossRef] [PubMed]

158. Kuzuya, A.; Sakai, Y.; Yamazaki, T.; Xu, Y.; Komiyama, M. Nanomechanical DNA origami 'single-molecule beacons' directly imaged by atomic force microscopy. *Nat. Commun.* **2011**, *2*, 449. [CrossRef] [PubMed]

159. Chen, A.L.; Yan, M.M.; Yang, S.M. Split aptamers and their applications in sandwich aptasensors. *Trac-Trends Anal. Chem.* **2016**, *80*, 581–593. [CrossRef]

160. Walter, H.K.; Bohlander, P.R.; Wagenknecht, H.A. Development of a wavelength-shifting fluorescent module for the adenosine aptamer using photostable cyanine dyes. *Chemistry* **2015**, *4*, 92–96. [CrossRef] [PubMed]

161. Holzhauser, C.; Wagenknecht, H.A. DNA and RNA "traffic lights": Synthetic wavelength-shifting fluorescent probes based on nucleic acid base substitutes for molecular imaging. *J. Org. Chem.* **2013**, *78*, 7373–7379. [CrossRef] [PubMed]

162. Bertrand, N.; Wu, J.; Xu, X.Y.; Kamaly, N.; Farokhzad, O.C. Cancer nanotechnology: The impact of passive and active targeting in the era of modern cancer biology. *Adv. Drug Deliv. Rev.* **2014**, *66*, 2–25. [CrossRef] [PubMed]

163. Strebhardt, K.; Ullrich, A. Paul Ehrlich's magic bullet concept: 100 years of progress. *Nat. Rev. Cancer* **2008**, *8*, 473–480. [CrossRef] [PubMed]

164. Liu, Q.L.; Jin, C.; Wang, Y.Y.; Fang, X.H.; Zhang, X.B.; Chen, Z.; Tan, W.H. Aptamer-conjugated nanomaterials for specific cancer cell recognition and targeted cancer therapy. *NPG Asia Mater.* **2014**, *6*, e95. [CrossRef]

165. Nair, R.R.; Wu, H.A.; Jayaram, P.N.; Grigorieva, I.V.; Geim, A.K. Unimpeded permeation of water through helium-leak-tight graphene-based membranes. *Science* **2012**, *335*, 442–444. [CrossRef] [PubMed]

166. Kim, J.; Park, S.J.; Min, D.H. Emerging approaches for graphene oxide biosensor. *Anal. Chem.* **2017**, *89*, 232–248. [CrossRef] [PubMed]

167. Nellore, B.P.V.; Kanchanapally, R.; Pramanik, A.; Sinha, S.S.; Chavva, S.R.; Hamme, A.; Ray, P.C. Aptamer-conjugated graphene oxide membranes for highly efficient capture and accurate identification of multiple types of circulating tumor cells. *Bioconjug. Chem.* **2015**, *26*, 235–242. [CrossRef] [PubMed]

168. Bahreyni, A.; Yazdian-Robati, R.; Hashemitabar, S.; Ramezani, M.; Ramezani, P.; Abnous, K.; Taghdisi, S.M. A new chemotherapy agent-free theranostic system composed of graphene oxide nano-complex and aptamers for treatment of cancer cells. *Int. J. Pharm.* **2017**, *526*, 391–399. [CrossRef] [PubMed]

169. Tang, Y.; Hu, H.; Zhang, M.G.; Song, J.; Nie, L.; Wang, S.; Niu, G.; Huang, P.; Lu, G.; Chen, X. An aptamer-targeting photoresponsive drug delivery system using "off-on" graphene oxide wrapped mesoporous silica nanoparticles. *Nanoscale* **2015**, *7*, 6304–6310. [CrossRef] [PubMed]

170. Yang, L.; Zhang, X.B.; Ye, M.; Jiang, J.H.; Yang, R.H.; Fu, T.; Chen, Y.; Wang, K.M.; Liu, C.; Tan, W.H. Aptamer-conjugated nanomaterials and their applications. *Adv. Drug Deliv. Rev.* **2011**, *63*, 1361–1370. [CrossRef] [PubMed]

171. Niu, W.; Chen, X.; Tan, W.; Veige, A.S. N-Heterocyclic Carbene–Gold(I) complexes conjugated to a Leukemia-Specific DNA aptamer for targeted drug delivery. *Angew. Chem. Int. Ed.* **2016**, *55*, 8889–8893. [CrossRef] [PubMed]

172. Siafaka, P.I.; Okur, N.U.; Karavas, E.; Bikiaris, D.N. Surface modified multifunctional and stimuli responsive nanoparticles for drug targeting: Current status and uses. *Int. J. Mol. Sci.* **2016**, *17*, 1440. [CrossRef] [PubMed]

173. Latorre, A.; Posch, C.; Garcimartin, Y.; Celli, A.; Sanlorenzo, M.; Vujic, I.; Ma, J.; Zekhtser, M.; Rappersberger, K.; Ortiz-Urda, S.; et al. DNA and aptamer stabilized gold nanoparticles for targeted delivery of anticancer therapeutics. *Nanoscale* **2014**, *6*, 7436–7442. [CrossRef] [PubMed]

174. Massich, M.D.; Giljohann, D.A.; Schmucker, A.L.; Patel, P.C.; Mirkin, C.A. Cellular response of polyvalent oligonucleotide-gold nanoparticle conjugates. *ACS Nano* **2010**, *4*, 5641–5646. [CrossRef] [PubMed]

175. Huang, S.-S.; Wei, S.-C.; Chang, H.-T.; Lin, H.-J.; Huang, C.-C. Gold nanoparticles modified with self-assembled hybrid monolayer of triblock aptamers as a photoreversible anticoagulant. *J. Control. Release* **2016**, *221*, 9–17. [CrossRef] [PubMed]

176. Swift, B.J.F.; Shadish, J.A.; DeForest, C.A.; Baneyx, F. Streamlined synthesis and assembly of a hybrid sensing architecture with solid binding proteins and click chemistry. *J. Am. Chem. Soc.* **2017**, *139*, 3958–3961. [CrossRef] [PubMed]

177. Geissler, D.; Linden, S.; Liermann, K.; Wegner, K.D.; Charbonniere, L.J.; Hildebrandt, N. Lanthanides and quantum dots as forster resonance energy transfer agents for diagnostics and cellular imaging. *Inorg. Chem.* **2014**, *53*, 1824–1838. [CrossRef] [PubMed]

178. Zhou, D.J. Quantum dot-nucleic acid/aptamer bioconjugate-based fluorimetric biosensors. *Biochem. Soc. Trans.* **2012**, *40*, 635–639. [CrossRef] [PubMed]

179. Elgqvist, J. Nanoparticles as theranostic vehicles in experimental and clinical applications-focus on prostate and breast cancer. *Int. J. Mol. Sci.* **2017**, *18*, 1102. [CrossRef] [PubMed]

180. Lin, Z.; Ma, Q.; Fei, X.; Zhang, H.; Su, X. A novel aptamer functionalized cuins2 quantum dots probe for daunorubicin sensing and near infrared imaging of prostate cancer cells. *Anal. Chim. Acta* **2014**, *818*, 54–60. [CrossRef] [PubMed]

181. Wu, Y.R.; Sefah, K.; Liu, H.P.; Wang, R.W.; Tan, W.H. DNA aptamer-micelle as an efficient detection/delivery vehicle toward cancer cells. *Proc. Natl. Acad. Sci. USA* **2010**, *107*, 5–10. [CrossRef] [PubMed]

182. Liu, H.P.; Zhu, Z.; Kang, H.Z.; Wu, Y.R.; Sefan, K.; Tan, W.H. DNA-based micelles: Synthesis, micellar properties and size-dependent cell permeability. *Chem. Eur. J.* **2010**, *16*, 3791–3797. [CrossRef] [PubMed]

183. Kim, J.; Kim, D.; Lee, J.B. DNA aptamer-based carrier for loading proteins and enhancing the enzymatic activity. *RSC Adv.* **2017**, *7*, 1643–1645. [CrossRef]

184. Xiong, X.L.; Liu, H.P.; Zhao, Z.L.; Altman, M.B.; Lopez-Colon, D.; Yang, C.J.; Chang, L.J.; Liu, C.; Tan, W.H. DNA aptamer-mediated cell targeting. *Angew. Chem. Int. Ed.* **2013**, *52*, 1472–1476. [CrossRef] [PubMed]

185. Lale, S.V.; Aswathy, R.G.; Aravind, A.; Kumar, D.S.; Koul, V. AS1411 aptamer and folic acid functionalized pH-responsive atrp fabricated pPEGMA-PCL-pPEGMA polymeric nanoparticles for targeted drug delivery in cancer therapy. *Biomacromolecules* **2014**, *15*, 1737–1752. [CrossRef] [PubMed]

186. Sun, P.C.; Zhang, N.; Tang, Y.F.; Yang, Y.N.; Chu, X.; Zhao, Y.X. Sl2b aptamer and folic acid dual-targeting DNA nanostructures for synergic biological effect with chemotherapy to combat colorectal cancer. *Int. J. Nanomed.* **2017**, *12*, 2657–2672. [CrossRef] [PubMed]

187. Dai, B.; Hu, Y.; Duan, J.; Yang, X.-D. Aptamer-guided DNA tetrahedron as a novel targeted drug delivery system for muc1-expressing breast cancer cells in vitro. *Oncotarget* **2016**, *7*, 38257–38269. [CrossRef] [PubMed]

188. Stiriba, S.E.; Frey, H.; Haag, R. Dendritic polymers in biomedical applications: From potential to clinical use in diagnostics and therapy. *Angew. Chem. Int. Ed.* **2002**, *41*, 1329–1334. [CrossRef]

189. MacEwan, S.R.; Chilkoti, A. From composition to cure: A systems engineering approach to anticancer drug carriers. *Angew. Chem. Int. Ed.* **2017**, *56*, 6712–6733. [CrossRef] [PubMed]

190. Gu, F.; Zhang, L.; Teply, B.A.; Mann, N.; Wang, A.; Radovic-Moreno, A.F.; Langer, R.; Farokhzad, O.C. Precise engineering of targeted nanoparticles by using self-assembled biointegrated block copolymers. *Proc. Natl. Acad. Sci. USA* **2008**, *105*, 2586–2591. [CrossRef] [PubMed]

191. Dhar, S.; Gu, F.X.; Langer, R.; Farokhzad, O.C.; Lippard, S.J. Targeted delivery of cisplatin to prostate cancer cells by aptamer functionalized Pt(IV) prodrug-PLGA-PEG nanoparticles. *Proc. Natl. Acad. Sci. USA* **2008**, *105*, 17356–17361. [CrossRef] [PubMed]

192. Dhar, S.; Kolishetti, N.; Lippard, S.J.; Farokhzad, O.C. Targeted delivery of a cisplatin prodrug for safer and more effective prostate cancer therapy in vivo. *Proc. Natl. Acad. Sci. USA* **2011**, *108*, 1850–1855. [CrossRef] [PubMed]

193. Zhuang, Y.Y.; Deng, H.P.; Su, Y.; He, L.; Wang, R.B.; Tong, G.S.; He, D.N.; Zhu, X.Y. Aptamer-functionalized and backbone redox-responsive hyperbranched polymer for targeted drug delivery in cancer therapy. *Biomacromolecules* **2016**, *17*, 2050–2062. [CrossRef] [PubMed]

194. Lao, Y.H.; Phua, K.K.L.; Leong, K.W. Aptamer nanomedicine for cancer therapeutics: Barriers and potential for translation. *ACS Nano* **2015**, *9*, 2235–2254. [CrossRef] [PubMed]

195. Taghavi, S.; Ramezani, M.; Alibolandi, M.; Abnous, K.; Taghdisi, S.M. Chitosan-modified plga nanoparticles tagged with 5tr1 aptamer for in vivo tumor-targeted drug delivery. *Cancer Lett.* **2017**, *400*, 1–8. [CrossRef] [PubMed]

196. Coles, D.J.; Rolfe, B.E.; Boase, N.R.B.; Veedu, R.N.; Thurecht, K.J. Aptamer-targeted hyperbranched polymers: Towards greater specificity for tumours in vivo. *Chem. Commun.* **2013**, *49*, 3836–3838. [CrossRef] [PubMed]

197. Yu, S.R.; Dong, R.J.; Chen, J.X.; Chen, F.; Jiang, W.F.; Zhou, Y.F.; Zhu, X.Y.; Yan, D.Y. Synthesis and self-assembly of arnphiphilic aptamer-functionalized hyperbranched multiarm copolymers for targeted cancer imaging. *Biomacromolecules* **2014**, *15*, 1828–1836. [CrossRef] [PubMed]

198. Xu, W.J.; Siddiqui, I.A.; Nihal, M.; Pilla, S.; Rosenthal, K.; Mukhtar, H.; Gong, S.Q. Aptamer-conjugated and doxorubicin-loaded unimolecular micelles for targeted therapy of prostate cancer. *Biomaterials* **2013**, *34*, 5244–5253. [CrossRef] [PubMed]

199. Torchilin, V.P. Recent advances with liposomes as pharmaceutical carriers. *Nat. Rev. Drug Discov.* **2005**, *4*, 145–160. [CrossRef] [PubMed]

200. Huwyler, J.; Wu, D.; Pardridge, W.M. Brain drug delivery of small molecules using immunoliposomes. *Proc. Natl. Acad. Sci. USA* **1996**, *93*, 14164–14169. [CrossRef] [PubMed]

201. Schnyder, A.; Huwyler, J. Drug transport to brain with targeted liposomes. *NeuroRX* **2005**, *2*, 99–107. [CrossRef] [PubMed]

202. Willis, M.C.; Collins, B.; Zhang, T.; Green, L.S.; Sebesta, D.P.; Bell, C.; Kellogg, E.; Gill, S.C.; Magallanez, A.; Knauer, S.; et al. Liposome anchored vascular endothelial growth factor aptamers. *Bioconjug. Chem.* **1998**, *9*, 573–582. [CrossRef] [PubMed]

203. Ara, M.N.; Matsuda, T.; Hyodo, M.; Sakurai, Y.; Hatakeyama, H.; Ohga, N.; Hida, K.; Harashima, H. An aptamer ligand based liposomal nanocarrier system that targets tumor endothelial cells. *Biomaterials* **2014**, *35*, 7110–7120. [CrossRef] [PubMed]

204. Alshaer, W.; Hillaireau, H.; Vergnaud, J.; Ismail, S.; Fattal, E. Functionalizing liposomes with anti-CD44 aptamer for selective targeting of cancer cells. *Bioconjug. Chem.* **2015**, *26*, 1307–1313. [CrossRef] [PubMed]

205. Plourde, K.; Derbali, R.M.; Desrosiers, A.; Dubath, C.; Vallée-Bélisle, A.; Leblond, J. Aptamer-based liposomes improve specific drug loading and release. *J. Control. Release* **2017**, *251*, 82–91. [CrossRef] [PubMed]

206. Wochner, A.; Menger, M.; Orgel, D.; Cech, B.; Rimmele, M.; Erdmann, V.A.; Glokler, J. A DNA aptamer with high affinity and specificity for therapeutic anthracyclines. *Anal. Biochem.* **2008**, *373*, 34–42. [CrossRef] [PubMed]

207. Barenholz, Y. Doxil®-the first FDA-approved nano-drug: Lessons learned. *J. Control. Release* **2012**, *160*, 117–134. [CrossRef] [PubMed]

208. Kang, S.; Hah, S.S. Improved ligand binding by antibody—Aptamer pincers. *Bioconjug. Chem.* **2014**, *25*, 1421–1427. [CrossRef] [PubMed]

209. Chu, T.C.; Twu, K.Y.; Ellington, A.D.; Levy, M. Aptamer mediated siRNA delivery. *Nucleic Acids Res.* **2006**, *34*, e73. [CrossRef] [PubMed]

210. Wang, R.W.; Zhu, G.Z.; Mei, L.; Xie, Y.; Ma, H.B.; Ye, M.; Qing, F.L.; Tan, W.H. Automated modular synthesis of aptamer-drug conjugates for targeted drug delivery. *J. Am. Chem. Soc.* **2014**, *136*, 2731–2734. [CrossRef] [PubMed]

211. Zhu, G.Z.; Niu, G.; Chen, X.Y. Aptamer-drug conjugates. *Bioconjug. Chem.* **2015**, *26*, 2186–2197. [CrossRef] [PubMed]

212. Huang, Y.-F.; Shangguan, D.; Liu, H.; Phillips, J.A.; Zhang, X.; Chen, Y.; Tan, W. Molecular assembly of an aptamer-drug conjugate for targeted drug delivery to tumor cells. *ChemBioChem* **2009**, *10*, 862–868. [CrossRef] [PubMed]

213. Zhu, G.; Zheng, J.; Song, E.; Donovan, M.; Zhang, K.; Liu, C.; Tan, W. Self-assembled, aptamer-tethered DNA nanotrains for targeted transport of molecular drugs in cancer theranostics. *Proc. Natl. Acad. Sci. USA* **2013**, *110*, 7998–8003. [CrossRef] [PubMed]

214. Shangguan, D.H.; Cao, Z.H.C.; Li, Y.; Tan, W.H. Aptamers evolved from cultured cancer cells reveal molecular differences of cancer cells in patient samples. *Clin. Chem.* **2007**, *53*, 1153–1155. [CrossRef] [PubMed]

215. Mallikaratchy, P. Evolution of complex target SELEX to identify aptamers against mammalian cell-surface antigens. *Molecules* **2017**, *22*, 215. [CrossRef] [PubMed]

216. Zhu, G.; Meng, L.; Ye, M.; Yang, L.; Sefah, K.; O'Donoghue, M.B.; Chen, Y.; Xiong, X.; Huang, J.; Song, E.; et al. Self-assembled aptamer-based drug carriers for bispecific cytotoxicity to cancer cells. *Chem. Asian J.* **2012**, *7*, 1630–1636. [CrossRef] [PubMed]

217. Zhang, P.; Zhao, N.X.; Zeng, Z.H.; Chang, C.C.; Zu, Y.L. Combination of an aptamer probe to CD4 and antibodies for multicolored cell phenotyping. *Am. J. Clin. Pathol.* **2010**, *134*, 586–593. [CrossRef] [PubMed]

218. Jo, N.; Mailhos, C.; Ju, M.H.; Cheung, E.; Bradley, J.; Nishijima, K.; Robinson, G.S.; Adarnis, A.P.; Shima, D.T. Inhibition of platelet-derived growth factor B signaling enhances the efficacy of anti-vascular enclothelial growth factor therapy in multiple models of ocular neovascularization. *Am. J. Pathol.* **2006**, *168*, 2036–2053. [CrossRef] [PubMed]

219. Heo, K.; Min, S.-W.; Sung, H.J.; Kim, H.G.; Kim, H.J.; Kim, Y.H.; Choi, B.K.; Han, S.; Chung, S.; Lee, E.S.; et al. An aptamer-antibody complex (oligobody) as a novel delivery platform for targeted cancer therapies. *J. Control. Release* **2016**, *229*, 1–9. [CrossRef] [PubMed]

220. Liu, Y.C.; Chen, Z.C.; He, A.B.; Zhan, Y.H.; Li, J.F.; Liu, L.; Wu, H.W.; Zhuang, C.L.; Lin, J.H.; Zhang, Q.X.; et al. Targeting cellular mRNAs translation by CRISPR-Cas9. *Sci. Rep.* **2016**, *6*, 29652. [CrossRef] [PubMed]

221. Wang, S.Y.; Su, J.H.; Zhang, F.; Zhuang, X.W. An RNA-aptamer-based two-color CRISPR labeling system. *Sci. Rep.* **2016**, *6*, 26857. [CrossRef] [PubMed]

222. Dassie, J.P.; Giangrande, P.H. Current progress on aptamer-targeted oligonucleotide therapeutics. *Ther. Deliv.* **2013**, *4*, 1527–1546. [CrossRef] [PubMed]

223. Kruspe, S.; Mittelberger, F.; Szameit, K.; Hahn, U. Aptamers as drug delivery vehicles. *Chem. Med. Chem.* **2014**, *9*, 1998–2011. [CrossRef] [PubMed]

224. Kruspe, S.; Giangrande, P. Aptamer-siRNA chimeras: Discovery, progress, and future prospects. *Biomedicines* **2017**, *5*, 45. [CrossRef] [PubMed]

225. McNamara, J.O.; Andrechek, E.R.; Wang, Y.; D Viles, K.; Rempel, R.E.; Gilboa, E.; Sullenger, B.A.; Giangrande, P.H. Cell type-specific delivery of siRNAs with aptamer-siRNA chimeras. *Nat. Biotechnol.* **2006**, *24*, 1005–1015. [CrossRef] [PubMed]

226. Dassie, J.P.; Liu, X.Y.; Thomas, G.S.; Whitaker, R.M.; Thiel, K.W.; Stockdale, K.R.; Meyerholz, D.K.; McCaffrey, A.P.; McNamara, J.O.; Giangrande, P.H. Systemic administration of optimized aptamer-siRNA chimeras promotes regression of PSMA-expressing tumors. *Nat. Biotechnol.* **2009**, *27*, 839–849. [CrossRef] [PubMed]

227. Liu, H.Y.; Yu, X.L.; Liu, H.T.; Wu, D.Q.; She, J.X. Co-targeting EGFR and survivin with a bivalent aptamer-dual siRNA chimera effectively suppresses prostate cancer. *Sci. Rep.* **2016**, *6*, 30346. [CrossRef] [PubMed]

228. Liu, H.Y.; Gao, X.H. A universal protein tag for delivery of siRNA-aptamer chimeras. *Sci. Rep.* **2013**, *3*, 3129. [CrossRef] [PubMed]

229. Jeong, H.; Lee, S.H.; Hwang, Y.; Yoo, H.; Jung, H.; Kim, S.H.; Mok, H. Multivalent aptamer-RNA conjugates for simple and efficient delivery of doxorubicin/siRNA into multidrug-resistant cells. *Macromol. Biosci.* **2017**, *17*, 1600343. [CrossRef] [PubMed]

230. Wilner, S.E.; Wengerter, B.; Maier, K.; Magalhaes, M.D.B.; Del Amo, D.S.; Pai, S.; Opazo, F.; Rizzoli, S.O.; Yan, A.; Levy, M. An RNA alternative to human transferrin: A new tool for targeting human cells. *Mol. Ther. Nucleic Acids* **2012**, *1*, e21. [CrossRef] [PubMed]

231. Breitz, H.B.; Weiden, P.L.; Beaumier, P.L.; Axworthy, D.B.; Seiler, C.; Su, F.M.; Graves, S.; Bryan, K.; Reno, J.M. Clinical optimization of pretargeted radioimmunotherapy with antibody-streptavidin conjugate and Y-90-DOTA-biotin. *J. Nucl. Med.* **2000**, *41*, 131–140. [PubMed]

232. Binzel, D.W.; Shu, Y.; Li, H.; Sun, M.Y.; Zhang, Q.S.; Shu, D.; Guo, B.; Guo, P.X. Specific delivery of miRNA for high efficient inhibition of prostate cancer by RNA nanotechnology. *Mol. Ther.* **2016**, *24*, 1267–1277. [CrossRef] [PubMed]

233. Serganov, A.; Patel, D.J. Ribozymes, riboswitches and beyond: Regulation of gene expression without proteins. *Nat. Rev. Genet.* **2007**, *8*, 776–790. [CrossRef] [PubMed]

234. Romero-Lopez, C.; Barroso-delJesus, A.; Puerta-Fernandez, E.; Berzal-Herranz, A. Interfering with hepatitis C virus ires activity using RNA molecules identified by a novel in vitro selection method. *Biol. Chem.* **2005**, *386*, 183–190. [CrossRef] [PubMed]

235. Romero-Lopez, C.; Diaz-Gonzalez, R.; Barroso-DelJesus, A.; Berzal-Herranz, A. Inhibition of hepatitis C virus replication and internal ribosome entry site-dependent translation by an RNA molecule. *J. Gen. Virol.* **2009**, *90*, 1659–1669. [CrossRef] [PubMed]

236. Romero-Lopez, C.; Lahlali, T.; Berzal-Herranz, B.; Berzal-Herranz, A. Development of optimized inhibitor RNAs allowing multisite-targeting of the HCV genome. *Molecules* **2017**, *22*, 861. [CrossRef] [PubMed]

237. Travascio, P.; Li, Y.F.; Sen, D. DNA-enhanced peroxidase activity of a DNA aptamer-hemin complex. *Chem. Biol.* **1998**, *5*, 505–517. [CrossRef]

238. Golub, E.; Albada, H.B.; Liao, W.C.; Biniuri, Y.; Willner, I. Nucleoapzymes: Hemin/G-quadruplex DNAzyme-aptamer binding site conjugates with superior enzyme-like catalytic functions. *J. Am. Chem. Soc.* **2016**, *138*, 164–172. [CrossRef] [PubMed]

239. Ni, S.; Yao, H.; Wang, L.; Lu, J.; Jiang, F.; Lu, A.; Zhang, G. Chemical modifications of nucleic acid aptamers for therapeutic purposes. *Int. J. Mol. Sci.* **2017**, *18*, 1683. [CrossRef] [PubMed]

240. Diafa, S.; Hollenstein, M. Generation of aptamers with an expanded chemical repertoire. *Molecules* **2015**, *20*, 16643–16671. [CrossRef] [PubMed]

241. Chen, T.; Hongdilokkul, N.; Liu, Z.X.; Thirunavukarasu, D.; Romesberg, F.E. The expanding world of DNA and RNA. *Curr. Opin. Chem. Biol.* **2016**, *34*, 80–87. [CrossRef] [PubMed]

242. Dellafiore, M.A.; Montserrat, J.M.; Iribarren, A.M. Modified nucleoside triphosphates for in vitro selection techniques. *Front. Chem.* **2016**, *4*, 18. [CrossRef] [PubMed]

243. Tolle, F.; Mayer, G. Dressed for success—Applying chemistry to modulate aptamer functionality. *Chem. Sci.* **2013**, *4*, 60–67. [CrossRef]

244. Dunn, M.R.; Jimenez, R.M.; Chaput, J.C. Analysis of aptamer discovery and technology. *Nat. Rev. Chem.* **2017**, *1*, 76. [CrossRef]

245. Gawande, B.N.; Rohloff, J.C.; Carter, J.D.; von Carlowitz, I.; Zhang, C.; Schneider, D.J.; Janjic, N. Selection of DNA aptamers with two modified bases. *Proc. Natl. Acad. Sci. USA* **2017**, *114*, 2898–2903. [CrossRef] [PubMed]

246. Lam, C.H.; Hipolito, C.J.; Hollenstein, M.; Perrin, D.M. A divalent metal-dependent self-cleaving DNAzyme with a tyrosine side chain. *Org. Biomol. Chem.* **2011**, *9*, 6949–6954. [CrossRef] [PubMed]

247. Renders, M.; Miller, E.; Lam, C.H.; Perrin, D.M. Whole cell-SELEX of aptamers with a tyrosine-like side chain against live bacteria. *Org. Biomol. Chem.* **2017**, *15*, 1980–1989. [CrossRef] [PubMed]

248. So, H.-M.; Park, D.-W.; Jeon, E.-K.; Kim, Y.-H.; Kim, B.S.; Lee, C.-K.; Choi, S.Y.; Kim, S.C.; Chang, H.; Lee, J.-O. Detection and Titer estimation of *Escherichia coli* using aptamer-functionalized single-walled carbon-nanotube field-effect transistors. *Small* **2008**, *4*, 197–201. [CrossRef] [PubMed]

249. Minagawa, H.; Onodera, K.; Fujita, H.; Sakamoto, T.; Akitomi, J.; Kaneko, N.; Shiratori, I.; Kuwahara, M.; Horii, K.; Waga, I. Selection, characterization and application of artificial DNA aptamer containing appended bases with sub-nanomolar affinity for a salivary biomarker. *Sci. Rep.* **2017**, *7*, 42716. [CrossRef] [PubMed]

250. Imaizumi, Y.; Kasahara, Y.; Fujita, H.; Kitadume, S.; Ozaki, H.; Endoh, T.; Kuwahara, M.; Sugimoto, N. Efficacy of base-modification on target binding of small molecule DNA aptamers. *J. Am. Chem. Soc.* **2013**, *135*, 9412–9419. [CrossRef] [PubMed]

251. Cho, E.J.; Lee, J.-W.; Ellington, A.D. Applications of aptamers as sensors. *Annu. Rev. Anal. Chem.* **2009**, *2*, 241–264. [CrossRef] [PubMed]

252. Vaish, N.K.; Larralde, R.; Fraley, A.W.; Szostak, J.W.; McLaughlin, L.W. A novel, modification-dependent ATP-binding aptamer selected from an RNA library incorporating a cationic functionality. *Biochemistry* **2003**, *42*, 8842–8851. [CrossRef] [PubMed]

253. Kabza, A.M.; Sczepanski, J.T. An L-RNA aptamer with expanded chemical functionality that inhibits microRNA biogenesis. *ChemBioChem* **2017**, *18*, 1824–1827. [CrossRef] [PubMed]

254. Kong, D.H.; Lei, Y.; Yeung, W.; Hili, R. Enzymatic synthesis of sequence-defined synthetic nucleic acid polymers with diverse functional groups. *Angew. Chem. Int. Ed.* **2016**, *55*, 13164–13168. [CrossRef] [PubMed]

255. Lei, Y.; Kong, D.H.; Hili, R. A high-fidelity codon set for the T4 DNA ligase-catalyzed polymerization of modified oligonucleotides. *ACS Comb. Sci.* **2015**, *17*, 716–721. [CrossRef] [PubMed]

256. Kong, D.; Yeung, W.; Hili, R. In vitro selection of diversely-functionalized aptamers. *J. Am. Chem. Soc.* **2017**, *139*, 13977–13980. [CrossRef] [PubMed]

257. Feldman, A.W.; Romesberg, F.E. In vivo structure–activity relationships and optimization of an unnatural base pair for replication in a semi-synthetic organism. *J. Am. Chem. Soc.* **2017**, *139*, 11427–11433. [CrossRef] [PubMed]

258. Kimoto, M.; Yamashige, R.; Matsunaga, K.-I.; Yokoyama, S.; Hirao, I. Generation of high-affinity DNA aptamers using an expanded genetic alphabet. *Nat. Biotechnol.* **2013**, *31*, 453–457. [CrossRef] [PubMed]

259. Kimoto, M.; Nakamura, M.; Hirao, I. Post-exSELEX stabilization of an unnatural-base DNA aptamer targeting VEGF(165) toward pharmaceutical applications. *Nucleic Acids Res.* **2016**, *44*, 7487–7494. [PubMed]

260. Matsunaga, K.-I.; Kimoto, M.; Hirao, I. High-affinity DNA aptamer generation targeting von willebrand factor A1-domain by genetic alphabet expansion for systematic evolution of ligands by exponential enrichment using two types of libraries composed of five different bases. *J. Am. Chem. Soc.* **2017**, *139*, 324–334. [CrossRef] [PubMed]

261. Hirao, I.; Kawai, G.; Yoshizawa, S.; Nishimura, Y.; Ishido, Y.; Watanabe, K.; Miura, K. Most compact hairpin-turn structure exerted by a short DNA fragment, d(GCGAAGC) in solution—An extraordinarily stable structure resistant to nucleases and heat. *Nucleic Acids Res.* **1994**, *22*, 576–582. [CrossRef] [PubMed]

262. Matsunaga, K.; Kimoto, M.; Hanson, C.; Sanford, M.; Young, H.A.; Hirao, I. Architecture of high-affinity unnatural-base DNA aptamers toward pharmaceutical applications. *Sci. Rep.* **2015**, *5*, 18478. [CrossRef] [PubMed]

263. Hirao, I.; Kimoto, M.; Lee, K.H. DNA aptamer generation by exSELEX using genetic alphabet expansion with a mini-hairpin DNA stabilization method. *Biochimie* **2017**. [CrossRef] [PubMed]

264. Yang, Z.Y.; Chen, F.; Chamberlin, S.G.; Benner, S.A. Expanded genetic alphabets in the polymerase chain reaction. *Angew. Chem. Int. Ed.* **2010**, *49*, 177–180. [CrossRef] [PubMed]

265. Benner, S.A.; Karalkar, N.B.; Hoshika, S.; Laos, R.; Shaw, R.W.; Matsuura, M.; Fajardo, D.; Moussatche, P. Alternative Watson-Crick synthetic genetic systems. *Cold Spring Harb. Perspect. Biol.* **2016**, *8*, a023770. [CrossRef] [PubMed]

266. Sefah, K.; Yang, Z.; Bradley, K.M.; Hoshika, S.; Jiménez, E.; Zhang, L.; Zhu, G.; Shanker, S.; Yu, F.; Turek, D.; et al. In vitro selection with artificial expanded genetic information systems. *Proc. Natl. Acad. Sci. USA* **2014**, *111*, 1449–1454. [CrossRef] [PubMed]

267. Zhang, L.; Yang, Z.; Le Trinh, T.; Teng, I.T.; Wang, S.; Bradley, K.M.; Hoshika, S.; Wu, Q.; Cansiz, S.; Rowold, D.J.; et al. Aptamers against cells overexpressing glypican 3 from expanded genetic systems combined with cell engineering and laboratory evolution. *Angew. Chem. Int. Ed.* **2016**, *55*, 12372–12375. [CrossRef] [PubMed]

268. Biondi, E.; Lane, J.D.; Das, D.; Dasgupta, S.; Piccirilli, J.A.; Hoshika, S.; Bradley, K.M.; Krantz, B.A.; Benner, S.A. Laboratory evolution of artificially expanded DNA gives redesignable aptamers that target the toxic form of anthrax protective antigen. *Nucleic Acids Res.* **2016**, *44*, 9565–9577. [CrossRef] [PubMed]

269. Kikin, O.; D'Antonio, L.; Bagga, P.S. Qgrs mapper: A web-based server for predicting G-quadruplexes in nucleotide sequences. *Nucleic Acids Res.* **2006**, *34*, W676–W682. [CrossRef] [PubMed]

270. Malyshev, D.A.; Dhami, K.; Lavergne, T.; Chen, T.J.; Dai, N.; Foster, J.M.; Correa, I.R.; Romesberg, F.E. A semi-synthetic organism with an expanded genetic alphabet. *Nature* **2014**, *509*, 385–388. [CrossRef] [PubMed]

271. Zhang, Y.; Lamb, B.M.; Feldman, A.W.; Zhou, A.X.; Lavergne, T.; Li, L.; Romesberg, F.E. A semisynthetic organism engineered for the stable expansion of the genetic alphabet. *Proc. Natl. Acad. Sci. USA* **2017**, *114*, 1317–1322. [CrossRef] [PubMed]

272. Rothlisberger, P.; Levi-Acobas, F.; Sarac, I.; Marliere, P.; Herdewijn, P.; Hollenstein, M. On the enzymatic incorporation of an imidazole nucleotide into DNA. *Org. Biomol. Chem.* **2017**, *15*, 4449–4455. [CrossRef] [PubMed]

273. Kaul, C.; Muller, M.; Wagner, M.; Schneider, S.; Carell, T. Reversible bond formation enables the replication and amplification of a crosslinking salen complex as an orthogonal base pair. *Nat. Chem.* **2011**, *3*, 794–800. [CrossRef] [PubMed]

274. Matyasovsky, J.; Perlikova, P.; Malnuit, V.; Pohl, R.; Hocek, M. 2-substituted dATP derivatives as building blocks for polymerase-catalyzed synthesis of DNA modified in the minor groove. *Angew. Chem. Int. Ed.* **2016**, *55*, 15856–15859. [CrossRef] [PubMed]

275. Hollenstein, M. Deoxynucleoside triphosphates bearing histamine, carboxylic acid, and hydroxyl residues—Synthesis and biochemical characterization. *Org. Biomol. Chem.* **2013**, *11*, 5162–5172. [CrossRef] [PubMed]

276. Eremeeva, E.; Abramov, M.; Margamuljana, L.; Herdewijn, P. Base-modified nucleic acids as a powerful tool for synthetic biology and biotechnology. *Chem. Eur. J.* **2017**, *23*, 9560–9576. [CrossRef] [PubMed]

277. Hollenstein, M. Synthesis of deoxynucleoside triphosphates that include proline, urea, or sulfamide groups and their polymerase incorporation into DNA. *Chem. Eur. J.* **2012**, *18*, 13320–13330. [CrossRef] [PubMed]

278. Röthlisberger, P.; Levi-Acobas, F.; Hollenstein, M. New synthetic route to ethynyl-dUTP: A means to avoid formation of acetyl and chloro vinyl base-modified triphosphates that could poison SELEX experiments. *Bioorg. Med. Chem. Lett.* **2017**, *27*, 897–900. [CrossRef] [PubMed]

279. Houlihan, G.; Arangundy-Franklin, S.; Holliger, P. Engineering and application of polymerases for synthetic genetics. *Curr. Opin. Biotechnol.* **2017**, *48*, 168–179. [CrossRef] [PubMed]

280. Chen, T.; Hongdilokkul, N.; Liu, Z.; Adhikary, R.; Tsuen, S.S.; Romesberg, F.E. Evolution of thermophilic DNA polymerases for the recognition and amplification of C2′-modified DNA. *Nat. Chem.* **2016**, *8*, 556–562. [CrossRef] [PubMed]

281. Thirunavukarasu, D.; Chen, T.; Liu, Z.; Hongdilokkul, N.; Romesberg, F.E. Selection of 2′-fluoro-modified aptamers with optimized properties. *J. Am. Chem. Soc.* **2017**, *139*, 2892–2895. [CrossRef] [PubMed]

282. Liu, Z.; Chen, T.; Romesberg, F.E. Evolved polymerases facilitate selection of fully 2′-OMe-modified aptamers. *Chem. Sci.* **2017**. [CrossRef]

283. Pinheiro, V.B.; Holliger, P. The XNA world: Progress towards replication and evolution of synthetic genetic polymers. *Curr. Opin. Chem. Biol.* **2012**, *16*, 245–252. [CrossRef] [PubMed]

284. Ferreira-Bravo, I.A.; Cozens, C.; Holliger, P.; DeStefano, J.J. Selection of 2′-deoxy-2′-fluoroarabinonucleotide (FANA) aptamers that bind HIV-1 reverse transcriptase with picomolar affinity. *Nucleic Acids Res.* **2015**, *43*, 9587–9599.

285. Yu, H.Y.; Zhang, S.; Chaput, J.C. Darwinian evolution of an alternative genetic system provides support for tna as an RNA progenitor. *Nat. Chem.* **2012**, *4*, 183–187. [CrossRef] [PubMed]

286. Diafa, S.; Evéquoz, D.; Leumann, C.J.; Hollenstein, M. Enzymatic synthesis of 7′,5′-bicyclo-DNA oligonucleotides. *Chem. Asian J.* **2017**, *12*, 1347–1352. [CrossRef] [PubMed]

287. Maiti, M.; Maiti, M.; Knies, C.; Dumbre, S.; Lescrinier, E.; Rosemeyer, H.; Ceulemans, A.; Herdewijn, P. Xylonucleic acid: Synthesis, structure, and orthogonal pairing properties. *Nucleic Acids Res.* **2015**, *43*, 7189–7200. [CrossRef] [PubMed]

288. Siegmund, V.; Santner, T.; Micura, R.; Marx, A. Screening mutant libraries of T7 RNA polymerase for candidates with increased acceptance of 2′-modified nucleotides. *Chem. Commun.* **2012**, *48*, 9870–9872. [CrossRef] [PubMed]

289. Inoue, N.; Shionoya, A.; Minakawa, N.; Kawakami, A.; Ogawa, N.; Matsuda, A. Amplification of 4′-thioDNA in the presence of 4′-thio-dTTP and 4′-thio-dCTP, and 4′-thioDNA-directed transcription in vitro and in mammalian cells. *J. Am. Chem. Soc.* **2007**, *129*, 15424–15425. [CrossRef] [PubMed]

290. Kojima, T.; Furukawa, K.; Maruyama, H.; Inoue, N.; Tarashima, N.; Matsuda, A.; Minakawa, N. PCR amplification of 4′-thioDNA using 2′-deoxy-4′-thionucleoside 5′-triphosphates. *ACS Synth. Biol.* **2013**, *2*, 529–536. [CrossRef] [PubMed]

291. Houlihan, G.; Arangundy-Franklin, S.; Holliger, P. Exploring the chemistry of genetic information storage and propagation through polymerase engineering. *Acc. Chem. Res.* **2017**, *50*, 1079–1087. [CrossRef] [PubMed]

292. Ghadessy, F.J.; Ramsay, N.; Boudsocq, F.; Loakes, D.; Brown, A.; Iwai, S.; Vaisman, A.; Woodgate, R.; Holliger, P. Generic expansion of the substrate spectrum of a DNA polymerase by directed evolution. *Nat. Biotechnol.* **2004**, *22*, 755–759. [CrossRef] [PubMed]

293. Higashimoto, Y.; Matsui, T.; Nishino, Y.; Taira, J.; Inoue, H.; Takeuchi, M.; Yamagishi, S. Blockade by phosphorothioate aptamers of advanced glycation end products-induced damage in cultured pericytes and endothelial cells. *Microvasc. Res.* **2013**, *90*, 64–70. [CrossRef] [PubMed]

294. Volk, D.; Lokesh, G. Development of phosphorothioate DNA and DNA thioaptamers. *Biomedicines* **2017**, *5*, 41. [CrossRef] [PubMed]

295. Yang, X.; Zhu, Y.; Wang, C.; Guan, Z.; Zhang, L.; Yang, Z. Alkylation of phosphorothioated thrombin binding aptamers improves the selectivity of inhibition of tumor cell proliferation upon anticoagulation. *Biochim. Biophys. Acta* **2017**, *1861*, 1864–1869. [CrossRef] [PubMed]

296. Padmanabhan, K.; Padmanabhan, K.P.; Ferrara, J.D.; Sadler, J.E.; Tulinsky, A. The structure of α-thrombin inhibited by a 15-mer single-stranded DNA aptamer. *J. Biol. Chem.* **1993**, *268*, 17651–17654. [PubMed]

297. Macaya, R.F.; Schultze, P.; Smith, F.W.; Roe, J.A.; Feigon, J. Thrombin-binding DNA aptamer forms a unimolecular quadruplex structure in solution. *Proc. Natl. Acad. Sci. USA* **1993**, *90*, 3745–3749. [CrossRef] [PubMed]

298. Lato, S.M.; Ozerova, N.D.S.; He, K.Z.; Sergueeva, Z.; Shaw, B.R.; Burke, D.H. Boron-containing aptamers to ATP. *Nucleic Acids Res.* **2002**, *30*, 1401–1407. [CrossRef] [PubMed]

299. Cheng, C.S.; Chen, Y.H.; Lennox, K.A.; Behlke, M.A.; Davidson, B.L. In vivo SELEX for identification of brain-penetrating aptamers. *Mol. Ther. Nucleic Acids* **2013**, *2*, e67. [CrossRef] [PubMed]

300. Monaco, I.; Camorani, S.; Colecchia, D.; Locatelli, E.; Calandro, P.; Oudin, A.; Niclou, S.; Arra, C.; Chiariello, M.; Cerchia, L.; et al. Aptamer functionalization of nanosystems for glioblastoma targeting through the blood-brain barrier. *J. Med. Chem.* **2017**, *60*, 4510–4516. [CrossRef] [PubMed]

301. Temme, J.S.; Drzyzga, M.G.; MacPherson, I.S.; Krauss, I.J. Directed evolution of 2g12-targeted nonamannose glycoclusters by SELMA. *Chem. Eur. J.* **2013**, *19*, 17291–17295. [CrossRef] [PubMed]

302. Larsen, A.C.; Dunn, M.R.; Hatch, A.; Sau, S.P.; Youngbull, C.; Chaput, J.C. A general strategy for expanding polymerase function by droplet microfluidics. *Nat. Commun.* **2016**, *7*, 11235. [CrossRef] [PubMed]

303. MacPherson, I.S.; Temme, J.S.; Krauss, I.J. DNA display of folded RNA libraries enabling RNA-SELEX without reverse transcription. *Chem. Commun.* **2017**, *53*, 2878–2881. [CrossRef] [PubMed]

304. Lai, J.C.; Hong, C.Y. Magnetic-assisted rapid aptamer selection (MARAS) for generating high-affinity DNA aptamer using rotating magnetic fields. *ACS Comb. Sci.* **2014**, *16*, 321–327. [CrossRef] [PubMed]

305. Renders, M.; Miller, E.; Hollenstein, M.; Perrin, D.M. A method for selecting modified DNAzymes without the use of modified DNA as a template in PCR. *Chem. Commun.* **2015**, *51*, 1360–1362. [CrossRef] [PubMed]

9

Development of an Impedimetric Aptasensor for the Detection of *Staphylococcus aureus*

Peggy Reich [1,*], **Regina Stoltenburg** [2], **Beate Strehlitz** [3], **Dieter Frense** [1] and **Dieter Beckmann** [1]

[1] Institut für Bioprozess- und Analysenmesstechnik e.V., 37308 Heilbad Heiligenstadt, Germany; Dieter.Frense@iba-heiligenstadt.de (D.F.); Dieter.Beckmann@iba-heiligenstadt.de (D.B.)

[2] UFZ – Helmholtz Centre for Environmental Research, 06120 Halle, Germany; regina.stoltenburg@ufz.de

[3] UFZ – Helmholtz Centre for Environmental Research, 04318 Leipzig, Germany; beate.strehlitz@ufz.de

[*] Correspondence: ReichPeggy@gmail.com

Abstract: In combination with electrochemical impedance spectroscopy, aptamer-based biosensors are a powerful tool for fast analytical devices. Herein, we present an impedimetric aptasensor for the detection of the human pathogen *Staphylococcus aureus*. The used aptamer targets protein A, a surface bound virulence factor of *S. aureus*. The thiol-modified protein A-binding aptamer was co-immobilized with 6-mercapto-1-hexanol onto gold electrodes by self-assembly. Optimization of the ratio of aptamer to 6-mercapto-1-hexanol resulted in an average density of $1.01 \pm 0.44 \times 10^{13}$ aptamer molecules per cm^2. As shown with quartz crystal microbalance experiments, the immobilized aptamer retained its functionality to bind recombinant protein A. Our impedimetric biosensor is based on the principle that binding of target molecules to the immobilized aptamer decreases the electron transfer between electrode and ferri-/ferrocyanide in solution, which is measured as an increase of impedance. Microscale thermophoresis measurements showed that addition of the redox probe ferri-/ferrocyanide has no influence on the binding of aptamer and its target. We demonstrated that upon incubation with various concentrations of *S. aureus*, the charge-transfer resistance increased proportionally. The developed biosensor showed a limit of detection of 10 CFU·mL^{-1} and results were available within 10 minutes. The biosensor is highly selective, distinguishing non-target bacteria such as *Escherichia coli* and *Staphylococcus epidermidis*. This work highlights the immense potential of impedimetric aptasensors for future biosensing applications.

Keywords: aptamer; staphylococcal protein A; label-free; biosensing techniques; rapid detection; self-assembly; limit of detection; protein binding; ferri-/ferrocyanide; gold electrode

1. Introduction

Staphylococcus aureus is a major pathogen for humans. It is a common cause of infections, from minor ones such as abscesses and sinusitis to life-threatening diseases such as bacteremia, endocarditis and sepsis [1]. Its antibiotic-resistant strains, e.g., methicillin-resistant *S. aureus* (MRSA) are a serious problem in healthcare [2]. Besides being the origin of hospital-acquired infections, *S. aureus* produces seven different toxins that cause food poisoning [3–5]. It is generally accepted, that 10^5 cells per g of food produce sufficient enterotoxins to cause food poisoning [6].

Since the relevance of this pathogen was discovered, many approaches in the development of rapid detection methods for infection control were investigated, as reviewed by Law et al. [7] and Zhao et al. [8]. According to these reviews, traditional methods, such as plate counts using selective agar, convince with their simplicity, low costs and high accuracy but take 4 to 6 days to yield results. Nevertheless, they are still regarded as the gold standard. One promising alternative method is polymerase chain reaction (PCR). The commercially available Xpert MRSA assay (Cepheid

International, Sunnyvale, CA, USA) for example requires 2 h from DNA extraction to assay result [9]. However, complex sample preparation by trained staff is needed.

According to Zhao et al., the most rapid detection methods are based on biosensor technology. Biosensors are devices, which use biological components as recognition elements to provide specific affinity to the desired target. The recognition element is coupled to a transducer, which transforms the biological into an electrical signal [10]. To be commercially successful, a biosensor has to meet several requirements, e.g., low cost, fast response and high sensitivity. Therefore, despite its complexity, many researchers recognize the high potential of electrochemical impedance spectroscopy (EIS).

EIS is a fast label-free technique to measure the properties of electrode surfaces and bulk electrolytes. Owed to the progress in engineering and electronics during the last decades, high performance miniaturized impedance instruments are available for a relatively low budget [11]. EIS was used successfully for biosensors with various recognition elements [12,13]. For example, Bekir et al., developed an electrochemical immunosensor using antibodies against S. aureus [14]. They report a detection limit of 10 CFU·mL^{-1} of S. aureus, exploiting the impedance change of the electrode surface caused by the affinity reaction of the immobilized antibodies.

To overcome the limitations of antibodies, such as high manufacturing costs, instability to high temperatures and short shelf life, aptasensors employ aptamers as recognition element [15]. Aptamers are synthetic, single-stranded nucleic acid molecules that can fold into complex three-dimensional structures allowing them to bind targets based on structure recognition with high affinity and specificity. They are selected using the SELEX procedure (systematic evolution of ligands by exponential enrichment), an iterative in vitro selection and amplification method [16].

Electrochemical aptasensors were reviewed by Willner et al. [17]: besides the well-known thrombin aptamer [18], other impedimetric aptasensors emerged ranging from the detection of potassium ions [19] and small molecules, such as ethanolamine [20], to whole cells, e.g., Salmonella typhimurium [21]. Shahdordizadeh et al., provided a review of recent advances in optical and electrochemical aptasensors for the detection of S. aureus [22]. They report on aptamers selected against staphylococcal toxins, staphylococcal teichoic acid, staphylococcal protein A and S. aureus as whole bacteria. The indirect detection of S. aureus via aptamers targeting the toxins excreted by the pathogen are limited due to the difficulty in correlation of the sensor signal to the presence of viable microorganisms. Therefore, direct detection is favored. In the field of optical aptasensors, fluorescence is most prominent, but also one colorimetric aptasensor was developed [23]. Using dielectrophoretic enrichment and fluorescent nanoparticles, Shangguan and coworkers developed an optical aptasensor with a limit of detection (LoD) of 93 CFU·mL^{-1} and an assay time of 2 h [24]. By the use of upconversion nanoparticles, the fluorescence intensity was increased and Duan et al., gained a LoD of 8 CFU·mL^{-1} [25]. Chang et al., developed an optical aptasensor for the single cell detection of S. aureus within 1.5 h [26]. The detection principle is based on resonance light scattering of modified gold nanoparticles. Optical sensors have the disadvantage that complex biological samples often interfere with the detection process. Furthermore, electrochemical methods are appreciated for their fast response time, higher sensitivity, low-cost fabrication, simple automation and lower sample volumes. In their review, Shahdordizadeh et al., described five electrochemical aptasensors for the detection of S. aureus [22]: Two are based on potentiometry with LoDs of 800 CFU·mL^{-1} [27] and single cell detection [28]. Another used voltammetry to reach a LoD of 1 CFU·mL^{-1} [29] and Lian et al., combined interdigital electrodes (IDE) with quartz crystal sensor to detect the bacteria as low as 12 CFU·mL^{-1} [30]. Jia et al., used a glassy carbon electrode with aptamer modified gold nanoparticles to impedimetric detect a lower limit of 10 CFU·mL^{-1} within 60 min [31].

All mentioned optical and electrochemical aptasensors used different aptamers, but have in common, that the aptamers were selected in a Cell-SELEX, wherein whole cells were used as target for aptamer generation. Although purposive, this has the disadvantage that it stays unknown, which part of the cell surface is targeted by the aptamer. Thus, it is also unknown, which S. aureus strains can be bound by these aptamers. S. aureus is known for its ability to adapt its genetics quickly to new environments. Nevertheless, the conserved sequence of the immune-evasive factor protein A shows only one mutation in 70 months [32]. The surface bound protein A enhances *S. aureus'* adhesion to wounds by binding to the von Willebrand factor (vWF) and prevents phagocytosis by binding to the Fc region of various immunoglobulins [33]. Protein A is bound to peptidoglycans on the cell wall of *S. aureus* and not found on other bacteria. Therefore, protein A is an excellent target for the detection of *S. aureus* cells. Also in PCR methods, the *spA* gene, encoding protein A, is used to distinguish between *S. aureus* and other bacteria.

A DNA aptamer targeting staphylococcal protein A was selected by the FluMag-SELEX procedure in 2015 [34,35]. This aptamer development aimed to detect intact bacterial cells of *S. aureus* via the protein A bound to its cell surface. Binding characteristics of the aptamer to protein A were studied intensively by different methods such as bead-based fluorescent binding assay, surface plasmon resonance (SPR), microscale thermophoresis (MST), and enzyme-linked oligonucleotide assay (ELONA) [35,36].

The structural features of an aptamer play a major role in biosensor development. In case of the protein A-binding aptamer, a combination of two structural elements is important for its functionality: First, an intact and free 5′-end, folding into an imperfect stem-loop motif, is crucial for binding to protein A. Second, the aptamer folds into a parallel G-quadruplex structure as demonstrated by circular dichroism spectroscopy [36].

In the present study, we developed a biosensor detecting *Staphylococcus aureus* by its surface bound protein A, which is highly conserved and only found on *S. aureus*. The protein A-binding aptamer served as biological recognition element. In combination with electrochemical impedance spectroscopy as measurement method, rapid and label-free detection was achieved. By immobilization of thiol-modified aptamer on gold electrodes by self-assembly, binding of *S. aureus* was detected in a flow-through chamber with a three-electrode setup in buffer solution containing ferri-/ferrocyanide. Upon binding of *S. aureus*, the impedance increased due to the hindrance of the electron transfer between ferri-/ferrocyanide and the electrode surface (Figure 1). Herein, we elucidate the development of an impedimetric aptasensor and present novel insights on the use of aptamer-based electrochemical biosensors for the rapid and selective detection of *S. aureus*.

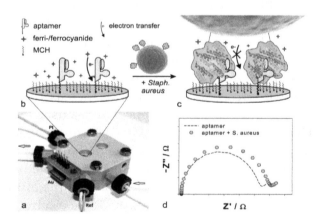

Figure 1. Flow-through measurement chamber and measurement principle. (**a**) Flow-through measurement chamber with liquid inlet and outlet (arrows) and a 3-electrode setup including a golden working electrode (Au), a platinum counter electrode (Pt) and a reference electrode (Ref); (**b**) Working electrode modified with aptamer (violet) and 6-mercapto-1-hexanol (MCH) in a buffer containing the redox probe ferri-/ferrocyanide (red stars); (**c**) Upon binding of protein A (green) on the surface of *S. aureus* (yellow) to the immobilized aptamer (violet), the impedance increased due to the hindrance of electron transfer (curved arrow) between ferri-/ferrocyanide (red stars) in solution and gold electrode surface (orange); (**d**) Characteristic Nyquist-plot of impedance spectra before and after incubation with *S. aureus* (based on data from measurements); Z′: real part of impedance and Z″: imaginary part of impedance.

2. Results and Discussion

2.1. Functionalization of the Gold Electrodes

Affinity of the protein A-binding aptamer (PAA) to its target has been intensively studied by Stoltenburg et al., using SPR-based measurements with the Biacore X100 [35]. They applied both, the protein A and PAA, respectively as biotinylated receptor, which was immobilized on a streptavidin-coated sensor surface. In the development of the impedimetric sensor, we modified the aptamer with C6-Spacer and thiol for immobilization via self-assembly. To enable high densities of the protein A-binding aptamer (PAA) on the surface, we used the co-immobilization strategy, described by Keighley et al. [37]. They found that in the presence of 6-mercapto-1-hexanol (MCH), oligonucleotides stand upright on the surface rather than lying down, thus, occupy less space and allow a higher density. For optimization studies of our sensor surface, we investigated the influence of different ratios of aptamer to total thiol (PAA + MCH) using chronocoulometry as described by Steel et al. [38]. They stated, that the reduction of hexaammineruthenium (III) chloride (RuHex), measured by chronocoulometry, is proportional to the number of oligonucleotides on the surface. Figure 2 shows the results of chronocoulometry measurements on co-immobilized PAA modified gold electrodes in 40 mM Tris buffer containing 200 µM RuHex. The highest density of PAA ($2.41 \pm 0.39 \times 10^{12}$ PAA/cm^2) was reached with a ratio of 1:5 (1 µM PAA and 4 µM MCH), thus, this ratio was used for further experiments.

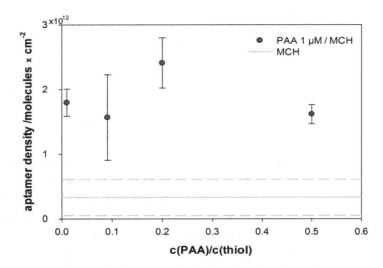

Figure 2. Aptamer density on the electrode surface depending on the protein A-binding aptamer (PAA)/ 6-mercapto-1-hexanol (MCH) molar ratio upon immobilization based on four experiments for each ratio; the grey lines show the average density and standard deviation determined for a MCH monolayer.

To ensure successful immobilization and aptamer functionality, we performed analysis using a quartz crystal microbalance as described in Section 3.3. In Figure 3a is shown the mass increase upon PAA immobilization, calculated accordingly from the measured frequency change. The formation of a self-assembled monolayer comprises two steps. The first is the initial attachment, which takes a few seconds, seen in the significant mass increase immediately after introduction of aptamer-solution in Figure 3a. The second step is the arrangement to an ordered monolayer, which takes more than 8 h [39]. Hence, to ensure an ordered monolayer, the electrode was incubated for at least 15 h. As seen in Figure 3a, upon PAA immobilization the mass increased by 500 ng/cm^2. Backfilling of gaps with MCH increased the mass slightly (~20 ng/cm^2). In comparison, an electrode covered with a pure MCH monolayer showed a mass increase of only 89 ng/cm^2. The difference between both provided the mass change due to immobilized PAA. Although immobilized aptamers hamper the formation of an entire layer of MCH, the error is negligible, because the aptamer mass is 142 times higher than the mass of a MCH-molecule. The average mass increase of immobilized PAA was 320 ± 139 ng/cm^2 (standard deviation obtained from three experiments). This correlates to $1.01 \pm 0.44 \times 10^{13}$ aptamers/cm^2. The aptamer density determined with chronocoulometry ($2.41 \pm 0.39 \times 10^{12}$ PAA/cm^2) is 4 times smaller than measured with QCM. This is due to the difference in the measurement techniques and surfaces. In the QCM measurements, not only the mass of the immobilized molecules, but also the adsorbed water and ions as well as the rigidity of the immobilized layer influence the resonance frequency [40]. Also, the different surfaces—a quartz crystal covered with gold and chrome as adhesion layer versus a glass test slide covered with gold and titanium as adhesion layer—contribute to the differences in surface coverage. Although their roughness, <1 nm RMS (= rough mean square) and 376 ± 74 pm RMS respectively, do not diverge significantly. However, both methods confirm successful immobilization and that a high density of aptamers was achieved.

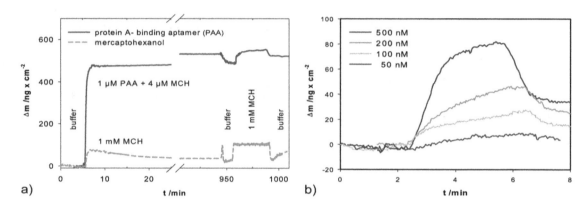

Figure 3. Quartz crystal microbalance measurements: (**a**) mass change Δm versus time of gold covered quartz crystals incubated with aptamer/6-mercapto-1-hexanol (MCH) (violet solid line) or MCH (orange dashed line) and subsequently blocked with 1 mM MCH; (**b**) Δm versus time of an aptamer-modified crystal incubated with different concentrations of protein A.

In general, high densities are desired to obtain a higher protein capture capacity resulting in higher sensitivity, but too high densities may lead to steric hindrance preventing correct aptamer folding and binding of target [41]. Assuming an even distribution, the obtained density results in an area of 9.9×10^{-14} cm^2 per aptamer. Assuming an aptamer occupies a squared area, the mean distance between two aptamers is 3.15 nm. Due to dimerization and the formation of quadruplexes [36], the true mean distance of the aptamers is likely >3.15 nm. According to Erickson et al., the partial specific volume of a protein can be calculated from its molecular mass [42]. For the recombinant protein A with a molecular mass of 45 kDa, this volume is 54.54 nm^3. Assuming protein A has the simplest shape, a sphere, its minimal diameter is 4.71 nm. Therefore, we assumed that the space around an aptamer (>6.3 nm) was sufficient for the binding of protein A.

To prove the functional binding of protein A to the immobilized PAA, we observed the mass change of modified electrodes upon incubation with protein A (Figure 3b). Concentrations of protein A in the range of 100 to 500 nM resulted in signals of 20 to 40 ng/cm^2, which correlates to 2.68×10^{11} and 5.35×10^{11} molecules/cm^2 respectively. Thus, we showed that the immobilized aptamer retained its functionality in binding protein A.

Figure 4 represents the impedance measurements during aptamer immobilization. It shows the Nyquist plots of a blank electrode and an aptamer-modified electrode. The impedance increased significantly after PAA immobilization.

2.2. Influence of Ferri-/Ferrocyanide

In faradaic impedimetric measurements, a redox probe for the transfer of electrons from the working electrode to the counter electrode is necessary. The ferri-/ferrocyanide couple is often used due to its fast electron transfer rate (2×10^{-2} cm/s) [43]. For biosensor development, it is important to examine, if the redox couple inhibits the affinity of the receptor to its target. To examine the influence of ferri-/ferrocyanide on binding of protein A to PAA, microscale thermophoresis (MST) experiments were performed. MST is a powerful technique to detect biomolecular interaction by quantifying directed movement of molecules along an induced microscopic temperature gradient. It is highly sensitive to changes in hydration shell, charge and size and therefore capable to detect many kinds of biomolecular interactions while both reaction partners remain in solution and no immobilization is required.

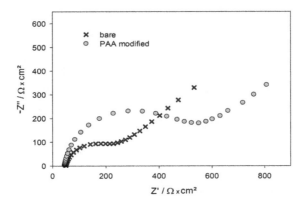

Figure 4. Nyquist plot of the impedance spectra of a blank gold electrode (black crosses) and a PAA/MCH-modified gold electrode (green circles); PAA = protein A-binding aptamer, MCH = 6-mercapto-1-hexanol.

In summary, we successfully immobilized PAA via self-assembly with a high density, whereas it retained its functionality of binding protein A.

Figure 5a shows characteristic curves of a MST experiment for a low and a high protein A concentration. The MST fluorescence signal was lower for the high protein A concentration, indicating that in solution, the aptamer-protein A-complex behaved differently than the free aptamer. Binding curves (Figure 5b) in buffer with and without 2 mM ferri-/ferrocyanide were measured as described in Section 3.4. Fitting to the Hill Equation (1) was performed to extract more information from the binding curves:

$$fraction\ bound = \frac{\Delta S}{\Delta S_{max}} = \frac{[T]^h}{K_D + [T]^h} \tag{1}$$

where ΔS = signal change, ΔS_{max} = maximal signal change, h = Hill coefficient, K_D = apparent binding constant, $[T]$ = target concentration (e.g., protein A).

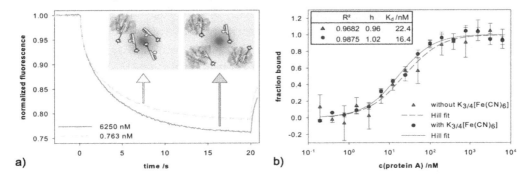

Figure 5. (a) Microscale thermophoresis (MST) curves for a low (0.763 nM, dashed line) and a high (6250 nM, solid line) protein A concentration with 38.5 nM labeled PAA (MST power 40%, LED power 100%); **(b)** Binding curves of recombinant protein A with aptamer in absence (red triangles) and presence (blue circles) of 2 mM ferri-/ferrocyanide measured in triplicates—the lines represent the Hill fit.

The determined K_D values in absence and presence of ferri-/ferrocyanide were 22.4 $\pm$ 5.8 nM and 16.4 $\pm$ 2.5 nM respectively. There is no statistically significant difference between the two curves (paired t-test, $p = 0.255$). Thus, we conclude, that ferri-/ferrocyanide has no significant influence on the binding of PAA and protein A.

The K_D value obtained by MST measurements in this work, 22.4 nM, differs from the value reported by Stoltenburg et al., 94.7 nM [35]. To investigate if this difference is due to the labeling-procedure or -site, MST measurements were repeated by 2bind GmbH (Regensburg, Germany). A similar analysis setup was applied using the maleimide-fluorophore on the 5′- and 3′-thiol-tagged PAA. The obtained K_D values were 115.6 nM and 110.8 nM, respectively. Hence, we concluded, that the binding behavior was not affected by the labeling site of the aptamer, for both partners in solution. However, the most distinct difference in the experiments was the concentration of Tween 20 (0.05% used by 2bind GmbH compared to 0.005% used in our first measurement), but the data of both experiments revealed no adhesion to the used capillaries. Thus, the variations in the K_D values in MST measurements are attributed to differences in buffer composition, amplified by the handling of very small volumes (10 μL).

2.3. Detection of Protein A by Impedance Spectroscopy

The gold electrodes modified with PAA were mounted in the flow-through chamber (see Figure 1) and exposed to different concentrations of protein A (2–700 nM). Impedance spectra were recorded in FeBB as described in Section 3.5. Every protein A concentration was measured in triplicates, i.e., on three different electrodes. Figure 6a shows the impedance spectra of PAA modified electrodes before and after exposure to 7–700 nM protein A. After incubation with protein A the impedance increased proportionally to the concentration of protein A.

To extract the relevant parameters, the impedance spectra were fitted to the modified Randles circuit (see Figure 6a), wherein R_{sol} is the solution resistance, CPE is the constant phase element for the double layer at the electrode surface, R_{ct} is the charge-transfer resistance (due to the interaction of ferri-/ferrocyanide with the electrode), and W is the Warburg impedance representing the diffusion of ions to the electrode surface. The fitting results are summarized in Table 1. As seen in Figure 6a and Table 1, the fits (lines) show good agreement with the experimental data (markers).

In Figure 6b, the fit parameter changes over all measured protein A concentrations are plotted. The charge transfer resistance R_{ct} showed a maximum increase of 33%, whereas the other parameters changed less than 4%, verifying that the charge transfer was influenced significantly by the binding of protein A. Thus R_{ct} was chosen as the significant parameter for further analysis.

The change of the extracted R_{ct} versus the logarithmic protein A concentration is plotted in Figure 6c. While only a slight increase of R_{ct} up to 10 nM protein A was observed, an exponential increase in R_{ct} from 10 nM to 100 nM was measured. Finally, above 100 nM protein A R_{ct} reached saturation. The resulting sigmoidal curve was fitted to the Hill Equation (2):

$$\Delta R_{ct} = \frac{\Delta R_{ct\_max} \times [T]^h}{K_D + [T]^h} \tag{2}$$

where ΔR_{ct} = change of the charge transfer resistance, ΔR_{ct_max} = maximal change, h = Hill coefficient, K_D = apparent binding constant, $[T]$ = target concentration (e.g., protein A).

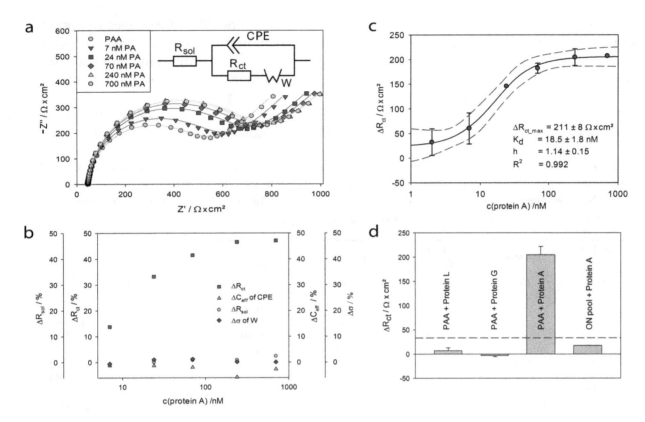

Figure 6. Electrochemical impedance spectroscopy measurements with protein A: (**a**) Nyquist plot of a PAA-modified electrode before (green circles) and after incubation with 7–700 nM protein A measured in FeBB—the fits to the equivalent circuit are shown as lines—modified Randles circuit with R_{sol}: solution resistance, CPE: constant phase element, R_{ct}: charge-transfer resistance and W: Warburg impedance; (**b**) the percentage change of the different fit parameters for all protein A concentrations, C_{eff}: the effective capacitance was calculated from the CPE under the assumption of a parallel distribution of time constants on the electrode surface as described in [44], σ is the parameter for the Warburg element; (**c**) binding curve: Change of the extracted charge transfer resistance ΔR_{ct} of aptamer-modified electrodes upon incubation of protein A measured in triplicate—dashed blue lines mark the 95% confidence interval; (**d**) unspecific signals: Extracted ΔR_{ct} of 1 μM protein G, protein L and protein A on PAA-modified electrodes as well as 1 μM protein A on electrodes modified with random oligonucleotides measured in triplicate—the dashed line represents the LoD.

Table 1. Results from the fitting of impedance data to the modified Randles circuit—SD = standard deviation, $X/\sqrt{N}$ = error of fitting normalized to the number of data points.

c(Protein A) /nM	R_{sol} /Ω·cm²	SD	CPE /μF·s$^{(\alpha-1)}$	SD	α	R_{ct} /Ω·cm²	SD	σ /Ω·s$^{-1/2}$	SD	$X/\sqrt{N}$
7	44.4	0.2	0.362	0.005	0.95	505.3	0.5	5769	2	0.0122
24	46.2	0.1	0.375	0.004	0.94	593.3	0.5	5847	2	0.0124
70	45.4	0.1	0.359	0.004	0.95	627.9	0.5	5861	2	0.0120
240	46.1	0.1	0.341	0.004	0.95	647.3	0.5	5834	2	0.0105
700	45.3	0.1	0.361	0.004	0.95	657.2	0.5	5811	2	0.0118

Thereby, an apparent K_D value of 18.5 ± 1.8 nM and Hill coefficient h of 1.14 ± 0.15 were obtained. The Hill coefficient h is slightly higher than 1, indicating a cooperative binding. Stoltenburg et al., already reported avidity effects [35]. They performed competitive experiments with protein A and immunoglobulin, which suggested that the protein A binding site for PAA overlaps with the known binding sites for immunoglobulin [15]. Hence, we can conclude that protein A provides more than one binding site for this aptamer. Reflective Interferometric Fourier Transform Spectroscopy (RIFTS) measurements of protein A binding to PAA, immobilized on porous silicon, resulted in an even higher h of 2.61 ± 0.69 [46]. Unlike the herein used planar gold electrodes, the rough and porous silicon surface increases the chance for aptamers being close enough to bind to the same protein A. Both observations substantiate the mentioned avidity affects.

Table 2 summarizes the apparent K_D values obtained with the same aptamer by different detection methods and setups. Affinities were found in the low nanomolar to micromolar range.

Table 2. Apparent dissociation constant, K_D, determined with different analysis methods for the protein A-binding aptamer—MST = microscale thermophoresis, SPR = surface plasmon resonance, ELONA = enzyme-linked oligonucleotide assay, EIS = electrochemical impedance spectroscopy, RIFTS = reflective interferometric Fourier transform spectroscopy.

Analysis Method	Aptamer	Aptamer Modification	Protein A	K_D/nM			Reference
MST	free	5'-fluorescence	free	94.7	±	64.6	[35]
MST	free	5'-fluorescence	free	115.6	±	26.9	this work
MST	free	3'-fluorescence	free	110.8	±	42.3	this work
MST	free	3'-fluorescence	free	22.4	±	5.8	this work
SPR	free	5'-fluorescence	immobilized	1920.0	±	250.0	[35]
SPR	immobilized	3'-biotin	free	287.0	±	16.2	[35]
ELONA	free	5'-biotin	immobilized	23.7	±	2.0	[36]
ELONA	free	3'-biotin	immobilized	11.3	±	1.4	[36]
EIS	immobilized	3'-thiol	free	18.5	±	1.8	this work
RIFTS	immobilized	3'-amino	free	13980.0	±	1540.0	[46]

Table 2 shows that each of these methods is capable of protein A detection utilizing PAA as receptor in a bioanalytical setup. It suggests that in the same analysis setup by the use of different designs, the measurement range can be adapted for the desired application. As shown in the example in the SPR experiments, immobilization of the aptamer instead of protein A, decreased the K_D almost by 14%. We want to emphasize that the K_D values are strongly dependent on the analysis method and setup used. Therefore, the performance of a biosensor cannot be judged based on the K_D value alone. Every method has to be evaluated for its purpose considering advantages and limitations. i.e., the strengths of EIS as detection method are fast, robust, label-free and non-destructive measurements.

Repeated measurements of PAA-modified electrodes in FeBB resulted in a standard deviation s of $11.10 \ \Omega\cdot cm^2$ of the R_{ct} value. As the limit of detection (LoD) is defined as the lowest target concentration at which the signal is higher than $3\cdot s$, a LoD of 2.99 ± 0.73 nM was determined using the approximated Hill equation.

Non-specific binding often presents a major challenge in biosensor development. Herein, we investigated the binding of the functionally similar proteins G and L to the aptamer-modified gold electrodes and observed that they neither bind to the surface nor to the aptamer (Figure 6d). Furthermore, the binding of protein A to an electrode, modified with random oligonucleotides, was determined to be $14.8 \pm 0.1 \ \Omega\cdot cm^2$ (Figure 4d), which is significantly below LoD ($3\cdot s$).

2.4. Detection of Staphylococcus aureus by Impedance Spectroscopy

Besides binding to the defined target protein A, the aptamer was also evaluated for its ability to recognize and bind to intact bacterial cells of *S. aureus*, expressing protein A on the cell surface [36]. Therefore, we performed experiments with our developed impedimetric biosensor and live *S. aureus* cells.

The gold electrodes modified with PAA were exposed to different concentrations of *S. aureus* (1 to 10^9 CFU·mL^{-1}). Figure 7a shows impedance spectra of a PAA modified electrode before and after exposure to *S. aureus*, recorded in FeBB. After incubation with *S. aureus*, the impedance increased proportionally to the cell concentration of *S. aureus*. To extract the relevant parameters, the impedance spectra were fitted to the modified Randles circuit (Figure 7a), which describes the phenomena occurring between the electrodes influencing the flow of current. As presented in Figure 7a and Table 3, the fits (lines) show good agreement with the experimental data (markers).

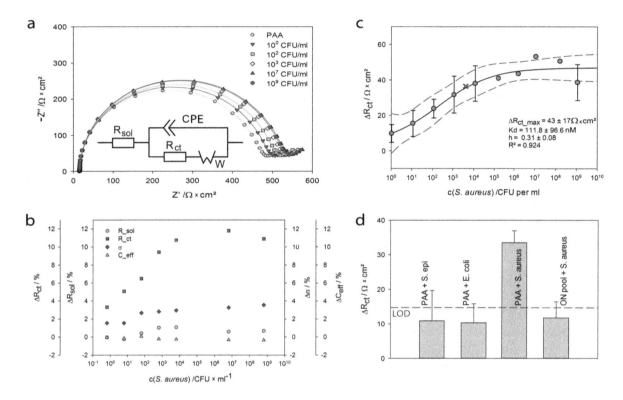

Figure 7. Electrochemical impedance spectroscopy (EIS) with *S. aureus* (**a**) Nyquist plot of a PAA-modified electrode before (cyan circles) and after incubation with 1 to 10^9 CFU·mL^{-1} measured in FeBB—the fits to the equivalent circuit are shown as lines modified Randles circuit with R_{sol}: solution resistance, *CPE*: constant phase element, R_{ct}: charge-transfer resistance and *W*: Warburg impedance; (**b**) percentage change of the different fit parameters for all *S. aureus* concentrations, C_{eff}: the effective capacitance was calculated from the CPE under the assumption of a parallel distribution of time constants on the electrode surface as described in [45], σ is the parameter for the Warburg element; (**c**) binding curve: Change of the extracted charge transfer resistance ΔR_{ct} of aptamer-modified electrodes upon incubation of 1 to 10^9 CFU·mL^{-1} *S. aureus* measured in triplicate—dashed blue lines mark the 95% confidence interval, green cross marks a sample of unknown concentration; (**d**) unspecific signals: 10^6 CFU·mL^{-1} of *Staphylococcus epidermidis*, *Escherichia coli* and *Staphylococcus aureus* on PAA-modified electrodes as well as 10^6 CFU·mL^{-1} *S. aureus* on electrodes modified with random oligonucleotides measured in triplicate—the dashed line represents the LoD.

Table 3. Results from the fitting of impedance data to the modified Randles circuit—SD = standard deviation, $X/\sqrt{N}$ = error of fitting normalized to the number of data points.

c(*S. aureus*) /CFU·mL^{-1}	R_{sol} /Ω·cm^2	SD	CPE /μF·s$^{(\alpha-1)}$	SD	α	R_{ct} /Ω·cm^2	SD	σ /Ω·s$^{-1/2}$	SD	$X/\sqrt{N}$
1E+01	14.8	0.1	1.230	0.02	0.99	459.8	0.3	310	3	0.0084
1E+02	14.8	0.1	1.232	0.02	0.99	467.8	0.3	311	3	0.0092
1E+03	14.9	0.1	1.229	0.02	0.99	474.7	0.3	312	3	0.0088
1E+04	15.0	0.1	1.224	0.02	0.99	485.7	0.3	315	3	0.0086
1E+05	15.0	0.1	1.232	0.02	0.99	492.2	0.3	312	3	0.0098
1E+08	14.9	0.1	1.231	0.02	0.99	497.9	0.3	312	3	0.0098
1E+10	14.9	0.1	1.226	0.02	0.99	491.8	0.3	317	3	0.0094

Figure 7b displays the fit parameter change for all measured bacteria concentrations. Only R_{ct} showed a significant increase of 12%, while the Warburg diffusion increased ~4% and R_{sol} as well as C_{eff}

changed <2%. This suggests that the bound *S. aureus* influences the electron transfer between electrode and the buffer containing ferri-/ferrocyanide. The R_{ct} parameter was chosen for further analysis.

In Figure 7c, the change of the extracted R_{ct} upon incubation with various *S. aureus* concentrations is plotted. Measurements were taken in triplicates, i.e., three electrodes were exposed to each *S. aureus* concentration. The standard deviation of these three measurements is shown as error bars. A diluted sample with 10 *S. aureus* cells per mL resulted in an average R_{ct} change of 35 Ohm. A sample with 10^5 cells per mL led to saturation of the sensor surface and the maximum change of 100 Ohm. The fit to the Hill Equation (1) resulted in an apparent K_D value of 111 ± 96 CFU·mL^{-1} and h of 0.31 ± 0.08. A measurement with a sample of unknown concentration showed good accordance to the data (green cross). With the approximated Hill equation a bacteria concentration of 5721 ± 2813 CFU·mL^{-1} was determined. Counting the sample under the microscope resulted in a value of 4150 CFU·mL^{-1}. The high standard deviation could be minimized by a higher amount of repetitions.

Whereas experiments with protein A resulted in a Hill coefficient slightly higher than 1, indicative for cooperative binding, biosensing experiments with whole cells resulted in a h significantly lower than 1, indicating negative cooperativity. This may be due to the size of *S. aureus* cells (~1 μm) and the spacing of protein A molecules on their surface not allowing for several aptamers to bind the same protein A. Another explanation may be the reduced flexibility of the cell surface bound protein A and therefore decreased accessibility of further binding sites for aptamers in the vicinity.

Repeated impedance measurements of a single PAA-modified electrode in FeBB resulted in a standard deviation of $s(R_{ct}) = 4.88$ Ω·cm^2. This corresponds to a calculated LoD of 10 CFU·mL^{-1}.

A comparison between previously reported *S. aureus* detection assays and the herein demonstrated biosensor is shown in Table 4. In summary, the achieved LoD is comparable to previous reports; however, our biosensor excels in simplicity, automation, low cost and rapid results.

Table 4. Comparison of several detection methods for *Staphylococcus aureus* (SA = *S. aureus* aptamer, LoD = limit of detection).

Detection Principle	Recognition Element	Assay Time	LoD/CFU·mL^{-1}	Reference
polymerase chain reaction	ssDNA	2 h	10	[47]
EIS immunosensor	anti-*S. aureus*-antibody	Not stated	10	[14]
resonance light scattering	SA 17 & SA 61 [26]	1.5 h	1	[26]
EIS	SA 43 [48]	1 h	10	[31]
fluorescent nanoparticles	SA 31 [48]	2 h	93	[24]
EIS	PA2#8[S1–58] [35]	10 min	10	this work

Lastly, we investigated the binding of protein A-deficient bacteria, such as *Staphylococcus epidermidis* and *Escherichia coli* to aptamer-modified gold electrodes and observed no cross-reactivity (Figure 7d). Similar results were found, if *S. aureus* was applied to electrodes modified with random oligonucleotides. Additionally, the binding of *S. aureus* (10^6 CFU·mL^{-1}) to a blank electrode and electrodes modified with a 6-mercapto-1-hexanol (MCH) monolayer, resulted in a negligible average R_{ct} change of 10.2 ± 2.2 Ω·cm^2 [49], a value below LoD (3·s).

These results are in agreement with results of Stoltenburg et al., reporting that protein A-deficient *S. aureus* strains and gram-negative bacteria, such as *E. coli*, were excluded from aptamer binding [36].

For the practical applicability of our sensor, we recommend that samples shall be extracted into tryptic soy broth or similar media and then diluted 1:100 in BB. The BB contains ions which are required for proper aptamer folding and thus binding to protein A. Simple sample preparation, such as a centrifugation step, showed improved results [49]. The influence of complex sample matrices, such as milk, on the performance of the developed sensor, still has to be examined.

For the food industry, detection of enterotoxins is more crucial than whole cell detection. However, *S. aureus* produces more than 7 different toxins, which all can cause illness. While an existing international standard operation [ISO 19020:2017] describes the screening of staphylococcal enterotoxins SEA, SEB, SECs, SED and SEE, others are excluded (SEG, SHE, SEI, SER, SES and SET). Extraction of the enterotoxins encompasses complex sample preparation, which requires fully equipped food testing laboratories. Furthermore, only enterotoxin assays with low detection limits, such as 0.05 ng/g food, fulfill the requirements [50].

Thus, the predominant approach, especially for small manufacturers, is the detection of *S. aureus* contamination. This approach, based on the Bacteriological Analytical Manual [51], requires culturing on selective agar plates for 2 days. Counting the colony forming units per g food in respect to the processing time is a good indicator for toxin formation. Our sensor could significantly reduce the time required for this approach.

In respect to clinical application, our sensor could improve tests by the rapid identification of bacteria type prior to antibiotic susceptibility testing and thus reduce the costs for isolating patients. Due to the microfluidic setup and the non-destructive impedance measurement, the combination with additional assays and steps will be realizable without great expense.

3. Materials and Methods

3.1. Reagents

The protein A-binding aptamer PA#2/8 (76 nt) was originally selected by Stoltenburg et al., using the FluMag-SELEX process [34]. A 3'-truncated variant of the full-length aptamer, PA#2/8[S1-58] (58 nt), was applied in the current study: 5'-ATACCAGCTTATTCAATTAGCAACATGAGGGGGAT AGAGGGGGTGGGTTCTCTCGGCT-3' [35]. This variant is herein referred to as PAA. A pool of randomized oligonucleotides (58 nt) was used as negative control (ON pool). Both were synthesized by Microsynth AG (Balgach, Switzerland), modified with C6-spacer and a thiol at the 3'-end, and purified with polyacrylamide gel electrophoresis.

Recombinant protein A (expressed in *Escherichia coli*, P7837) and 6-mercapto-1-hexanol (MCH, 99%) were purchased from Sigma-Aldrich Chemie GmbH (Taufkirchen, Germany). *Staphylococcus aureus* (DSM 20231), *Staphylococcus epidermidis* (DSM 3269) and *Escherichia coli* (DSM 498) were purchased from Leibniz Institut DSMZ GmbH (Braunschweig, Germany). Potassium-hexacyanoferrate (II) and (III) ($K_{3/4}[Fe(CN)_6]$) were purchased from Merck Chemicals GmbH (Darmstadt, Germany). Recombinant protein G and protein L (21193 and 21189) were purchased from Pierce Biotechnology (Rockford, USA). Tris-(2-carboxyethyl)-phosphine-hydrochloride (TCEP) was purchased from Carl Roth GmbH + Co. KG (Karlsruhe, Germany). All reagents were of analytical grade and used without further purification. All working solutions were prepared in water purified with a Milli-Q Type-1-system (EMD Millipore Corporation, Billerica, MA, USA; 18 MΩ·cm).

For the aptamer experiments, the binding buffer (BB) as for aptamer selection [35] was used. It consisted of 100 mM NaCl, 20 mM Tris, 10 mM $MgCl_2$, 5 mM KCl, 1 mM $CaCl_2$ (adjusted with HCl to a pH of 7.6) and was autoclaved, sterile filtered before further use. Protein A was directly dissolved and diluted in BBT, BB containing 0.005% Tween 20, to reduce unspecific binding to surfaces. *S. aureus*, *S. epidermidis* and *E. coli* cells were cultured in tryptic soy broth (TSB) and washed twice in BB. A fraction was dyed with SYTO and counted in a cell chamber of 0.02 mm depth. Washed cell suspensions were diluted to desired concentration in BB. For electrochemical measurements, 2 mM of $K_{3/4}[Fe(CN)_6]$ (equimolar) were added to BB (FeBB).

3.2. Preparation of Electrodes

The thiol-modified PAA was preconditioned with TCEP (200 μM/μM thiol) for 20 min to reduce disulfides and heated to 95 °C in BB for 5 min to enable proper folding.

The gold electrodes and gold covered quartz crystals were exposed to ultraviolet light for 5 min and subsequently incubated in hot alkaline piranha solution (5:1:1 water:NH_3:H_2O_2; CAUTION: this acidic mixture reacts violently with organic solvents and must be handled with care!) in an ultrasonic bath for 5 min. After rigorous rinsing with ultrapure water, the electrodes were dried in nitrogen and immediately covered by 1 µM preconditioned PAA and 4 µM MCH in BB followed by incubation at room temperature overnight. Next day, the surface was washed intensively with BB and exposed to 1 mM MCH for 30 min, followed by another wash with BB. The modified electrodes were stored in BB until use.

3.3. QCM Measurements

A quartz crystal microbalance (QCM, Q-Sense Analyzer E4, Biolin Scientific Holding AB, Stockholm, Sweden) was used for verification of aptamer immobilization and binding of protein A to immobilized aptamers. For measurements, the gold covered quartz crystals (QSX301, resonance frequency 4.95 MHz ± 50 kHz, Biolin Scientific) were cleaned as described above and mounted into the flow module QFM401. First, a baseline in BB was established, and then the chamber was flushed with a solution of 1 µM preconditioned PAA and 4 µM MCH in BB and incubated overnight while the frequency change was continuously measured. The temperature was held at 21 °C. The next day, the crystal was incubated with 1 mM MCH for 30 min and washed with BB. The modified crystals were incubated for 4 min with different concentrations of protein A (50–500 nM). Unbound protein A was washed away with BB. The relative frequency change—difference of frequency change measured in BB before and after incubation with aptamer or protein A—was used to determine the mass change by the Sauerbrey equation [52]. Thereby, most of the influences of viscosity and density of the fluids on the measurements were compensated.

3.4. MST Measurements

To determine the influence of ferri-/ferrocyanide on aptamer-target-binding, microscale thermophoresis (MST) measurements were performed with the Monolith NT.115 (NanoTemper Technologies GmbH, Munich, Germany) and standard capillaries. The fluorescence dye NT-547 was bound to the thiol group of PAA using the labeling kit MO-L005 Monolith™ by NanoTemper. A fluorophore-per-aptamer-ratio of 0.5 was determined by measuring the absorbance at 260 and 546 nm with a nanophotometer™ (Implen GmbH, Munich, Germany). The MST power was set to 40% and the LED power was set to 100% (no bleaching was observed). 38.5 nM labeled aptamer (77 nM in total, including the unlabeled aptamer) were incubated with 0.2 nM to 6.25 µM recombinant protein A. The analyses were performed in BBT at 25 °C. Normalization of the fluorescence signal and fitting to the Hill equation were performed using the software MO Affinity Analysis v2.1.3 (NanoTemper). To investigate if the labeling procedure or labeling site influences the aptamer affinity, additional MST measurements were conducted by commercial analysis service of 2bind GmbH (Regensburg, Germany). The same labeling procedure as described above was applied to the 3'- and 5'-end, and 30 nM labeled PAA with 0.354 nM to 11.6 µM protein A in BB with 0.05% Tween 20 were used.

3.5. EIS Measurements

Electrochemical impedance spectroscopy (EIS) measurements were performed using the SP-300 potentiostat/galvanostat with impedance analyzer (Bio-Logic Science Instruments SAS, Claix, France). A custom flow-through chamber (Figure 1) made of polyether-ether-ketone was designed and manufactured. Flow-through was realized by a peristaltic pump (IPC-16, Ismatec, Cole-Parmer GmbH, Wertheim, Germany) and a Teflon tube of 0.5 mm inner diameter. The fluid chamber with a volume of 100 µL was covered with glass. It consists of a three-electrode system with a working electrode cut from gold coated test slides (TA134-(Ti/Au), EMF Corporation, Ithaca, NY, USA), a Pt-black wire as counter electrode and a leak-free Ag/AgCl reference electrode (LF-1 Ø1 mm, Innovative Instruments Inc., Tampa, FL, USA). The electrochemical active area (A_{true}) of the working electrode was determined by cyclic voltammetry in 0.5 M H_2SO_4 with 100 mV/s from 0.2 to 1.5 V. By integration of the reduction peak at ~800 mV, the required charge for reduction of a gold oxide monolayer was obtained (215 µC). Division by the theoretical value, 482 µC/cm^2, calculated by [53], resulted in A_{true} = 0.444 ± 0.039 cm^2.

The geometric surface area A_{geo} was 0.246 cm^2 and thus, the roughness factor $R = A_{true}/A_{geo}$ was 1.8. The flow-through chamber and the solutions were kept in a temperature cabinet at 21 °C while the pump and impedance analyzer were positioned outside.

EIS measurements were performed with the aptamer-modified gold electrode mounted in the flow-through measurement chamber (Figure 1). Different concentrations of protein A (2–700 nM), 6 mL of each, were pumped (50 µL/s) through the chamber and incubated for 5 min while flow-through was paused. Then 6 mL FeBB was pumped (50 µL/s) through the system. Finally, the impedance was measured from 1 Hz to 200 kHz with 7 logarithmic spaced frequencies per decade. The sinusoidal alternating voltage with an amplitude of 10 mV was applied at the equilibrium potential of ferri-/ferrocyanide (E_{eq} ~140 mV). The impedance measurement was repeated four times while the average of the last three cycles was used for fitting and analysis. Fitting was performed with the simplex algorithm implemented in EC-Lab®software (v11.00, Bio-Logic Science Instruments SAS, Claix, France). The same procedure was applied to control samples, protein G and L, of which 1 µM were used and bacteria cell suspensions.

4. Conclusions

This study provides the proof of principle for an impedimetric biosensor for the rapid detection of S. aureus, based on the protein A-binding aptamer. Successful co-immobilization of protein A-binding aptamers and 6-mercapto-1-hexanol on gold electrodes resulted in an average density of $1.01 \pm 0.44 \times 10^{13}$ aptamers per cm^2. The immobilization density can be influenced by the ratio of aptamer to 6-mercapto-1-hexanol (MCH) as shown with chronocoulometry. We showed with MST measurements that ferri-/ferrocyanide, necessary as redox couple for faradaic impedance measurements, has no significant influence on the binding of the aptamer to its target. The biosensor displayed sensitive binding to protein A with a K_D of 18.5 ± 1.8 nM and a LoD of 3 nM. Our results also showed the excellent selectivity of the developed sensor, with signals below LoD upon exposure to high concentrations of the functionally similar proteins G and L.

When exposed to live S. aureus cells, our developed aptamer-based biosensor showed a K_D of 111 ± 96 CFU·mL^{-1} and a LoD of 10 CFU·mL^{-1}, which is in good agreement with other reported assays or sensors. Our results also prove the high selectivity of the aptamer, distinguishing between S. aureus and protein A- deficient bacteria, such as E. coli and S. epidermidis.

For application in a clinical setting, an additional step for the evaluation of the different detectable S. aureus strains and their possible antibiotic resistance (e.g., by PCR) may have to be considered. Furthermore, the influence of different ionic strength buffers and sample matrices on the biosensor performance have to be investigated closely.

This work demonstrated that the protein A-binding aptamer can be used as recognition element in impedimetric aptasensors for successful, rapid, sensitive and selective detection of S. aureus in buffer. It contributes to the deeper understanding of impedimetric aptasensors and their development. We provided a fundamental base for inexpensive and robust biosensing, utilizing aptamer receptors. The advantages of using gold electrodes are their robustness, enabling regeneration and subsequent reuse of the biosensor. The simplicity of our design enables easy reproduction and the developed microfluidic system can be easily automated. Furthermore, combination with electrode patterning may enable the parallel measurement of multiple analytes when functionalized with different aptamers, in the future.

Acknowledgments: We want to thank Tobias Pflüger for his excellent experimental assistance with MST experiments. Our special thanks go to Katharina Urmann for her help in reviewing the manuscript.

Author Contributions: Peggy Reich designed the experimental concept, performed experiments and data analysis. Dieter Frense helped with experiments. Regina Stoltenburg and Beate Strehlitz helped profoundly with advice and consultation and provided the aptamer, Dieter Beckmann supervised the project. Peggy Reich wrote the manuscript with help of Regina Stoltenburg and Beate Strehlitz. All authors reviewed the manuscript.

Abbreviations

BB	Binding buffer
BBT	Binding buffer and 0.005% Tween 20
CFU	Colony-forming unit
CPE	Constant phase element
DNA	Deoxyribonucleic acid
EIS	Electrochemical impedance spectroscopy
ELONA	Enzyme-linked oligonucleotide assay
FeBB	Binding buffer and 2 mM equimolar ferri-/ferrocyanide
h	Hill coefficient
K_D	Apparent binding constant
kDa	kiloDalton
LoD	Limit of detection
MCH	Mercaptohexanol
MDPI	Multidisciplinary Digital Publishing Institute
MRSA	Methicillin-resistant *Staphylococcus aureus*
MST	Microscale thermophoresis
ON pool	Random oligonucleotide pool
PAA	Protein A-binding Aptamer
PCR	Polymerase chain reaction
QCM	Quartz crystal microbalance
$R_{ct}/\Delta R_{ct}$	Charge transfer resistance/change of charge transfer resistance
RIFTS	Reflective interferometric fourier transform spectroscopy
RMS	Rough mean square
R_{sol}	Solution resistance
RuHex	hexaammineruthenium(III) chloride
SELEX	Systematic Evolution of Ligands by Exponential Enrichment
spA	Gene encoding protein A
SPR	Surface plasmon resonance spectroscopy
SD	Standard deviation
TSB	Tryptic soy broth
vWF	Von Willebrand factor
W	Warburg element

References

1. Tong, S.Y.; Davis, J.S.; Eichenberger, E.; Holland, T.L.; Fowler, V.G., Jr. *Staphylococcus aureus* infections: Epidemiology, pathophysiology, clinical manifestations, and management. *Clin. Microbiol. Rev.* **2015**, *28*, 603–661. [CrossRef] [PubMed]
2. Carroll, K.C. Rapid diagnostics for methicillin-resistant *Staphylococcus aureus*: Current status. *Mol. Diagn. Ther.* **2008**, *12*, 15–24. [CrossRef] [PubMed]
3. Jin, W.; Yamada, K. Staphylococcal enterotoxins in processed dairy products A2. In *Food Hygiene and Toxicology in Ready to Eat Foods*; Kotzekidou, P., Ed.; Academic Press: San Diego, CA, USA, 2016; pp. 241–258.
4. Hibnick, H.E.; Bergdoll, M.S. Staphylococcal enterotoxin. II. Chemistry. *Arch. Biochem. Biophys.* **1959**, *85*, 70–73. [CrossRef]
5. DeDent, A.C.; McAdow, M.; Schneewind, O. Distribution of protein A on the surface of *Staphylococcus aureus*. *J. Bacteriol.* **2007**, *189*, 4473–4484. [CrossRef] [PubMed]
6. Schelin, J.; Wallin-Carlquist, N.; Cohn, M.T.; Lindqvist, R.; Barker, G.C.; Radstrom, P. The formation of *Staphylococcus aureus* enterotoxin in food environments and advances in risk assessment. *Virulence* **2011**, *2*, 580–592. [CrossRef] [PubMed]
7. Law, J.W.; Ab Mutalib, N.S.; Chan, K.G.; Lee, L.H. Rapid methods for the detection of foodborne bacterial pathogens: Principles, applications, advantages and limitations. *Front. Microbiol.* **2014**, *5*, 770. [CrossRef] [PubMed]

8. Zhao, X.; Wei, C.; Zhong, J.; Jin, S. Research advance in rapid detection of foodborne *Staphylococcus aureus*. *Biotechnol. Biotechnol. Equip.* **2016**, *30*, 827–833. [CrossRef]

9. Oh, A.C.; Lee, J.K.; Lee, H.N.; Hong, Y.J.; Chang, Y.H.; Hong, S.I.; Kim, D.H. Clinical utility of the Xpert MRSA assay for early detection of methicillin-resistant *Staphylococcus aureus*. *Mol. Med. Rep.* **2013**, *7*, 11–15. [CrossRef] [PubMed]

10. Mohanty, S.P.; Kougianos, E. Biosensors: A tutorial review. *IEEE Potentials* **2006**, *25*, 35–40. [CrossRef]

11. Hoja, J.; Lentka, G. A family of new generation miniaturized impedance analyzers for technical object diagnostics. *Metrol. Meas. Syst.* **2013**, *20*. [CrossRef]

12. Lisdat, F.; Schafer, D. The use of electrochemical impedance spectroscopy for biosensing. *Anal. Bioanal. Chem.* **2008**, *391*, 1555–1567. [CrossRef] [PubMed]

13. Bahadir, E.B.; Sezginturk, M.K. A review on impedimetric biosensors. *Artif. Cells Nanomed. Biotechnol.* **2016**, *44*, 248–262. [CrossRef] [PubMed]

14. Bekir, K.; Barhoumi, H.; Braiek, M.; Chrouda, A.; Zine, N.; Abid, N.; Maaref, A.; Bakhrouf, A.; Ouada, H.B.; Jaffrezic-Renault, N.; et al. Electrochemical impedance immunosensor for rapid detection of stressed pathogenic *Staphylococcus aureus* bacteria. *Environ. Sci. Pollut. Res. Int.* **2015**, *22*, 15796–15803. [CrossRef] [PubMed]

15. O'Sullivan, C.K. Aptasensors—The future of biosensing? *Anal. Bioanal. Chem.* **2002**, *372*, 44–48. [CrossRef] [PubMed]

16. Gopinath, S.C. Methods developed for SELEX. *Anal. Bioanal. Chem.* **2007**, *387*, 171–182. [CrossRef] [PubMed]

17. Willner, I.; Zayats, M. Electronic aptamer-based sensors. *Angew. Chem. Int. Ed. Engl.* **2007**, *46*, 6408–6418. [CrossRef] [PubMed]

18. Radi, A.E.; Acero Sanchez, J.L.; Baldrich, E.; O'Sullivan, C.K. Reusable impedimetric aptasensor. *Anal. Chem.* **2005**, *77*. [CrossRef] [PubMed]

19. Zhu, B.; Booth, M.A.; Woo, H.Y.; Hodgkiss, J.M.; Travas-Sejdic, J. Label-Free, electrochemical quantitation of potassium ions from femtomolar levels. *Chem. Asian. J.* **2015**, *10*, 2169–2175. [CrossRef] [PubMed]

20. Liang, G.; Man, Y.; Jin, X.; Pan, L.; Liu, X. Aptamer-based biosensor for label-free detection of ethanolamine by electrochemical impedance spectroscopy. *Anal. Chim. Acta.* **2016**, *936*, 222–228. [CrossRef] [PubMed]

21. Labib, M.; Zamay, A.S.; Kolovskaya, O.S.; Reshetneva, I.T.; Zamay, G.S.; Kibbee, R.J.; Sattar, S.A.; Zamay, T.N.; Berezovski, M.V. Aptamer-based viability impedimetric sensor for bacteria. *Anal. Chem.* **2012**, *84*, 8966–8969. [CrossRef] [PubMed]

22. Shahdordizadeh, M.; Taghdisi, S.M.; Ansari, N.; Langroodi, F.A.; Abnous, K.; Ramezani, M. Aptamer based biosensors for detection of *Staphylococcus aureus*. *Sens. Actuators B Chem.* **2017**, *241*, 619–635. [CrossRef]

23. Yuan, J.; Wu, S.; Duan, N.; Ma, X.; Xia, Y.; Chen, J.; Ding, Z.; Wang, Z. A sensitive gold nanoparticle-based colorimetric aptasensor for *Staphylococcus aureus*. *Talanta* **2014**, *127*, 163–168. [CrossRef] [PubMed]

24. Shangguan, J.; Li, Y.; He, D.; He, X.; Wang, K.; Zou, Z.; Shi, H. A combination of positive dielectrophoresis driven on-line enrichment and aptamer-fluorescent silica nanoparticle label for rapid and sensitive detection of *Staphylococcus aureus*. *Analyst* **2015**, *140*, 4489–4497. [CrossRef] [PubMed]

25. Duan, N.; Wu, S.; Zhu, C.; Ma, X.; Wang, Z.; Yu, Y.; Jiang, Y. Dual-color upconversion fluorescence and aptamer-functionalized magnetic nanoparticles-based bioassay for the simultaneous detection of *Salmonella typhimurium* and *Staphylococcus aureus*. *Anal. Chim. Acta* **2012**, *723*, 1–6. [CrossRef] [PubMed]

26. Chang, Y.C.; Yang, C.Y.; Sun, R.L.; Cheng, Y.F.; Kao, W.C.; Yang, P.C. Rapid single cell detection of *Staphylococcus aureus* by aptamer-conjugated gold nanoparticles. *Sci. Rep.* **2013**, *3*. [CrossRef] [PubMed]

27. Zelada-Guillen, G.A.; Sebastian-Avila, J.L.; Blondeau, P.; Riu, J.; Rius, F.X. Label-free detection of *Staphylococcus aureus* in skin using real-time potentiometric biosensors based on carbon nanotubes and aptamers. *Biosens. Bioelectron.* **2012**, *31*, 226–232. [CrossRef] [PubMed]

28. Hernandez, R.; Valles, C.; Benito, A.M.; Maser, W.K.; Rius, F.X.; Riu, J. Graphene-based potentiometric biosensor for the immediate detection of living bacteria. *Biosens. Bioelectron.* **2014**, *54*, 553–557. [CrossRef] [PubMed]

29. Abbaspour, A.; Norouz-Sarvestani, F.; Noori, A.; Soltani, N. Aptamer-conjugated silver nanoparticles for electrochemical dual-aptamer-based sandwich detection of *Staphylococcus aureus*. *Biosens. Bioelectron.* **2015**, *68*. [CrossRef] [PubMed]

30. Lian, Y.; He, F.; Wang, H.; Tong, F. A new aptamer/graphene interdigitated gold electrode piezoelectric sensor for rapid and specific detection of *Staphylococcus aureus*. *Biosens. Bioelectron.* **2015**, *65*, 314–319. [CrossRef] [PubMed]

31. Jia, F.; Duan, N.; Wu, S.; Ma, X.; Xia, Y.; Wang, Z.; Wei, X. Impedimetric aptasensor for *Staphylococcus aureus* based on nanocomposite prepared from reduced graphene oxide and gold nanoparticles. *Microchim. Acta* **2014**, *181*, 967–974. [CrossRef]

32. Kahl, B.C.; Mellmann, A.; Deiwick, S.; Peters, G.; Harmsen, D. Variation of the polymorphic region X of the protein A gene during persistent airway infection of cystic fibrosis patients reflects two independent mechanisms of genetic change in *Staphylococcus aureus*. *J. Clin. Microbiol.* **2005**, *43*, 502–505. [CrossRef] [PubMed]

33. O'Seaghdha, M.; van Schooten, C.J.; Kerrigan, S.W.; Emsley, J.; Silverman, G.J.; Cox, D.; Lenting, P.J.; Foster, T.J. *Staphylococcus aureus* protein A binding to von Willebrand factor A1 domain is mediated by conserved IgG binding regions. *FEBS J.* **2006**, *273*, 4831–4841. [CrossRef] [PubMed]

34. Stoltenburg, R.; Reinemann, C.; Strehlitz, B. FluMag-SELEX as an advantageous method for DNA aptamer selection. *Anal. Bioanal. Chem.* **2005**, *383*, 83–91. [CrossRef] [PubMed]

35. Stoltenburg, R.; Schubert, T.; Strehlitz, B. In vitro selection and interaction studies of a DNA aptamer targeting protein A. *PLoS ONE* **2015**, *10*. [CrossRef] [PubMed]

36. Stoltenburg, R.; Krafcikova, P.; Viglasky, V.; Strehlitz, B. G-quadruplex aptamer targeting protein A and its capability to detect *Staphylococcus aureus* demonstrated by ELONA. *Sci. Rep.* **2016**, *6*, 33812. [CrossRef] [PubMed]

37. Keighley, S.D.; Li, P.; Estrela, P.; Migliorato, P. Optimization of DNA immobilization on gold electrodes for label-free detection by electrochemical impedance spectroscopy. *Biosens. Bioelectron.* **2008**, *23*, 1291–1297. [CrossRef] [PubMed]

38. Steel, A.B.; Herne, T.M.; Tarlov, M.J. Electrochemical quantitation of DNA immobilized on gold. *Anal. Chem.* **1998**, *70*, 4670–4677. [CrossRef] [PubMed]

39. Debono, R.F.; Loucks, G.D.; Manna, D.D.; Krull, U.J. Self-assembly of short and long-chain *n*-alkyl thiols onto gold surfaces: A real-time study using surface plasmon resonance techniques. *Can. J. Chem.* **1996**, *74*, 677–688. [CrossRef]

40. Martin, S.J.; Granstaff, V.E.; Frye, G.C. Characterization of a quartz crystal microbalance with simultaneous mass and liquid loading. *Anal. Chem.* **2002**, *63*, 2272–2281. [CrossRef]

41. Urmann, K.; Modrejewski, J.; Scheper, T.; Walter Johanna, G. Aptamer-modified nanomaterials: Principles and applications. *BioNanoMaterials* **2017**, *18*. [CrossRef]

42. Erickson, H.P. Size and shape of protein molecules at the nanometer level determined by sedimentation, gel filtration, and electron microscopy. *Biol. Proced. Online* **2009**, *11*, 32–51. [CrossRef] [PubMed]

43. Angell, D.H.; Dickinson, T. The kinetics of the ferrous/ferric and ferro/ferricyanide reactions at platinum and gold electrodes. *J. Electroanal. Chem. Interfacial Electrochem.* **1972**, *35*, 55–72. [CrossRef]

44. Hirschorn, B.; Orazem, M.E.; Tribollet, B.; Vivier, V.; Frateur, I.; Musiani, M. Determination of effective capacitance and film thickness from constant-phase-element parameters. *Electrochim. Acta* **2010**, *55*, 6218–6227. [CrossRef]

45. Sjödahl, J. Structural studies on the four repetitive Fc-binding regions in protein A from *Staphylococcus aureus*. *Eur. J. Biochem.* **1977**, *78*, 471–490. [CrossRef] [PubMed]

46. Urmann, K.; Reich, P.; Walter, J.G.; Beckmann, D.; Segal, E.; Scheper, T. Rapid and label-free detection of protein A by aptamer-tethered porous silicon nanostructures. *J. Biotechnol.* **2017**, *257*, 171–177. [CrossRef] [PubMed]

47. Banada, P.P.; Chakravorty, S.; Shah, D.; Burday, M.; Mazzella, F.M.; Alland, D. Highly sensitive detection of *Staphylococcus aureus* directly from patient blood. *PLoS ONE* **2012**, *7*, e31126. [CrossRef] [PubMed]

48. Cao, X.; Li, S.; Chen, L.; Ding, H.; Xu, H.; Huang, Y.; Li, J.; Liu, N.; Cao, W.; Zhu, Y.; et al. Combining use of a panel of ssDNA aptamers in the detection of *Staphylococcus aureus*. *Nucleic Acids Res.* **2009**, *37*, 4621–4628. [CrossRef] [PubMed]

49. Reich, P. Entwicklung eines impedimetrischen Aptasensor zur Detektion von *Staphylococcus aureus*. Ph.D. Thesis, Leibniz Universität Hannover, Hannover, Germany, 2017. in preparation.

50. Tallent, S.M.; Bennett, R.W.; Hait, J.M. Chapter 13 B Staphylococcal Enterotoxins Detection Methods. 2017. U.S. Food and Drug Administration. Available online: https://www.fda.gov/Food/FoodScienceResearch/ LaboratoryMethods/ucm564359.htm (accessed on 13 November 2017).

51. Bennett, R.W.; Lancette, G.A. Chapter 12 *Staphylococcus aureus*. 2016; U.S. Food and Drug Administration. Available online: https://www.fda.gov/Food/FoodScienceResearch/LaboratoryMethods/ucm071429.htm (accessed on 13 November 2017).

52. Sauerbrey, G. Verwendung von Schwingquarzen zur Wägung dünner Schichten und zur Mikrowägung. *Zeitschrift für Physik* **1959**, *155*, 206–222. [CrossRef]

53. Oesch, U.; Janata, J. Electrochemical study of gold electrodes with anodic oxide films—I. Formation and reduction behaviour of anodic oxides on gold. *Electrochim. Acta* **1983**, *28*, 1237–1246. [CrossRef]

10

Aptamers and Glioblastoma: Their Potential use for Imaging and Therapeutic Applications

Emma M. Hays, Wei Duan and Sarah Shigdar *

Centre for Molecular and Medical Research, School of Medicine, Deakin University, 75 Pigdons Road, Waurn Ponds, Victoria 3216, Australia; ehays@deakin.edu.au (E.M.H.); wei.duan@deakin.edu.au (W.D.)
* Correspondence: sarah.shigdar@deakin.edu.au

Abstract: Glioblastoma is a highly aggressive primary brain tumour, renowned for its infiltrative growth and varied genetic profiles. The current treatment options are insufficient, and their off-target effects greatly reduce patient quality of life. The major challenge in improving glioblastoma diagnosis and treatment involves the development of a targeted imaging and drug delivery platform, capable of circumventing the blood brain barrier and specifically targeting glioblastoma tumours. The unique properties of aptamers demonstrate their capability of bridging the gap to the development of successful diagnosis and treatment options, where antibodies have previously failed. Aptamers possess many characteristics that make them an ideal novel imaging and therapeutic agent for the treatment of glioblastoma and other brain malignancies, and are likely to provide patients with a better standard of care and improved quality of life. Their target sensitivity, selective nature, ease of modification and low immunogenicity make them an ideal drug-delivery platform. This review article summarises the aptamers previously generated against glioblastoma cells or its identified biomarkers, and their potential application in diagnosis and therapeutic targeting of glioblastoma tumours.

Keywords: glioblastoma; aptamers; SELEX; cancer; targeted therapies; imaging; diagnosis; biomarkers; blood brain barrier

---

1. Introduction

Glioblastoma or grade IV glioma is a highly malignant primary brain tumour with a particularly poor median survival duration of 14 months [1]. Whilst the incidence rate of glioblastoma currently stands markedly low at 3.2 per 100,000 population, its five-year survival rate is below 6% [2]. These tumours can arise spontaneously from glial cells, or can develop from the progression of a lower grade glioma and are defined as primary or secondary glioblastomas, respectively (see Figure 1) [3]. Histological features of glioblastoma include increased cellularity and angiogenesis, vascular proliferation, hemorrhage, necrosis, and cystic regions throughout the tumours [4,5]. Cell populations within these tumours tend to vary greatly, with the presence of both small, undifferentiated cells and large cells with multiple nuclei reported [4].

According to the World Health Organization, glioblastomas are grouped into three categories based on isocitrase dehydrogenase (IDH) status: glioblastoma IDH-wildtype, which is designated as primary or to have originated de novo; glioblastoma IDH-mutant, which is identified as a secondary tumour, arising from a lower grade glioma; and glioblastoma not otherwise specified (NOS) for tumours when IDH evaluation cannot be performed [6]. An estimated 90% of glioblastoma arise de novo, with the remaining 10% developing from a lower grade glioma [7]. Mutations of IDH lead to hypermethylation of histones and DNA, altering gene expression, promoting the activation of oncogenes and blocking tumour suppressing mechanisms [8]. A variety of genetic mutations are implicated in glioblastoma development including the *epidermal growth factor receptor* (*EGFR*), *human epidermal growth factor receptor two* (*ERBB2*), *isocitrase dehydrogenase one* (*IDH1*), *neurofibromin one* (*NF1*), *phosphoinositide three-kinase (PI3K), phosphatase and tensin homolog (PTEN), retinoblastoma protein (RB1)* and *tumour suppressor p53 (TP53)* [9]. The gene expression patterns can be used to further categorise glioblastomas into four sub-categories: proneural; neural; mesenchymal; and classical [10].

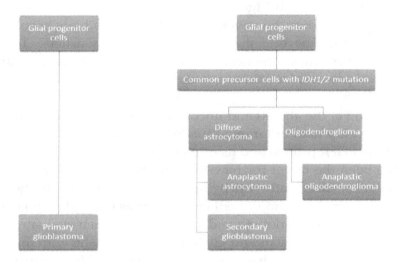

Figure 1. Glioblastomas can develop through two different pathways; arising as a primary malignancy; or through the progression of a lower grade glioma [11].

The classical subtype is associated with amplification of chromosome 7 teamed with loss of chromosome 10, EGFRoverexpression and mutations. Mesenchymal glioblastomas maintain a high expression of *CH13L1, MET,* and genes associated with tumour necrosis factor and nuclear factor-κB pathways, along with mutations and deletions of NF1. The proneural subclass parallels secondary glioblastoma and lower-grade glioma with mutations in *IDH1* and *TP53,* and modification of platelet-derived growth factor receptor A (PDGFR-A) [10]. Finally, neural glioblastomas are similar to normal brain tissue; however, they do overexpress EGFR [10,12].The current treatment modality includes surgery, radiotherapy, and chemotherapy with the DNA-alkylating agent temozolomide [1,13]. Surgery is performed to debulk the tumour thereby reducing mass effect symptoms in the patient, while also allowing for the collection of tissue specimens for histologic analysis [14]. Glioblastomas are renowned for their heterogeneity and infiltrative growth; complete surgical resection is near impossible as a result, and further complicated by the inability to differentiate the tumour from normal brain tissue [15]. Radiation and chemotherapy form the next line of treatment, aiming to destroy the cancer cells that were missed or could not be removed during surgery [1]. Temozolomide is a DNA-alkylating agent capable of inducing single- and double-stranded breaks in DNA, resulting in senescence and cell death [16]. Whilst surgery and the addition of temozolomide during and post-radiotherapy has led to an increase in patient survival times, they are responsible for various adverse effects and poor quality of life in patients [13,17]. In addition, chemo-resistance and glioblastoma recurrence are inevitable for the majority of patients, highlighting the necessity of improved treatment options [16,18]. One of the greatest challenges facing modern medicine is the development of tumour targeting molecules, capable of specifically binding to their target with no adverse effects to normal cells and tissues within the body. Whilst the discovery and subsequent development of antibodies have become an integral part of scientific research, disease diagnosis and therapies, these molecules possess undesired characteristics and many pose significant risks to patients, limiting their clinical efficacy [19,20]. Antibody generation occurs in vivo, in both a time consuming and costly process, with incidences of great batch-to-batch variability [21]. The high cost of the complicated antibody production process limits their clinical use, particularly as large doses are needed for effective responses in patients [22]. Their large size hinders their ability to reach the desired targets due to poor tissue penetration, and can be irreversibly denatured by small changes in temperature, thereby limiting their shelf life and transport options [23]. Attempts have been made to humanize antibodies; however, complement dependent toxicity (CDC), antibody-dependent cellular cytotoxicity (ADCC), and cytokine storms still occur in patients [24–26]. Despite these known issues, antibodies remain in development and clinical trials, while the development of alternatives with improved safety profiles are being investigated [27,28]. Targeted peptides and aptamers may bridge the gaps of diagnostic and therapeutic applications that are currently unfilled by antibodies; however, targeted peptides have been reviewed extensively elsewhere (see [29–31]) and fall beyond the scope of this review article. Targeted therapies will pave the way for personalized cancer medicine, ensuring patients receive treatment based on their tumour's gene expression profiles, for effective treatment against glioblastoma and other currently incurable conditions.

Further understanding of the altered and uncontrolled signaling pathways in glioblastoma has potential to aid in the development of targeted treatments against the disease. Gene amplification and overexpression of the EGFR protein and its mutant variant EGFRvIII contributes to glioblastoma tumorigenesis and has been the target for new therapeutics [32,33]. Tyrosine kinase inhibitors of EGFR block downstream signaling pathways by inhibiting ATP binding to the intracellular domain, thereby impeding with responses leading to cell growth, invasion, and angiogenesis [34]. The use of the EGFR tyrosine kinase inhibitors erlotinib and gefitinib have been evaluated in numerous clinical trials, alone and in combination with standard glioblastoma treatment, with minimal effect on patient survival [35–38]. Monoclonal antibodies have also been investigated for the treatment of glioblastoma. In particular, bevacizumab was generated to inhibit vascular endothelial growth factor (VEGF) in order to prevent angiogenesis, survival and migration of glioblastoma tumour cells [39]. Two large phase III trials determined that the use of bevacizumab did not lead to increased survival compared to standard treatment, despite this, it has remained in use to treat recurrent glioblastoma when temozolomide rechallenge fails [40–42]. The unsuccessful development and clinical translation of more effective treatments for glioblastoma are hindered by the restrictive nature of the blood brain barrier.

The blood brain barrier's role is to control the passage of molecules and cells into the brain, in order to protect this vital organ [43]. Tight junctions formed by a uniform monolayer of endothelial cells maintain an almost impermeable barrier, increasing the difficulty of conveying chemotherapeutics into the brain [44]. Great efforts have been made to develop mechanisms capable of delivering therapeutic agents across this barrier, including both invasive and non-invasive strategies. The proposed invasive strategies include temporary disruption of the blood brain barrier and intrathecal drug delivery [45–48]. A temporary disruption of the barrier can be achieved using focused ultrasound and its interaction with microbubbles to alter the structure of the endothelial layer [49]. However, this technique can lead to necrosis in the target tissue area, vasculitis, and seizures [50]. Intrathecal drug delivery introduces therapeutics via direct injection into the cerebrospinal fluid [48]. While this is a simple method of bypassing the blood brain barrier, the drugs are often filtered from the cerebrospinal fluid before they reach the targeted area, and the use of a catheter often results in infection [47,51]. A less invasive method involves chemically altering drugs, potentially by the addition of a lipid-like structure, or modifying the drug to suit specific receptors, where it would be converted within the target organ to the active state [52]. Modifying drugs may result in considerable changes from the parental drug, in turn altering the pharmacodynamics and biological efficacy [44]. Finally, biological delivery systems can be used to increase the uptake of drugs by transporters that would normally be used for essential nutrients [53].

The presence of transport molecules on the surface of the blood brain barrier have recently proven to be an effective mechanism to utilise for the delivery of substances into the brain via receptor-mediated endocytosis. Yu and colleagues successfully utilised the transferrin receptor to transport an antibody across the endothelial layer into the brain itself [54]. On the contrary, the use of antibodies is associated with an immune response, therefore the development of a novel targeting agent similar to antibodies is required for the safe treatment of brain malignancies [55]. The recent development of a bi-functional aptamer capable of transcytosing across the blood brain barrier aided by the transferrin receptor in vivo, can be conjugated to a glioblastoma targeting aptamer (Figure 2), and potentially improve its treatment modality [56].

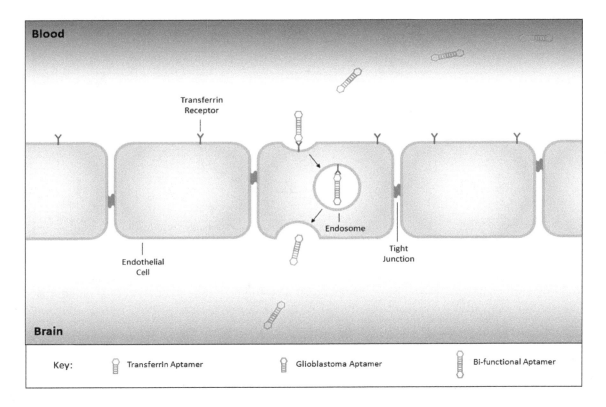

Figure 2. The fusion of a transferrin aptamer with a glioblastoma targeting aptamer would create a bi-functional aptamer capable of crossing the blood brain barrier to target glioblastoma tumour tissues specifically. This targeted mechanism may be utilised for the safe delivery of both imaging and therapeutic agents, sparing healthy cells throughout the body and the surrounding brain tissues from off-target effects.

2. Aptamers

Aptamers, also referred to as 'chemical antibodies' are single stranded oligonucleotides capable of binding to a target via shape recognition, in a similar fashion to antibodies [57]. Aptamers are generated against specific targets via the process of the systematic evolution of ligands by exponential enrichment (SELEX) and their sequence can be modified to optimise binding affinity and selectivity (see Figure 3) [58–61]. The selection of aptamers occurs in vitro, as does their subsequent manufacture, leading to little to no variation between batches, low immunogenicity, and are a considerably smaller size than antibodies, further increasing their ease of tissue penetration [62]. The enhanced binding properties of aptamers can saturate the available binding sites on the tumour surface, and therefore lead to improved imaging signals and intratumoural delivery of therapeutic agents, comparative to antibodies [22,63]. Chemotherapeutic agents can be attached to aptamers in order to deliver them to tumour cells, thereby reducing unwanted off-target effects, a particularly important factor for the development of new therapeutic agents for brain tumours [57]. Specificity of aptamer binding highlights their suitability for tumour imaging and targeted drug delivery, particularly in an organ as important as the brain. The attachment of imaging agents to aptamers has previously been achieved, and may lead to the development of more effective tumour imaging strategies for both histologic analysis of tissues as well as determining tumour location within the patient [64–67]. The use of aptamers for both imaging and therapeutic delivery have been reviewed extensively elsewhere [68–71].

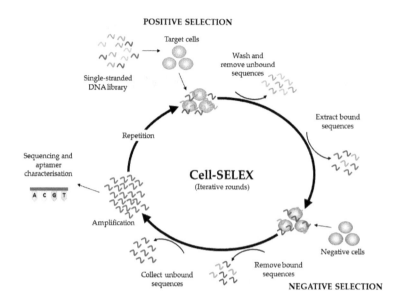

Figure 3. Schematic representation of a Cell-SELEX (systematic evolution of ligands by exponential enrichment) protocol. Initially, a single-stranded DNA library is incubated with the target cells. The cells are washed to remove the unbound sequences, and the bound sequences are collected, prior to incubation with negative (control) cells. The bound sequences are removed and discarded, whilst the unbound sequences are amplified and used to begin the next round of SELEX. This cycle is repeated a number of times, before the pool is sequenced and characterised. The bold arrows denote one iterative round of SELEX, with the small arrows depicting the different steps of the protocol: incubation of the single-stranded DNA library with the target cells; washing and removal of the unbound sequences, incubation with negative cells; removal of bound sequences; collection of bound sequences; amplification and cycle repetition prior to sequencing and aptamer characterization.

3. Glioblastoma Targeting Aptamers

To date, a multitude of aptamers have been generated against cell membrane proteins expressed on glioblastoma cells (see Tables 1 and 2). These aptamers have potential to be used to identify new biomarkers, and to specifically deliver imaging or therapeutic agents to revolutionise glioblastoma diagnosis and treatment (see Figure 4).

Table 1. Glioblastoma aptamers and their SELEX methods.

Aptamer	Target	SELEX Method	Positive Selection	Negative Selection	Reference(s)
A3, A4	Unknown	Differential cell-SELEX	CD133+ TIC	CD133− cells; human neural progenitor cells	[72]
TTA1	Tenascin-C	Crossover-SELEX	U251 cells; human Tenascin-C	No negative selection	[66,73]
GBI-10	Tenascin-C	Cell-SELEX	U251 cells	No negative selection	[74,75]
Gint4.T	PDGFRβ	Cell-SELEX	U87MG cells	No negative selection	[76,77]
GL21.T	Axl	Differential cell-SELEX	U87MG cells	T98G cells	[77–80]
U2	EGFRvIII	Differential cell-SELEX	U87MG-EGFRvIII cells	U87MG cells	[81]
Aptamer 32	EGFRvIII	Differential cell-SELEX	U87MG-EGFRvIII cells	U87MG cells	[82–84]
E07	EGFR; EGFRvIII	Protein-SELEX	Human EGFR	No negative selection	[85–87]
GMT 3–9	Unknown	Differential cell-SELEX	A172 cells	No negative selection	[88]
GBM128, GBM131	Unknown	Differential cell-SELEX	U118-MG cells	SVGp12 cells	[89]

Table 2. Aptamers generated against potential biomarkers for glioblastoma targeting.

Aptamer	Target	SELEX Method	Positive Selection	Negative Selection	Reference(s)
CD133-A15, CD133-B19	CD133	Differential cell-SELEX	HEK293T-CD133 cells	HEK293T cells	[90]
CL4	EGFR; EGFRvIII	Differential cell-SELEX	A549 cells	H460 cells	[91–93]

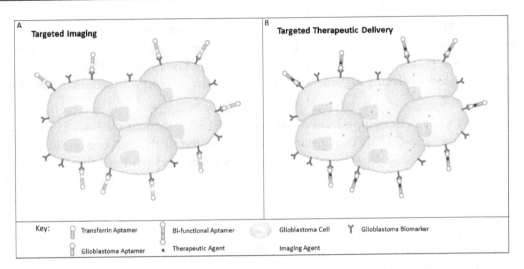

Figure 4. Schematic representation of aptamers for targeting glioblastoma tissue. (**A**) Targeted delivery of imaging agents to glioblastomas with the use of a bi-functional aptamer conjugated with an imaging agent such as radionuclides, may improve tumour detection; (**B**) Therapeutic agents can be conjugated to the bi-functional aptamer for the specific delivery to the glioblastoma cells, thereby reducing unwanted off-target effects and sparing the healthy surrounding tissues and throughout the rest of the body.

3.1. Aptamers Targeting Tumour Intitiating Cells

A relatively new concept in cancer biology is the discovery of tumour initiating cells (TIC), referred to as cancer stem cells, thought to be a driving force in tumour development and therapeutic resistance [94]. While a population of TIC in glioblastomas can be identified by the glycoprotein CD133, some CD133-negative cells in these tumours have the ability to self-renew, thus this marker may not be indicative of all stem cells within glioblastoma tumours [95]. It should be noted that the CD133 receptor, usually detected with the AC133 epitope monoclonal antibody, can become truncated, thereby hindering the antibody's access to its binding region and leading to false-negative CD133 expression results in cancer cell populations [96]. As such, CD133 expression should be confirmed with multiple techniques. Despite these controversies, CD133 is still an attractive therapeutic target that requires further investigation due to its presence within these tumours.

To develop aptamers against CD133, Shigdar et al. utilised differential cell-SELEX with HEK293T-CD133 transfected cells and normal HEK293T cells as a negative control, to ensure the aptamers were binding to the desired epitope [90]. The generation of RNA containing 2′-F-pyrimidine aptamers capable of binding to CD133 was confirmed using both CD133 positive, and negative cell lines, via flow cytometry and confocal microscopy. The CD133-A15 aptamer binds to the AC133 epitope, similar to that of the standard antibody used for this receptor's detection. In contrast, CD133-B19 does not bind to the AC133 epitope and therefore has potential to be used to detect cancer stem cells within glioblastoma tumours, as well as potentially deliver therapeutics to these target cells.

Differential cell-SELEX was also utilised to generate the aptamers A3 and A4 against glioblastoma TIC, by using cells extracted from human glioblastoma tissue xenografts [72]. The collected cells were sorted based on CD133 expression and used for positive and negative selection in all SELEX rounds, with the addition of human neural progenitor cells as an extra counter selection step to promote binding to a cancer surface marker. This selection process developed aptamers capable of specific binding and uptake to glioblastoma TIC, generating a potential method for their detection and treatment. Further characterisation of the identified aptamers is required to determine their molecular targets and determine their efficacy as diagnostic or therapeutic delivery agents.

3.2. Aptamers Targeting Tenascin-C

Tenascin-C is a large glycoprotein found in the extracellular matrix [73]. While normal adult tissues have little or no expression, higher levels are present during foetal development, wound healing, atherosclerosis, psoriasis, and tumour growth [97,98]. The high levels of tenascin-C expression in tumours is associated with angiogenesis, and may be an effective biomarker for diagnosis and treatment [73].

The aptamer TTA1 was developed via a crossover-SELEX protocol with U251 glioblastoma cells and purified human tenascin-C [73]. This aptamer was generated with RNA containing 2'-F-pyrimidine to induce resistance to nucleases in the blood, and was truncated prior to further modifications to improve aptamer stability and half-life with the substitution of 2'-OCH_3 purines, the addition of a thymidine cap at the 3' end, and a 5' amine as a conjugation site. In vitro testing determined the aptamer to bind to human tenascin-C with high affinity, with a 20-fold reduction in affinity to the mouse protein. The addition of the conjugate site on TTA1 enables the attachment of various molecules, ensuring its adaptability for various clinical applications. ^{99m}Tc radiolabelled-TTA1 administered intravenously to nude mice successfully targeted tumour xenografts of U251 glioblastoma cells and MDA-MB-435 breast cancer cells, and its rapid clearance from the blood and uptake into the tumours indicates TTA1 as a viable tumour imaging modality [66]. It should be noted that there are limitations with the use of animals as human disease models, as xenografts do not wholly replicate the disease that occurs in humans [99]. The use of cultured cells in xenografts can lead to tumours with genetic profiles differing greatly from patient tumours and therefore do not always accurately represent treatment outcomes in patients. The use of xenograft models for glioblastoma tumours completely fails to replicate the cancer's microenvironment, and indeed the ability of these aptamers to cross the blood brain barrier, as these xenografts are heterotopic. Despite these drawbacks, animal models still provide important information for translational drug development prior to experimental clinical trials, and give great insight into efficacy of newly developed imaging agents. It is vital that the TTA1 aptamer is tested in healthy animal models to determine if it is capable of specifically targeting tumour tissue and also asses its ability to cross the blood brain barrier. If TTA1 is not able to do so, this aptamer may be attached to a transferrin aptamer to ensure its successful transition into the brain. TTA1 must also be tested in animal models that provide a relevant clinical representation of glioblastoma, to ensure its effectiveness as both a diagnostic and therapeutic delivery agent.

Another aptamer capable of specifically binding to tenascin-C is GBI-10, a DNA aptamer generated against U251 glioblastoma cells. Radiolabeling GBI-10 with positron emission tomography (PET) isotopes, ^{18}F and ^{64}Cu, developed aptamers capable of imaging tenascin-C within U87MG glioblastoma and MDA-MB-435 breast cancer xenografts in mice [74]. The aptamer's rapid clearance hindered intratumoural uptake; although these properties are ideal for PET imaging agents, they would need to be modified for effective therapeutic delivery in in vivo systems. To improve GBI-10's stability and affinity, D-/L-isonucleotides and 2'-dI phosphoramidite were incorporated into the aptamer structure, and successfully increased the binding affinity to U251 cells in vitro [75]. Aptamer binding affinity and serum stability are vital characteristics that determine their efficacy as diagnostic or drug-delivery platforms. Modifications to aptamers to improve their half-life and targeting ability may be vital to ensure their success in clinical applications. Further testing is warranted in more appropriate glioblastoma animal models to prove GBI-10's capability to specifically target glioblastoma tissues in a clinically relevant in vivo system.

3.3. Aptamers Targeting Platelet Derived Growth Factor Receptor β

Overexpression and mutations of the platelet derived growth factor receptor (PDGFR) are associated with gliomagenesis and tumour progression [100]. In glioblastoma, deregulation of the PDGFR signaling cascade and overexpression of this receptor is frequently displayed, indicating it as a potential therapeutic target [101].

Cell-SELEX was employed to develop a 2′-F-pyrimidine containing RNA aptamer capable of recognising platelet derived growth factor receptor β (PDGFRβ) [76]. This aptamer, Gint4.T, was determined to internalise into the U87MG glioblastoma cells, indicating its ability to be developed as a targeted delivery platform of therapeutics to the tumour tissues. Binding of this aptamer to the PDGFRβ was found to interfere with the receptor's signaling cascade by preventing phosphorylation of the tyrosine kinase domains, thereby reducing cell proliferation and migration both in vitro and in athymic CD-1 nude mice tumour xenografts in vivo. To prove Gint4.T's success as a therapeutic delivery platform, this aptamer was conjugated with anti-microRNA-10b to counteract expression of the oncogenic microRNA-10b seen in glioblastoma cancer stem cells, and induced cellular differentiation, thereby preventing tumoursphere growth and development when combined with GL21.T conjugated with microRNA-37 [77]. Interestingly, Gint4.T was capable of crossing the blood brain barrier in a physiologically relevant in vitro model previously developed by Kumar and associates (see [102]), likely via receptor mediated transcytosis, although this is yet to be replicated in an in vivo model. The ability of this aptamer to cross the blood brain barrier and specifically target glioblastoma cancer stem cells may lead to an effective therapeutic against this normally evasive cell population, thereby improving prognosis and increasing survival times for glioblastoma patients.

3.4. Aptamers Targeting Axl

Axl is a receptor tyrosine kinase commonly overexpressed in glioblastomas and other systemic tumours, associated with cancer cell survival, angiogenesis, proliferation, invasion, and motility [103–105]. The GL21.T aptamer is capable of binding and inhibiting Axl signaling, and may serve as a novel therapeutic agent to control tumour cell growth and invasion driven by this oncogenic protein [78]. This aptamer reduced migration and invasion of A549 lung cancer cells and U87MG glioblastoma cells in vitro and in athymic CD-1 nude mice xenografts in vivo. As previously mentioned, GL21.T conjugated with microRNA-37 used in combination with microRNA-10b-conjugated Gint4.T, transformed tumourspheres into adherent-like cells, inhibiting their ability to grow in spheroid form [77]. GL21.T was also able to traverse the in vitro blood brain barrier model, if this aptamer is capable of crossing this barrier in vivo, it could potentially be developed as a therapeutic targeting agent in the clinical setting. However, previous attempts to inhibit PDGFR in clinical trials have failed due to tumour resistance and pathway compensatory mechanisms [79,80]. Therefore, GL21.T should be used as a therapeutic delivery platform, and used in combination with other targeted therapies to ensure complete tumour eradication.

3.5. Aptamers Targeting the Epidermal Growth Factor Receptor

EGFR is involved with important cellular functions including cell growth, differentiation, survival, and migration, and deregulation of its signaling cascade is a driving force of glioblastoma tumorigenesis [32,106]. An aptamer capable of targeting EGFR and the mutant variant EGFRvIII would serve to be an ideal candidate for use in glioblastoma therapy, as their overexpression and mutation are prevalent in the majority of glioblastoma tumours [32].

An anti-EGFR RNA aptamer, CL4, capable of binding to both the wildtype and mutant variant has previously been generated via differential cell-SELEX with lung cancer cell lines [91]. Upon binding to the target receptor, CL4 prevented tyrosine-phosphorylation in a time-dependent manner, decreasing cell viability, proliferation, migration, and invasion in vitro [91,92]. Whilst minimal in vivo work has been undertaken with CL4, promising results from a triple-negative breast cancer xenograft model in athymic CD-1 nude mice determined this aptamer was capable of reducing tumour size, without inducing toxicity in the treated mice [93]. CL4 displays a therapeutic effect against EGFR activation; however, EGFR inhibitors have failed to prove success as monotherapies for glioblastoma, as such, CL4 should be used as a shuttle for therapeutic or diagnostic agents to these tumours in order to have clinical success [107].

Differential cell-SELEX developed an aptamer, U2, with U87MG cells expressing EGFRvIII or wildtype EGFR as positive and negative selectors, respectively [81]. This aptamer's high affinity for the membrane receptors indicated its potential use as a diagnostic agent. Therefore, U2 was radiolabelled with [188]Re to investigate its ability to function as a molecular probe in EGFRvIII expressing glioblastoma xenografts in male BALB/c nude mice. This aptamer successfully accumulated in the target tumours, highlighting its potential to be used as an effective glioblastoma imaging tool. There is little evidence to date that suggests U2 can penetrate the blood brain barrier in these mice, as such, conjugation with a transferrin receptor aptamer would ensure effective transport into the brain, though this would need to be extensively investigated both in vitro and in vivo. Tan et al. created aptamer 32 via differential cell-SELEX, utilising U87MG-EGFRvIII cells as the target for aptamer selection and U87MG cells as a negative control [82]. Binding assays were performed to evaluate aptamer specificity, with aptamer 32 showing binding to the U87MG-EGFRvIII target cells, and no binding to U87MG cells or the kidney cell line HEK293. Pull-down assays and confocal microscopy concluded the aptamer's target was EGFRvIII, and the aptamer was translocated to the nucleus of target cells, indicating its potential to be used for the targeted delivery of therapeutics within glioblastoma tumours. A follow-on study conjugated c-Met small interfering RNA (siRNA) to aptamer 32, and successfully delivered the siRNA to target U87MG-EGFRvIII cells, inhibiting the target gene's expression and cell proliferation whilst inducing apoptosis [83]. Conjugation of aptamer 32 to quantum dots could verify the expression of EGFRvIII in various glioblastoma tissues, the presence of this receptor was confirmed with immunohistochemistry [84]. This novel imaging probe was also capable of crossing the blood brain barrier in vivo, and accumulated in tumour tissues within the brains of male nude mice. Analysis of mouse body weight and major organs determined that the quantum dot-aptamer conjugates did not induce toxicity in the animals, further highlighting this aptamer's potential for use in the clinical setting. Overall, these results indicate aptamer 32 as a candidate tumour imaging probe and as a therapeutic delivery platform for the treatment of glioblastoma, though further investigation is required to determine aptamer 32's mechanism of crossing the blood brain barrier, and to further determine its stability and specificity to glioblastoma tumours within the brain.

An alternative SELEX method was employed to develop a 2′-F-pyrimidine containing RNA aptamer against human EGFR protein [85]. This aptamer, E07, was determined to competitively bind to both the wildtype and mutant variant three of EGFR, hindering ligand binding and autophosphorylation of these receptors in vitro. E07 has since been utilised to successfully capture and isolate circulating tumour cells with the use of biochips in vitro, with potential to be used for cancer diagnosis and evaluation of metastasis in patients [86,87].

3.6. Targeting Nucleolin with a Guanine-Rich Oligonucleotide

Nucleolin is associated with ribosome biogenesis, and the overexpression of this protein in cancer cells is linked to cell division, angiogenesis and inhibiting apoptosis, and therefore, is a viable target for delivery of therapeutic agents to glioblastomas, or other forms of cancer [108,109]. The development and use of guanine-rich oligonucleotides (GRO) stems from their intriguing biological properties, notably, their enhanced uptake that could aid in the delivery of therapeutic agents to target tissues, although they are known to display non-specific effects [110,111]. AS1411 is a GRO, utilised for its antiproliferative effects and nucleolin targeting ability in vitro [112]. The efficacy of AS1411 as a cancer therapy was replicated in vivo, as intraperitoneal injections were able to significantly reduce pancreatic cancer xenograft growth in mice. In terms of using AS1411 as a treatment for glioblastoma, Luo et al. successfully delivered AS1411-conjugated paclitaxel nanoconjugates to U87MG-PMT48 orthotopic glioblastoma xenografts in BALB/c nude mice, significantly increasing the median survival time for the treated mice [113]. Despite these promising results, the clinical translation of AS1411 was not so successful, with evaluation as a therapeutic in a phase II trial for metastatic renal cell carcinoma, only one patient out of 35 responded to the treatment [114]. Rigorous evaluation of AS1411 as a glioblastoma treatment in physiologically relevant in vitro and in vivo models is vital to ensure successful transition into clinical trials, and the potential off-target uptake must be evaluated if AS1411 is to be used for targeted drug delivery in the future.

3.7. Glioblastoma Aptamers with Unknown Targets

Cell-SELEX can be advantageous for aptamer development as whole cells have a myriad of surface markers that aptamers could bind to, many which remain unidentified. Therefore, aptamers can be generated against unknown surface markers that can later be identified, improving our knowledge of glioblastoma and other cancers.

Cell-SELEX was employed to develop DNA aptamers using the A172 glioblastoma cell line [88]. The sequenced aptamers GMT 3, GMT 4, GMT 5, GMT 8, and GMT 9 bound to glioblastoma cells with high affinity, and were further characterised to determine if binding was specific to glioblastoma cell lines. GMT 4 and 8 bound to the CEM leukemia cells and breast cancer cells GMBJ1, indicating that these cells share a surface marker recognised by these aptamers. Whilst these aptamers did bind to the glioblastoma cells, they had increased affinities to one of each leukemic and breast cancer cell line by more than two-fold. As such, these aptamers may have potential for use as a targeted imaging or therapeutic delivery strategy in these cancers as well. Treatment with proteinase K and trypsin was performed to determine if the aptamer binding targets were membrane proteins. All aptamers lost binding affinity after proteinase K treatment, indicating all of the aptamers likely bind to proteins. Trypsin application lead to a slight decrease in GMT 3 and 5 binding, demonstrating their targets are resistant to trypsin cleavage. These results show that negative selection is vital to cell-SELEX to identify aptamers specific to a tumour type, and to date, no publications have indicated the targets of these aptamers.

Differential cell-SELEX generated aptamers GBM128 and 131 using U118-MG cells for positive selection and astroglial cells SVGp12 for negative selection [89]. These aptamers were found to bind to the U118-MG glioblastoma cells in vitro, and did not recognise the astroglial cell line used in the SELEX process. Further testing with paraffin embedded tissues determined these aptamer's ability to bind to glioblastoma and glioma tissues, with no binding observed in the astroglial tissues. Whilst these aptamers could not differentiate between glioblastoma and other glioma tumours, they did not bind to normal brain tissue, indicating their potential use in glioma diagnosis. Further investigation is required to determine the aptamers' targets and prove their efficacy for glioma diagnosis.

3.8. Circumventing the Blood Brain Barrier

Whilst there are a number of aptamers that have been generated against cancer cells or their surface markers, the majority have only been investigated in vitro. Although the results show good selectivity, specificity, and efficacy, the challenge will be to see if these aptamers can cross the blood brain barrier in clinical models. One way of rescuing these aptamers, should they fail, is to use a transferrin receptor aptamer for receptor-mediated transcytosis in order to give these aptamers access to the brain. This mechanism has been successfully utilised by Macdonald et al. with a bi-functional aptamer capable of transcytosing through an in vitro blood brain barrier model, and also confirmed to reach the brain of a healthy mouse model following intravenous injection [56]. The conjugation of aptamers seen in the previous paper serves as a proof of concept that combining a cancer-targeting aptamer to a transferrin receptor aptamer, can develop a bi-functional aptamer capable of transcytosing through the blood brain barrier, to specifically target tumours within the brain.

4. Conclusions

Brain cancers, particularly glioblastomas, have very poor five-year survival rates and there is an urgent need to develop new therapeutic strategies with superior mechanisms of action to the current available therapies. The issue of getting new therapies across the blood brain barrier has been investigated in a number of studies, and there are now several strategies available to actively transport therapeutics through the barrier in order to reach the cancer cells within the brain. Whilst some of these transport mechanisms are unlikely to provide the patient with a better standard of care and improved quality of life, there are some with potential to revolutionise the treatment of glioblastoma, and other brain disorders.

Aptamers possess many characteristics that make them an ideal novel therapeutic agent for the treatment of glioblastoma and other brain malignancies. Their target sensitivity, selective nature, ease of modification and low immunogenicity make them an ideal drug-delivery platform. The development and optimisation of aptamers capable of transcytosing through the blood brain barrier to specifically deliver imaging and chemotherapeutic agents to glioblastoma tissues have potential to revolutionise brain tumour treatment.

In conclusion, aptamers show great promise as novel agents for tumour detection and treatment, although ongoing investigations are required to ensure their effective clinical translation. The research performed to date indicates aptamers are an innovative diagnostic and therapeutic platform for glioblastomas. Furthermore, these novel nucleic acids are improving our understanding of cancer and glioblastoma biology.

Acknowledgments: We would like to thank the Australian Government for the support through an Australian Government Research Training Program Scholarship.

Author Contributions: Emma M. Hays, Wei Duan, and Sarah Shigdar conceived the idea, Emma Hays performed the literature search. Emma M. Hays and Sarah Shigdar wrote the manuscript, and Wei Duan and Sarah Shigdar edited the manuscript.

Abbreviations

ADCC	Antibody-dependent cellular cytotoxicity
ATP	Adenosine triphosphate
CDC	Complement-dependent cytotoxicity
DNA	Deoxyribose nucleic acid
EGFR	Epidermal growth factor receptor
EGFRvIII	Epidermal growth factor receptor variant three
ERBB2	Human epidermal growth factor receptor two
NF1	Neurofibromin one
NOS	Not otherwise specified
PDGFRA	Platelet derived growth factor receptor A
PDGFRβ	Platelet derived growth factor receptor β
PI3K	Phosphoinositide three-kinase
PTEN	Phosphatase and tensin homolog
SELEX	Systematic evolution of ligands by exponential enrichment
PET	Positron emission tomography
RB1	Retinoblastoma protein
RNA	Ribonucleic acid
siRNA	Small interfering RNA
TIC	Tumour initiating cells
TP53	Tumour suppressor p53
VEGF	Vascular endothelial growth factor

References

1. Hanif, F.; Muzaffar, K.; Perveen, K.; Malhi, S.M.; Simjee, S.U. Glioblastoma multiforme: A review of its epidemiology and pathogenesis through clinical presentation and treatment. *Asian Pac. J. Cancer Prev.* **2017**, *18*, 3–9. [PubMed]

2. Ostrom, Q.T.; Gittleman, H.; Xu, J.; Kromer, C.; Wolinsky, Y.; Kruchko, C.; Barnholtz-Sloan, J.S. Cbtrus statistical report: Primary brain and other central nervous system tumors diagnosed in the united states in 2009–2013. *Neuro Oncol.* **2016**, *18*, v1–v75. [CrossRef] [PubMed]

3. Cohen, A.L.; Holmen, S.L.; Colman, H. Idh1 and Idh2 mutations in gliomas. *Curr. Neurol. Neurosci. Rep.* **2013**, *13*, 345. [CrossRef] [PubMed]

4. Smith, C.; Ironside, J.W. Diagnosis and pathogenesis of gliomas. *Curr. Diagn. Pathol.* **2007**, *13*, 180–192. [CrossRef]

5. Louis, D.N.; Ohgaki, H.; Wiestler, O.D.; Cavenee, W.K. *Who Classification of Tumours of the Central Nervous System*, 4th ed.; IARC Press: Lyon, France, 2016.

6. Louis, D.N.; Perry, A.; Reifenberger, G.; von Deimling, D.; Figarella-Branger, W.; Cavenee, K.; Ohgaki, H.; Wiestler, O.D.; Kleihues, P.; Ellison, D.W. The 2016 world health organization classification of tumors of the central nervous system: A summary. *Acta Neuropathol.* **2016**, *131*, 803–820. [CrossRef] [PubMed]

7. Ohgaki, H.; Kleihues, P. Genetic pathways to primary and secondary glioblastoma. *Am. J. Pathol.* **2007**, *170*, 1445–1453. [CrossRef] [PubMed]

8. Molenaar, R.J.; Radivoyevitch, T.; Maciejewski, J.P.; van Noorden, C.J.F.; Bleeker, F.E. The driver and passenger effects of isocitrate dehydrogenase 1 and 2 mutations in oncogenesis and survival prolongation. *Biochim. Biophys. Acta Rev. Cancer* **2014**, *1846*, 326–341. [CrossRef] [PubMed]

9. Liu, A.; Hou, C.; Chen, H.; Zong, X.; Zong, P. Genetics and epigenetics of glioblastoma: Applications and overall incidence of Idh1 mutation. *Front. Oncol.* **2016**, *6*, 16. [CrossRef] [PubMed]

10. Szopa, W.; Burley, T.A.; Kramer-Marek, G.; Kaspera, W. Diagnostic and therapeutic biomarkers in glioblastoma: Current status and future perspectives. *BioMed Res. Int.* **2017**, *2017*. [CrossRef] [PubMed]

11. Ohgaki, H.; Kleihues, P. The definition of primary and secondary glioblastoma. *Clin. Cancer Res.* **2013**, *19*, 764–772. [CrossRef] [PubMed]

12. Eder, K.; Kalman, B. Molecular heterogeneity of glioblastoma and its clinical relevance. *Pathol. Oncol. Res.* **2014**, *20*, 777–787. [CrossRef] [PubMed]

13. Stupp, R.; Mason, W.P.; van den Bent, M.J.; Weller, M.; Fisher, B.; Taphoorn, M.J.B.; Belanger, K.; Brandes, A.A.; Marosi, C.; Bogdahn, U.; et al. Mirimanoff radiotherapy plus concomitant and adjuvant temozolomide for glioblastoma. *N. Engl. J. Med.* **2005**, *352*, 987–996. [CrossRef] [PubMed]

14. Oppenlander, M.E.; Wolf, A.B.; Snyder, L.A.; Bina, R.; Wilson, J.R.; Coons, S.W.; Ashby, L.S.; Brachman, D.; Nakaji, P.; Porter, R.W.; et al. An extent of resection threshold for recurrent glioblastoma and its risk for neurological morbidity. *J. Neurosurg.* **2014**, *120*, 846–853. [CrossRef] [PubMed]

15. Urbanska, K.; Sokolowska, J.; Szmidt, M.; Sysa, P. Glioblastoma multiforme—An overview. *Contemp. Oncol.* **2014**, *18*, 307–312.

16. Thomas, R.P.; Recht, L.; Nagpal, S. Advances in the management of glioblastoma: The role of temozolomide and mgmt testing. *Clin. Pharmacol.* **2013**, *5*, 1–9. [PubMed]

17. Jakola, A.S.; Gulati, S.; Weber, C.; Unsgard, G.; Solheim, O. Postoperative Deterioration in health related quality of life as predictor for survival in patients with glioblastoma: A prospective study. *PLoS ONE* **2011**, *6*, e28592. [CrossRef] [PubMed]

18. Omuro, A.; DeAngelis, L.M. Glioblastoma and other malignant gliomas: A clinical review. *JAMA* **2013**, *310*, 1842–1850. [CrossRef] [PubMed]

19. Silverstein, A.M. Paul Ehrlich's passion: The origins of his receptor immunology. *Cell. Immunol.* **1999**, *194*, 213–221. [CrossRef] [PubMed]

20. Carvalho, S.; Levi-Schaffer, F.; Sela, M.; Yarden, Y. Immunotherapy of cancer: From monoclonal to oligoclonal cocktails of anti-cancer antibodies: Iuphar review 18. *Br. J. Pharmacol.* **2016**, *173*, 1407–1424. [CrossRef] [PubMed]

21. Baker, M. Reproducibility crisis: Blame it on the antibodies. *Nature* **2015**, *521*, 274–276. [CrossRef] [PubMed]

22. Chames, P.; van Regenmortel, M.; Weiss, E.; Baty, D. Therapeutic antibodies: Successes, limitations and hopes for the future. *Br. J. Pharmacol.* **2009**, *157*, 220–233. [CrossRef] [PubMed]

23. Radom, F.; Jurek, P.M.; Mazurek, M.P.; Otlewski, J.; Jeleń, F. Aptamers: Molecules of great potential. *Biotechnol. Adv.* **2013**, *31*, 1260–1274. [CrossRef] [PubMed]

24. Maggi, E.; Vultaggio, A.; Matucci, A. Acute infusion reactions induced by monoclonal antibody therapy. *Expert Rev. Clin. Immunol.* **2011**, *7*, 55–63. [CrossRef] [PubMed]

25. Suntharalingam, G.; Perry, M.R.; Ward, S.; Brett, S.J.; Castello-Cortes, A.; Brunner, M.D.; Panoskaltsis, N. Cytokine storm in a phase 1 trial of the Anti-CD28 monoclonal antibody Tgn1412. *N. Engl. J. Med.* **2006**, *355*, 1018–1028. [CrossRef] [PubMed]

26. Descotes, J. Immunotoxicity of monoclonal antibodies. *mABs* **2009**, *1*, 104–111. [CrossRef] [PubMed]

27. Gan, H.K.; van den Bent, M.; Lassman, A.B.; Reardon, D.A.; Scott, A.M. Antibody–drug conjugates in glioblastoma therapy: The right drugs to the right cells. *Nat. Rev. Clin. Oncol.* **2017**, *14*, 695. [CrossRef] [PubMed]

28. Reichert, J.M. Antibodies to Watch in 2017. *mABs* **2017**, *9*, 167–181. [CrossRef] [PubMed]

29. Sarafraz-Yazdi, E.; Pincus, M.R.; Michl, J. Tumor-targeting peptides and small molecules as anti-cancer agents to overcome drug resistance. *Curr. Med. Chem.* **2014**, *21*, 1618–1630. [CrossRef] [PubMed]

30. Le Joncour, V.; Laakkonen, P. Seek & destroy, use of targeting peptides for cancer detection and drug delivery. *Bioorg. Med. Chem.* **2017**. [CrossRef]

31. Mousavizadeh, A.; Jabbari, A.; Akrami, M.; Bardania, H. Cell targeting peptides as smart ligands for targeting of therapeutic or diagnostic agents: A systematic review. *Colloids Surf. B Biointerfaces* **2017**, *158*, 507–517. [CrossRef] [PubMed]

32. Gan, H.K.; Cvrljevic, A.N.; Johns, T.G. The epidermal growth factor receptor variant III (EGFRvIII): Where wild things are altered. *FEBS J.* **2013**, *280*, 5350–5370. [CrossRef] [PubMed]

33. Xu, H.; Zong, H.; Ma, C.; Ming, X.; Shang, M.; Li, K.; He, X.; Du, H.; Cao, L. Epidermal growth factor receptor in glioblastoma. *Oncol. Lett.* **2017**, *14*, 512–516. [CrossRef] [PubMed]

34. Raymond, E.; Faivre, S.; Armand, J.P. Epidermal growth factor receptor tyrosine kinase as a target for anticancer therapy. *Drugs* **2000**, *60*, 15–23. [CrossRef] [PubMed]

35. Prados, M.D.; Chang, S.M.; Butowski, N.; DeBoer, R.; Parvataneni, R.; Carliner, H.; Kabuubi, P.; Ayers-Ringler, J.; Rabbitt, J.; Page, M.; et al. Phase II study of erlotinib plus temozolomide during and after radiation therapy in patients with newly diagnosed glioblastoma multiforme or gliosarcoma. *J. Clin. Oncol.* **2009**, *27*, 579–584. [CrossRef] [PubMed]

36. Van den Bent, M.J.; Brandes, A.A.; Rampling, R.; Kouwenhoven, M.C.; Kros, J.M.; Carpentier, A.F.; Clement, P.M.; Frenay, M.; Campone, M.; Baurain, J.F.; et al. Randomized phase II trial of erlotinib versus temozolomide or carmustine in recurrent glioblastoma: Eortc brain tumor group study. *J. Clin. Oncol.* **2009**, *27*, 1268–1274. [CrossRef] [PubMed]

37. Rich, J.N.; Reardon, D.A.; Peery, T.; Dowell, J.M.; Quinn, J.A.; Penne, K.L.; Wikstrand, C.J.; van Duyn, L.B.; Dancey, J.E.; McLendon, R.E.; et al. Phase II trial of gefitinib in recurrent glioblastoma. *J. Clin. Oncol.* **2004**, *22*, 133–142. [CrossRef] [PubMed]

38. Franceschi, E.; Cavallo, G.; Lonardi, S.; Magrini, E.; Tosoni, A.; Grosso, D.; Scopece, L.; Blatt, V.; Urbini, B.; Pession, A.; et al. Gefitinib in patients with progressive high-grade gliomas: A multicentre phase II study by gruppo italiano cooperativo di neuro-oncologia (GICNO). *Br. J. Cancer* **2007**, *96*, 1047–1051. [CrossRef] [PubMed]

39. Narita, Y. Drug review: Safety and efficacy of bevacizumab for glioblastoma and other brain tumors. *Jpn. J. Clin. Oncol.* **2013**, *43*, 587–595. [CrossRef] [PubMed]

40. Gilbert, M.R.; Dignam, J.J.; Armstrong, T.S.; Wefel, J.S.; Blumenthal, D.T.; Vogelbaum, M.A.; Colman, H.; Chakravarti, A.; Pugh, S.; Won, M.; et al. A randomized trial of bevacizumab for newly diagnosed glioblastoma. *N. Engl. J. Med.* **2014**, *370*, 699–708. [CrossRef] [PubMed]

41. Chinot, O.L.; Wick, W.; Mason, W.; Henriksson, R.; Saran, F.; Nishikawa, R.; Carpentier, A.F.; Hoang-Xuan, K.; Kavan, P.; Cernea, D.; et al. Bevacizumab plus radiotherapy-temozolomide for newly diagnosed glioblastoma. *N. Engl. J. Med.* **2014**, *370*, 709–722. [CrossRef] [PubMed]

42. Tipping, M.; Eickhoff, J.; Robins, H.I. Clinical outcomes in recurrent glioblastoma with bevacizumab therapy: An analysis of the literature. *J. Clin. Neurosci.* **2017**, *44*, 101–106. [CrossRef] [PubMed]

43. Abbott, N.J. Blood-brain barrier structure and function and the challenges for CNS drug delivery. *J. Inherit. Metab. Dis.* **2013**, *36*, 437–449. [CrossRef] [PubMed]

44. Gabathuler, R. Approaches to transport therapeutic drugs across the blood-brain barrier to treat brain diseases. *Neurobiol. Dis.* **2010**, *37*, 48–57. [CrossRef] [PubMed]

45. Patel, M.M.; Goyal, B.R.; Bhadada, S.V.; Bhatt, J.S.; Amin, A.F. Getting into the brain: Approaches to enhance brain drug delivery. *CNS Drugs* **2009**, *23*, 35–58. [CrossRef] [PubMed]

46. Baseri, B.; Choi, J.J.; Tung, Y.S.; Konofagou, E.E. Multi-modality safety assessment of blood-brain barrier opening using focused ultrasound and definity microbubbles: A short-term study. *Ultrasound Med. Biol.* **2010**, *36*, 1445–1459. [CrossRef] [PubMed]

47. Calias, P.; Banks, W.A.; Begley, D.; Scarpa, M.; Dickson, P. Intrathecal delivery of protein therapeutics to the brain: A critical reassessment. *Pharmacol. Ther.* **2014**, *144*, 114–122. [CrossRef] [PubMed]

48. Benjamin, S.B.; Kohman, R.E.; Feldman, R.E.; Ramanlal, S.; Han, X. Permeabilization of the blood-brain barrier via mucosal engrafting: Implications for drug delivery to the brain. *PLoS ONE* **2013**, *8*, e61694.

49. Sheikov, N.; McDannold, N.; Vykhodtseva, N.; Jolesz, F.; Hynynen, K. Cellular mechanisms of the blood-brain barrier opening induced by ultrasound in presence of microbubbles. *Ultrasound Med. Biol.* **2004**, *30*, 979–989. [CrossRef] [PubMed]

50. Miyake, M.M.; Bleier, B.S. The blood-brain barrier and nasal drug delivery to the central nervous system. *Am. J. Rhinol. Allergy* **2015**, *29*, 124–127. [CrossRef] [PubMed]

51. Aprili, D.; Bandschapp, O.; Rochlitz, C.; Urwyler, A.; Ruppen, W. Serious complications associated with external intrathecal catheters used in cancer pain patients: A systematic review and meta-analysis. *Anesthesiology* **2009**, *111*, 1346–1355. [CrossRef] [PubMed]

52. Chen, Y.; Dalwadi, G.; Benson, H.A. Drug delivery across the blood-brain barrier. *Curr. Drug Deliv.* **2004**, *1*, 361–376. [CrossRef] [PubMed]

53. Jones, A.R.; Shusta, E.V. Blood-brain barrier transport of therapeutics via receptor-mediation. *Pharm. Res.* **2007**, *24*, 1759–1771. [CrossRef] [PubMed]

54. Yu, Y.J.; Zhang, Y.; Kenrick, M.; Hoyte, K.; Luk, W.; Lu, Y.; Atwal, J.; Elliott, J.M.; Prabhu, S.; Watts, R.J.; et al. Boosting brain uptake of a therapeutic antibody by reducing its affinity for a transcytosis target. *Sci. Transl. Med.* **2011**, *3*, 84ra44. [CrossRef] [PubMed]

55. Pestourie, C.; Tavitian, B.; Duconge, F. Aptamers against extracellular targets for in vivo applications. *Biochimie* **2005**, *87*, 921–930. [CrossRef] [PubMed]

56. Macdonald, J.; Henri, J.; Goodman, L.; Xiang, D.; Duan, W.; Shigdar, S. Development of a bifunctional aptamer targeting the transferrin receptor and epithelial cell adhesion molecule (EPCAM) for the treatment of brain cancer metastases. *ACS Chem. Neurosci.* **2017**, *8*, 777–784. [CrossRef] [PubMed]

57. Yang, Y.; Yang, D.; Schluesener, H.J.; Zhang, Z. Advances in SELEX and application of aptamers in the central nervous system. *Biomol. Eng.* **2007**, *24*, 583–592. [CrossRef] [PubMed]

58. Tuerk, C.; Gold, L. Systematic evolution of ligands by exponential enrichment: RNA ligands to bacteriophage T4 DNA polymerase. *Science* **1990**, *249*, 505–510. [CrossRef] [PubMed]

59. Ellington, A.D.; Szostak, J.W. In vitro selection of RNA molecules that bind specific ligands. *Nature* **1990**, *346*, 818–822. [CrossRef] [PubMed]

60. Hasegawa, H.; Savory, N.; Abe, K.; Ikebukuro, K. Methods for improving aptamer binding affinity. *Molecules* **2016**, *21*, 421. [CrossRef] [PubMed]

61. Sefah, K.; Shangguan, D.; Xiong, X.; O'Donoghue, M.B.; Tan, W. Development of DNA aptamers using cell-SELEX. *Nat. Protoc.* **2010**, *5*, 1169–1185. [CrossRef] [PubMed]

62. Bunka, D.H.J.; Stockley, P.G. Aptamers come of age—At last. *Nat. Rev. Microbiol.* **2006**, *4*, 588–596. [CrossRef] [PubMed]

63. Hicke, B.J.; Stephens, A.W. Escort aptamers: A delivery service for diagnosis and therapy. *J. Clin. Investig.* **2000**, *106*, 923–928. [CrossRef] [PubMed]

64. Dougherty, C.A.; Cai, W.; Hong, H. Applications of aptamers in targeted imaging: State of the art. *Curr. Top. Med. Chem.* **2015**, *15*, 1138–1152. [CrossRef] [PubMed]

65. Keshtkar, M.; Shahbazi-Gahrouei, D.; Khoshfetrat, S.M.; Mehrgardi, M.A.; Aghaei, M. Aptamer-conjugated magnetic nanoparticles as targeted magnetic resonance imaging contrast agent for breast cancer. *J. Med. Signals Sens.* **2016**, *6*, 243–247. [PubMed]

66. Hicke, B.J.; Stephens, A.W.; Gould, T.; Chang, Y.F.; Lynott, C.K.; Heil, J.; Borkowski, S.; Hilger, C.S.; Cook, G.; Warren, S.; et al. Tumor targeting by an aptamer. *J. Nucl. Med.* **2006**, *47*, 668–678. [PubMed]

67. Rockey, W.M.; Huang, L.; Kloepping, K.C.; Baumhover, N.J.; Giangrande, P.H.; Schultz, M.K. Synthesis and radiolabeling of chelator-RNA aptamer bioconjugates with copper-64 for targeted molecular imaging. *Bioorg. Med. Chem.* **2011**, *19*, 4080–4090. [CrossRef] [PubMed]

68. Shigdar, S. What potential do aptamers hold in therapeutic delivery? *Ther. Deliv.* **2017**, *8*, 53–55. [CrossRef] [PubMed]

69. Chandola, C.; Kalme, S.; Casteleijn, M.G.; Urtti, A.; Neerathilingam, M. Application of aptamers in diagnostics, drug-delivery and imaging. *J. Biosci.* **2016**, *41*, 535–561. [CrossRef] [PubMed]

70. Keefe, A.D.; Pai, S.; Ellington, A. Aptamers as therapeutics. *Nat. Rev. Drug Discov.* **2010**, *9*, 537–550. [CrossRef] [PubMed]

71. Gijs, M.; Aerts, A.; Impens, N.; Baatout, S.; Luxen, A. Aptamers as radiopharmaceuticals for nuclear imaging and therapy. *Nucl. Med. Biol.* **2016**, *43*, 253–271. [CrossRef] [PubMed]

72. Kim, Y.; Wu, Q.; Hamerlik, P.; Hitomi, M.; Sloan, A.E.; Barnett, G.H.; Weil, R.J.; Leahy, P.; Hjelmeland, A.B.; Rich, J.N. Aptamer identification of brain tumor-initiating cells. *Cancer Res.* **2013**, *73*, 4923–4936. [CrossRef] [PubMed]

73. Hicke, B.J.; Marion, C.; Chang, Y.F.; Gould, T.; Lynott, C.K.; Parma, D.; Schmidt, P.G.; Warren, S. Tenascin-C Aptamers are generated using tumor cells and purified protein. *J. Biol. Chem.* **2001**, *276*, 48644–48654. [CrossRef] [PubMed]

74. Jacobson, O.; Yan, X.; Niu, G.; Weiss, I.D.; Ma, Y.; Szajek, L.P.; Shen, B.; Kiesewetter, D.O.; Chen, X. Pet imaging of tenascin-C with a radiolabeled single-stranded DNA aptamer. *J. Nucl. Med.* **2015**, *56*, 616–621. [CrossRef] [PubMed]

75. Li, K.; Deng, J.; Jin, H.; Yang, X.; Fan, X.; Li, L.; Zhao, Y.; Guan, Z.; Wu, Y.; Zhang, L.; et al. Chemical modification improves the stability of the DNA aptamer GBI-10 and its affinity towards tenascin-c. *Org. Biomol. Chem.* **2017**, *15*, 1174–1182. [CrossRef] [PubMed]

76. Camorani, S.; Esposito, C.L.; Rienzo, A.; Catuogno, S.; Iaboni, M.; Condorelli, G.; de Franciscis, V.; Cerchia, L. Inhibition of receptor signaling and of glioblastoma-derived tumor growth by a novel PDGFRβ aptamer. *Mol. Ther.* **2014**, *22*, 828–841. [CrossRef] [PubMed]

77. Esposito, C.L.; Nuzzo, S.; Kumar, S.A.; Rienzo, A.; Lawrence, C.L.; Pallini, R.; Shaw, L.; Alder, J.E.; Ricci-Vitiani, L.; Catuogno, S.; et al. A combined microrna-based targeted therapeutic approach to eradicate glioblastoma stem-like cells. *J. Control Release* **2016**, *238*, 43–57. [CrossRef] [PubMed]

78. Cerchia, L.; Esposito, C.L.; Camorani, S.; Rienzo, A.; Stasio, L.; Insabato, L.; Affuso, A.; de Franciscis, V. Targeting Axl with an high-affinity inhibitory aptamer. *Mol. Ther.* **2012**, *20*, 2291–2303. [CrossRef] [PubMed]

79. Jayson, G.C.; Parker, G.J.; Mullamitha, S.; Valle, J.W.; Saunders, M.; Broughton, L.; Lawrance, J.; Carrington, B.; Roberts, C.; Issa, B.; et al. Blockade of platelet-derived growth factor receptor-β by CDP860, a humanized, pegylated DI-FAB', leads to fluid accumulation and is associated with increased tumor vascularized volume. *J. Clin. Oncol.* **2005**, *23*, 973–981. [CrossRef] [PubMed]

80. Chee, C.E.; Krishnamurthi, S.; Nock, C.J.; Meropol, N.J.; Gibbons, J.; Fu, P.F.; Bokar, J.; Teston, L.; O'Brien, T.; Gudena, V.; et al. Phase II study of dasatinib (BMS-354825) in patients with metastatic adenocarcinoma of the pancreas. *Oncologist* **2013**, *18*, 1091–1092. [CrossRef] [PubMed]

81. Wu, X.; Liang, H.; Tan, Y.; Yuan, C.; Li, S.; Li, X.; Li, G.; Shi, Y.; Zhang, X. Cell-SELEX aptamer for highly specific radionuclide molecular imaging of glioblastoma in vivo. *PLoS ONE* **2014**, *9*, e90752. [CrossRef] [PubMed]

82. Tan, Y.; Shi, Y.S.; Wu, X.D.; Liang, H.Y.; Gao, Y.B.; Li, S.J.; Zhang, X.M.; Wang, F.; Gao, T.M. DNA aptamers that target human glioblastoma multiforme cells overexpressing epidermal growth factor receptor variant iii in vitro. *Acta Pharmacol. Sin.* **2013**, *34*, 1491–1498. [CrossRef] [PubMed]

83. Zhang, X.; Liang, H.; Tan, Y.; Wu, X.; Li, S.; Shi, Y. A U87-EGFRvIII cell-specific aptamer mediates small interfering RNA delivery. *Biomed. Rep.* **2014**, *2*, 495–499. [CrossRef] [PubMed]

84. Tang, J.; Huang, N.; Zhang, X.; Zhou, T.; Tan, Y.; Pi, J.; Pi, L.; Cheng, S.; Zheng, H.; Cheng, Y. Aptamer-conjugated pegylated quantum dots targeting epidermal growth factor receptor variant III for fluorescence imaging of glioma. *Int. J. Nanomed.* **2017**, *12*, 3899–3911. [CrossRef] [PubMed]

85. Li, N.; Nguyen, H.H.; Byrom, M.; Ellington, A.D. Inhibition of cell proliferation by an anti-EGFR aptamer. *PLoS ONE* **2011**, *6*, e20299. [CrossRef] [PubMed]

86. Wan, Y.; Liu, Y.; Allen, P.B.; Asghar, W.; Mahmood, M.A.I.; Tan, J.; Duhon, H.; Kim, Y.; Ellington, A.D.; Iqbal, S.M. Capture, isolation and release of cancer cells with aptamer-functionalized glass bead array. *Lab Chip* **2012**, *12*, 4693–4701. [CrossRef] [PubMed]

87. Wan, Y.; Mahmood, M.A.I.; Li, N.; Allen, P.B.; Kim, Y.; Bachoo, R.; Ellington, A.D.; Iqbal, S.M. Nanotextured substrates with immobilized aptamers for cancer cell isolation and cytology. *Cancer* **2012**, *118*, 1145–1154. [CrossRef] [PubMed]

88. Bayrac, A.T.; Sefah, K.; Parekh, P.; Bayrac, C.; Gulbakan, B.; Oktem, H.A.; Tan, W. In vitro selection of dna aptamers to glioblastoma multiforme. *ACS Chem. Neurosci.* **2011**, *2*, 175–181. [CrossRef] [PubMed]

89. Kang, D.; Wang, J.; Zhang, W.; Song, Y.; Li, X.; Zou, Y.; Zhu, M.; Zhu, Z.; Chen, F.; Yang, C.J. Selection of DNA aptamers against glioblastoma cells with high affinity and specificity. *PLoS ONE* **2012**, *7*, e42731. [CrossRef] [PubMed]

90. Shigdar, S.; Qiao, L.; Zhou, S.F.; Xiang, D.; Wang, T.; Li, Y.; Lim, L.Y.; Kong, L.; Li, L.; Duan, W. RNA aptamers targeting cancer stem cell marker CD133. *Cancer Lett.* **2013**, *330*, 84–95. [CrossRef] [PubMed]

91. Esposito, C.L.; Passaro, D.; Longobardo, I.; Condorelli, G.; Marotta, P.; Affuso, A.; de Franciscis, V.; Cerchia, L. A neutralizing RNA aptamer against EGFR causes selective apoptotic cell death. *PLoS ONE* **2011**, *6*, e24071. [CrossRef] [PubMed]

92. Camorani, S.; Crescenzi, E.; Colecchia, D.; Carpentieri, A.; Amoresano, A.; Fedele, M.; Chiariello, M.; Cerchia, L. Aptamer targeting EGFRvIII mutant hampers its constitutive autophosphorylation and affects migration, invasion and proliferation of glioblastoma cells. *Oncotarget* **2015**, *6*, 37570–37587. [CrossRef] [PubMed]

93. Camorani, S.; Crescenzi, E.; Gramanzini, M.; Fedele, M.; Zannetti, A.; Cerchia, L. Aptamer-mediated impairment of EGFR-integrin Avβ3 complex inhibits vasculogenic mimicry and growth of triple-negative breast cancers. *Sci. Rep.* **2017**, *7*, 46659. [CrossRef] [PubMed]

94. Singh, S.K.; Hawkins, C.; Clarke, I.D.; Squire, J.A.; Bayani, J.; Hide, T.; Henkelman, R.M.; Cusimano, M.D.; Dirks, P.B. Identification of human brain tumour initiating cells. *Nature* **2004**, *432*, 396–401. [CrossRef] [PubMed]

95. Brescia, P.; Ortensi, B.; Fornasari, L.; Levi, D.; Broggi, G.; Pelicci, G. CD133 is essential for glioblastoma stem cell maintenance. *Stem Cells* **2013**, *31*, 857–869. [CrossRef] [PubMed]

96. Schmohl, J.U.; Vallera, D.A. CD133, selectively targeting the root of cancer. *Toxins* **2016**, *8*, 165. [CrossRef] [PubMed]

97. Erickson, H.P.; Bourdon, M.A. Tenascin: An extracellular matrix protein prominent in specialized embryonic tissues and tumors. *Annu. Rev. Cell Biol.* **1989**, *5*, 71–92. [CrossRef] [PubMed]

98. Tiitta, O.; Virtanen, I.; Sipponen, P.; Gould, V. Tenascin expression in inflammatory, dysplastic and neoplastic lesions of the human stomach. *Virchows Arch.* **1994**, *425*, 369–374. [CrossRef] [PubMed]

99. Denayer, T.; Stöhr, T.; van Roy, M. Animal models in translational medicine: Validation and prediction. *New Horiz. Transl. Med.* **2014**, *2*, 5–11. [CrossRef]

100. Östman, A. PDGF receptors-mediators of autocrine tumor growth and regulators of tumor vasculature and stroma. *Cytokine Growth Factor Rev.* **2004**, *15*, 275–286. [CrossRef] [PubMed]

101. Kim, Y.; Kim, E.; Wu, Q.; Guryanova, O.; Hitomi, M.; Lathia, J.D.; Serwanski, D.; Sloan, A.E.; Weil, R.J.; Lee, J.; et al. Platelet-derived growth factor receptors differentially inform intertumoral and intratumoral heterogeneity. *Genes Dev.* **2012**, *26*, 1247–1262. [CrossRef] [PubMed]

102. Kumar, S.; Shaw, L.; Lawrence, C.; Lea, R.; Alder, J. P50developing a physiologically relevant blood brain barrier model for the study of drug disposition in glioma. *Neuro Oncol.* **2014**, *16*, vi8. [CrossRef]

103. Shieh, Y.-S.; Lai, C.-Y.; Kao, Y.-R.; Shiah, S.-G.; Chu, Y.-W.; Lee, H.-S.; Wu, C.-W. Expression of Axl in lung adenocarcinoma and correlation with tumor progression. *Neoplasia* **2005**, *7*, 1058–1064. [PubMed]

104. Zhang, Y.X.; Knyazev, P.G.; Cheburkin, Y.V.; Sharma, K.; Knyazev, Y.P.; Orfi, L.; Szabadkai, I.; Daub, H.; Keri, G.; Ullrich, A. Axl is a potential target for therapeutic intervention in breast cancer progression. *Cancer Res.* **2008**, *68*, 1905–1915. [CrossRef] [PubMed]

105. Hutterer, M.; Knyazev, P.; Abate, A.; Reschke, M.; Maier, H.; Stefanova, N.; Knyazeva, T.; Barbieri, V.; Reindl, M.; Muigg, A.; et al. Axl and *growth arrest-specific gene 6* are frequently overexpressed in human gliomas and predict poor prognosis in patients with glioblastoma multiforme. *Clin. Cancer Res.* **2008**, *14*, 130 138. [CrossRef] [PubMed]

106. Voldborg, B.R.; Damstrup, L.; Spang-Thomsen, M.; Poulsen, H.S. Epidermal growth factor receptor (EGFR) and EGFR mutations, function and possible role in clinical trials. *Ann. Oncol.* **1997**, *8*, 1197–1206. [CrossRef] [PubMed]

107. Padfield, E.; Ellis, H.P.; Kurian, K.M. Current therapeutic advances targeting EGFR and EGFRvIII in glioblastoma. *Front Oncol.* **2015**, *5*, 5. [CrossRef] [PubMed]

108. Ginisty, H.; Sicard, H.; Roger, B.; Bouvet, P. Structure and functions of nucleolin. *J. Cell Sci.* **1999**, *112 Pt 6*, 761–772. [PubMed]

109. Berger, C.M.; Gaume, X.; Bouvet, P. The roles of nucleolin subcellular localization in cancer. *Biochimie* **2015**, *113*, 78–85. [CrossRef] [PubMed]

110. Wu, C.C.N.; Castro, J.E.; Motta, M.; Cottam, H.B.; Kyburz, D.; Kipps, T.J.; Corr, M.; Carson, D.A. Selection of oligonucleotide aptamers with enhanced uptake and activation of human leukemia B cells. *Hum. Gene Ther.* **2003**, *14*, 849. [CrossRef] [PubMed]

111. Stein, C.A. The experimental use of antisense oligonucleotides: A guide for the perplexed. *J. Clin. Investig.* **2001**, *108*, 641–644. [CrossRef] [PubMed]

112. Bates, P.J.; Laber, D.A.; Miller, D.M.; Thomas, S.D.; Trent, J.O. Discovery and development of the G-rich oligonucleotide AS1411 as a novel treatment for cancer. *Exp. Mol. Pathol.* **2009**, *86*, 151–164. [CrossRef] [PubMed]

113. Luo, Z.; Yan, Z.; Jin, K.; Pang, Q.; Jiang, T.; Lu, H.; Liu, X.; Pang, Z.; Yu, L.; Jiang, X. Precise glioblastoma targeting by AS1411 aptamer-functionalized poly (L-γ-glutamylglutamine)-paclitaxel nanoconjugates. *J. Colloid Interface Sci.* **2017**, *490*, 783–796. [CrossRef] [PubMed]

114. Rosenberg, J.E.; Bambury, R.M.; van Allen, E.M.; Drabkin, H.A.; Lara, P.N., Jr.; Harzstark, A.L.; Wagle, N.; Figlin, R.A.; Smith, G.W.; Garraway, L.A.; et al. A phase II trial of AS1411 (a novel nucleolin-targeted DNA aptamer) in metastatic renal cell carcinoma. *Investig. New Drugs* **2014**, *32*, 178–187. [CrossRef] [PubMed]

11

Aptamers for DNA Damage and Repair

Maureen McKeague

Department of Health Sciences and Technology, ETH Zürich, Schmelzbergstrasse 9, 8092 Zurich, Switzerland; maureen.mckeague@hest.ethz.ch

Abstract: DNA is damaged on a daily basis, which can lead to heritable mutations and the activation of proto-oncogenes. Therefore, DNA damage and repair are critical risk factors in cancer, aging and disease, and are the underlying bases of most frontline cancer therapies. Much of our current understanding of the mechanisms that maintain DNA integrity has been obtained using antibody-based assays. The oligonucleotide equivalents of antibodies, known as aptamers, have emerged as potential molecular recognition rivals. Aptamers possess several ideal properties including chemical stability, in vitro selection and lack of batch-to-batch variability. These properties have motivated the incorporation of aptamers into a wide variety of analytical, diagnostic, research and therapeutic applications. However, their use in DNA repair studies and DNA damage therapies is surprisingly un-tapped. This review presents an overview of the progress in selecting and applying aptamers for DNA damage and repair research.

Keywords: aptamer; DNA damage; DNA repair; in vitro selection; SELEX; mutation; therapeutics

1. Introduction

While DNA was originally considered an extremely stable molecule, Tomas Lindahl determined that the nucleobases of DNA react slowly with water. This insight led him to discover the molecular machinery by which cells repair damaged DNA and won him, along with Aziz Sancar and Paul Modrich, the Nobel Prize in Chemistry in 2015 [1–3]. Indeed, our DNA is damaged on a daily basis by radiation, ultraviolet light and contaminants in our food and in our environment. DNA damage can lead to heritable mutations and the activation of proto-oncogenes. DNA damage is therefore a critical risk factor in cancer, aging and heritable diseases [4], and is the underlying basis for most frontline cancer therapies [5,6] (Figure 1).

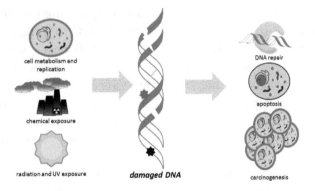

Figure 1. Our DNA is damaged by normal cell processes, contaminants in our food and environment, radiation and ultraviolet light. Damage may include strand-breaks, crosslinks (yellow line in DNA), or adducts (black and red stars). If not repaired, DNA damage can lead to cell death or heritable mutations and cancer.

Cells have a number of strategies to detect (e.g., damage checkpoints) and deal with (i.e., repair pathways) damage to DNA. If such detection and repair mechanisms are impaired, cells may experience genomic instability, apoptosis or senescence, and furthermore, predispose organisms to immunodeficiency, neurological disorders and cancer. Antibodies have been widely used as the molecular recognition platform of choice for the detection of DNA adducts and repair signaling and activation [7,8]. Over the past several decades, their use in DNA damage and repair research has facilitated the discovery of and insight into the mechanisms by which cells respond to DNA damage and initiate repair.

More recently, the nucleic acid analogues to antibodies, known as aptamers, have emerged. Since their discovery, aptamers have been compared to antibodies due to their similar ability to bind to specific targets. However, aptamers offer several broad advantages over antibodies as molecular recognition molecules (reviewed elsewhere [9,10]). Antibodies must be developed in vivo, whereas aptamers are composed of oligonucleotides which can be selected in vitro (using a process termed the systematic evolution of ligands by exponential enrichment, SELEX). As a result, aptamers can be selected against DNA-damaging toxins or the resulting lethal DNA adducts that are challenging for antibody generation. Furthermore, the nucleic acid composition affords aptamers with the ability to reversibly change conformation, making it possible to develop aptamers in conditions with varying pH, temperatures and ionic strengths that would cause antibodies to be irreversibly denatured. Finally, aptamers can be chemically modified during their synthesis to increase shelf life and nuclease resistance, as well as impart different chemical functionality (e.g., fluorescence and electrochemical properties) [11].

The unique combination of aptamer qualities listed above has led to a surge in the application of aptamers for analytical and diagnostic detection, therapeutics and drug delivery, intracellular imaging, and for gene regulation and control. With this increased interest in aptamers, coupled to a growing concern of the quality of commercial research antibodies [12], it is surprising that the application of aptamers to the study and detection of DNA damage and repair processes is limited. This review covers the aptamers that have been selected for DNA damage and repair proteins to-date and discusses the challenges in their selection and potential applications.

2. Aptamers for Damaged DNA

Damage to DNA includes oxidation, depurination or depyrimidation, single-strand or double-strand DNA breaks, deamination or alkylation. In this section, aptamers that can be potentially used to detect these types of DNA damage are highlighted.

2.1. Guanine Oxidation

The most abundant oxidatively-damaged base is 8-oxoguanine (8-oxoG.) When occurring in the genome, mismatched pairing of 8-oxoG with adenine results in G to T transversion mutations [13]. G to T mutations are often observed in oxidative stress-associated diseases, such as cancer, atherosclerosis, diabetes and pathologies of the central nervous system, as well as during aging. Therefore, it is not surprising that this damaged-nucleobase has been the focus of recent detection strategies [14], and the target of many aptamer-related selections. 8-oxoG can be further oxidized, leading to the formation of a variety of products, including the spiroiminodihydantoin (Sp) lesion. Unlike the 8-oxoG lesion, the Sp lesion is not planar. Instead, it is shaped like a propeller. Given this unusual shape, both Sp lesions strongly destabilize the DNA duplex [15] (Figure 2).

The first example of an aptamer that recognized DNA damage was reported in 1998 by Rink et al. Following ten rounds of conventional SELEX using an 8-oxodG affinity matrix, a specific RNA aptamer (Clone R10-B35) that exhibited highly specific binding to 8-oxodG compared to dG and other nucleosides was selected [16]. Using the electrophoretic mobility-shift assay (EMSA), the authors demonstrated that the aptamer bound to a single-strand DNA sequence containing a 3' terminal

8-oxodG with an apparent dissociation constant (K_d) of 270 nM. Surprisingly, the RNA aptamer also recognized 8-oxodG present in the center of a 19 nt ssDNA with an apparent K_d of 2.8 μM.

Almost ten years later, in 2009, the first DNA aptamer that recognized the oxidative lesion 8-oxodG was reported. To select these aptamers, Miyachi and co-workers used guanosine-monophosphate as an analog of 8-oxodG (due to the difficulty in immobilizing sufficiently high concentrations of 8-oxodG). The highest affinity aptamer that emerged from their selection was capable of binding to the free nucleoside with a K_d of 0.1 μM [17].

In 2012, an aptamer for 8-oxoG (the free base) was rationally designed. To achieve this, the thermal stability of nine 8-oxodG-containing hairpin DNA triplexes were compared. Next, the 8-oxoG moiety was removed from the two most stable triplexes. As a result, the abasic site allowed for the free oxidized base to bind in a specific manner [18]; thus creating a "rationally designed aptamer". The same group also tested several aliphatic side chain modifications at the abasic site to improve binding specificity. Introduction of a β-alanine side chain allowed selective binding of the 8-oxoG nucleobase with a dissociation constant of 5.5 μM [19].

Finally, in 2015, the Burrows group employed the recently described "structure-switching" SELEX [20] to isolate DNA aptamers for several products of guanine oxidation, including 8-oxodG and its nucleobase (8-oxo-G), the dSp nucleoside diastereomers: (−),-(R)-dSp and (+),-(S)-dSp and one of the Sp nucleobase enantiomers: (−),-(R)-Sp. The DNA aptamers resulting from this work bound to their respective targets with K_d values in the low nanomolar range, with the exception of the aptamer for 8-oxo-dG, which bound in the micromolar range (Table 1) [21].

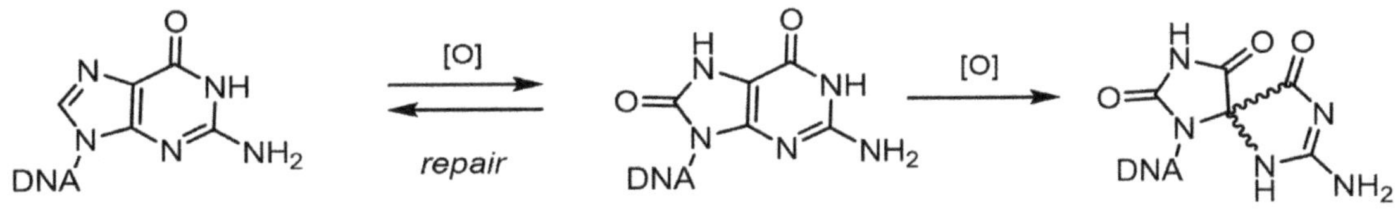

Figure 2. When guanine is oxidized, forming 8-oxoguanine (8-oxo-G), the resulting preferential basepairing to A ultimately leads to a G to T tranversion mutation. Further oxidation results in the spiroiminodihydantoin (Sp) adduct diastereomers, for example, which are highly destabilizing to the DNA duplex.

Table 1. Aptamers for DNA damage and repair targets.

Target Class	Target	Nucleic Acid	K_d [1]	Reference
	8-oxodG	RNA	270 nM	[16]
	8-oxodG	DNA	100 nM	[17]
	8-oxoG	DNA [2]	5.5 μM	[19]
	8-oxoG	DNA	3 nM	[21]
DNA	8-oxodG	DNA	25 μM	[21]
Adducts	(−),-(R)-dSp	DNA	28 nM	[21]
	(+),-(S)-dS	DNA	76 nM	[21]
	(−),-(R)-Sp	DNA	12 nM	[21]
	m⁷-GTP	RNA	500 nM	[22]
	benzylguanine	RNA	200 nM	[23]
	homopurine/pyrimidine duplex	RNA	1 μM	[24]
Strand	20 bp duplex	DNA [3]	43.9 nM	[25]
Breaks	3'LTR	RNA	300 nM	[26]
	Ku protein	RNA	2 nM	[27]

Table 1. *Cont.*

Target Class	Target	Nucleic Acid	K_d [1]	Reference
Repair Proteins	Fpg (DNA glycosylase)	RNA	2.5 nM	[28]
	Polβ/polκ	RNA	290 nM	[29]
	MutS	DNA	3.6 nM	[30]
	AlkB	DNA	20 nM	[31]
	AlkB homologue 2	DNA	85 nM	[32]
Mutated Gene	KRAS^{V12}	RNA	4.04 nM	[33]

[1] Only aptamers with K_d values are reported; for each, the best K_d is included. [2] With a β-alanine side chain. [3] Presence of benzoindoloquinolin required.

2.2. Guanine Alkylation

Adduct formation is the result of covalent binding between reactive electrophilic substances and nucleophilic sites (ring nitrogens and exocyclic oxygen atoms) of DNA bases. The N7 atom of guanine is the most vulnerable site for attack by alkylating agents [34]. Other common sites of DNA alkylation include the N^3 and N^1 positions of adenine, as well as the N^3 position of cytosine. Furthermore, even though O-alkyl lesions are generated to a much lesser extent than N-alkyl adducts, the induction of O^6-alkyl-G lesions is of significant interest because O^6-alkyl-G can readily mispair with thymine during DNA replication to cause mutagenic and cytotoxic biological effects.

While there are no reports of aptamers that directly bind to the adduct N^7-methylguanine (m^{7}dG), there are a few potential examples that might be explored in the future. As one example, Haller et al. reported the selection of RNA aptamers that bind to the methylated ribose, 7-methyl-guanosine (m^{7}G) which is typically part of the 5′ cap at the ends of mRNA transcripts in eukaryotes. The authors performed conventional SELEX with the goal of isolating aptamers capable of inhibiting translation of capped mRNA transcripts. The resulting aptamers bound to m^{7}G with modest affinity (0.5 μM) and high specificity, discriminating between non-methylated nucleotides by over 2000 fold. Interestingly, the presence of phosphate groups, or the identity of the purine group (e.g., adenine vs. guanine) had little effect on binding [22]. Therefore, it is possible that these aptamers may recognize the m^{7}dG adduct.

As another example, Larguinho and co-workers used an RNA aptamer that binds to xanthine [35] to develop nanoprobes that preferentially detected glycidamide (GA) adducts. GA is an epoxide metabolite of the genotoxic carcinogen acrylamide, and alkylates the N^7 position of guanine. Surprisingly, the authors were able to preferentially detect GA adducts compared to similar compounds (e.g., dGTP and glycidamide metabolites). While the sensor was capable of detecting the GA adducts, it was unfortunately ineffective in the presence of high concentrations of nucleotides [36].

Finally, Xu et al. developed an RNA aptamer that binds benzylguanine [23]. While benzylguanine is not a biologically-relevant DNA adduct, it is frequently used in place of O^6-methylguanine for studies [37]. The best selected aptamer displayed high affinity for the target (~200 nM), but most importantly, it exhibited a 20,000 fold selectivity compared to other guanine metabolites.

Together, these examples suggest that aptamers could be selected to recognize and bind to biologically-relevant alkylated guanine adducts in the future.

2.3. Double-Strand Breaks

The most dangerous forms of DNA damage are double-strand breaks (DSB). DNA DSBs occur when the two complementary stands of the double helix are simultaneously broken at locations that are close enough to one another, so that base pairing and chromatin structure cannot keep the two DNA ends together. As a consequence, repair is difficult and detrimental recombination with other sites in the genome may occur [38].

Despite their importance, there have been no reports of selecting aptamers to specifically recognize the termini of DSBs. This is expected to be difficult due to the electrostatic repulsion of the sugar–phosphate backbone of DNA that would prevent single strand aptamer binding to the DNA duplex. However, there have been several efforts in the past two decades to overcome the challenges presented by binding oligonucleotides to duplex DNA.

The Maher group reported the first example in 1996. The authors applied conventional SELEX against a 21 nt homopurine/homopyrimidine duplex DNA target. Following 26 rounds of selection under conditions that increased from pH 5.0–7.4, several aptamers emerged that bound to the duplex target with modest affinity (less than 1 μM) at pH 6, and approximately 10 μM at physiological pH [24]. Importantly, these selected aptamers approach the binding affinity of a 21 nt RNA oligonucleotide that forms a canonical triple helix with the duplex. Building on the success of this SELEX approach, selection of oligonucleotides to a duplex was performed again 13 years later, this time at neutral pH and in the presence of the triplex stabilizing agent, benzoindoloquinoline (BIQ). Following only seven rounds of selection, aptamers were capable of binding a 20 bp duplex with a K_d of 43.9 nM. However, in the absence of BIQ, no binding was observed [25]. Together, these examples indicate that there are strategies to reduce the electrostatic repulsion between DNA and potential aptamers; thus, potentially enabling future approaches for aptamer-based detection of DSBs.

As a final exciting example, Srisawat et al. performed SELEX using the 3′ long terminal repeat (LTR) of human immunodeficiency virus type 1. This 325 bp DNA duplex lacks a long polypurine/polypyrimiding tract and is therefore unlikely to favor triplex formation. The goal of this work was to identify aptamers that bind to the internal region of the LTR and thereby regulate its transcription. As such, several efforts were made to promote binding of the aptamer library to the internal region. The authors found that the selected RNA aptamers had a tendency to bind to the "ends" of the dsDNA; some aptamers were specific for the 3′ end while others recognized the 5′ end of the LTR [26]. This result suggests that aptamers might be selected to recognize the damaged ends of double-strand breaks.

3. Aptamers for Repair Proteins

Cells have unique molecular pathways to correct common types of DNA damage. The major pathways include non-homologous end joining, homologous recombination, mismatch repair, nucleotide excision repair, base excision repair and direct repair [39]. Below, the aptamers that bind to the repair proteins that mediate these repair pathways are described. The repair mechanisms and proteins involved in each pathway have been extensively reviewed by others (see for example [40,41]); however, a short overview is also provided. To our knowledge, there are no aptamers available for the components of homologous recombination repair pathways. Furthermore, there are no aptamers that specifically interact with proteins from transcription-coupled repair (TC-NER) involving the transcription factor TFIIH. However, a recent review describes many aptamers that bind to the transcription factor, TFIIA [42]. This suggests that the TC-NER pathway may be a suitable target for future aptamer development.

3.1. Non-Homologous End Joining

Double-strand breaks (DSBs) are repaired by the non-homologous end joining (NHEJ) and homologous recombination repair (HR) pathways. While the HR pathways require a homologous template, the NHEJ pathway repairs DSBs by directly ligating the ends. There are at least two genetically distinct sub-pathways of NHEJ: the classical-NHEJ (C-NHEJ) and alternative-NHEJ (A-NHEJ). The C-NHEJ pathway requires at least seven different proteins [43]. Two of these proteins, Ku70 and Ku86, form a heterodimer that functions as a molecular scaffold for the other NHEJ proteins to bind and initiate repair [44,45].

The first example of aptamer selection for a repair protein was performed in 1998 to the important scaffolding protein Ku. Here, Yoo et al. performed electrophoretic mobility shift assay (EMSA)–SELEX using an RNA library against the dimeric Ku protein purified from HeLa cell extracts. Their selection yielded 18 individual aptamers, each binding to the Ku protein with dissociation constants below

2 nM [27]. Excitingly, four of the aptamers were sufficiently selective that they were able to bind to the Ku protein in crude Hela cell extracts. Furthermore, these aptamers inhibited the binding and catalytic activity of the DNA dependent protein kinase catalytic subunit (DNA-PK), which is normally recruited to broken ends of DNA by the Ku protein [27].

3.2. Base Excision Repair

The base excision repair (BER) pathway is responsible for repairing small, non-helix-distorting damaged bases. BER is initiated by DNA glycosylases that recognize and remove the damaged base, creating an abasic site (AP site). Next, an AP endonuclease cleaves the AP site, and DNA polymerase β (polβ) removes the resulting 5'-deoxyribose phosphate via its 5' to 3' nuclease activity. Finally, the gap is filled by either short-patch (polβ replaces a single nucleotide) or long-patch BER (2–10 nucleotides are newly synthesized) [46,47].

There are currently two known examples of aptamers selected to BER-related proteins. First, as a proof-of-concept study, the Beal lab selected RNA aptamers to formamidopyrimidine DNA glycosylase (Fpg). Fpg (also known as 8-oxoguanine DNA glycosylase) is found in bacteria and repairs a wide range of oxidized purines. In this work, Vuyisich et al. used conventional SELEX with Fpg isolated from *E. coli* as the target. However, the authors were interested in isolating ligand-induced binding aptamers (i.e., those that only bind to the target in certain conditions). Therefore, the selections included a range of neomycin concentrations. As a result, the emerging aptamers could only bind to the target, Fpg, in the presence of the antibiotic, neomycin. Regardless, the best aptamer displayed high affinity to this repair protein, with a reported K_d of 7.5 nM [28]. Next, in 2006, Gening and co-workers performed seven rounds of selection to uncover RNA aptamers that bind to polβ isolated from *E. coli*. Upon further characterization, these aptamers bound to polβ with K_d values as low as 290 nM. Unfortunately, the aptamers did not display high specificity, binding also to polκ with similar affinity (K_d = 410 nM). More impressively, the aptamers were able to inhibit the activity of both polymerases in primer extension assays [29], highlighting potential applications for these aptamers.

3.3. Mismatch Repair

Mismatches in the genome can occur due to mis-incorporation during the replication process, or as a result of chemical damage to a complementary nucleobase. If the mismatches are not removed by the proofreading activities of the replisome, then the post-replicative "Mismatch Repair System" (MMRS) is activated. For simplicity, the process in *E. coli* is described; however, homologues of all these proteins are found in eukaryotes. This process is initiated by MutS, a protein that recognizes and binds to mispaired nucleotides. MutS then works together with MutL to direct the excision of the newly synthesized DNA strand by MutH [48]. This is followed by removal of the mismatch and subsequent re-synthesis by DNA polymerases [49].

The Krylov group has been using non-equilibrium capillary electrophoresis of equilibrium mixtures (NECEEM) SELEX [50] to identify many aptamers to various repair proteins. In 2006, NECEEM was first used to select aptamers to MutS from *Thermus aquaticus*. The isolated aptamers displayed dissociation constants as low as 3.6 nM to the isolated protein [30]. Other selections by the Krylov group have focused on DNA dealkylating proteins (see below).

3.4. Direct Repair

The simplest form of repairing DNA damage is direct repair, because cleavage of the phosphodiester backbone is not required. As a result, highly specialized proteins are involved for each type of damage [51]. Direct reversal is primarily used for correcting damage caused by DNA alkylating agents. For example, O^6-alkylguanine DNA alkyltransferase (AGT) is known to specifically reverse O^6-methylguanine back to guanine. As another example, the repair protein AlkB directly repairs N^1-methyladenine and N^3-methylcytosine base lesions [52,53].

In 2011, the Krylov group selected DNA aptamers using the NECEEM SELEX platform to the AlkB protein isolated from *E. coli*. The resulting aptamers displayed K_d values in the nanomolar range (as low as 20 nM) [31]. Later, these aptamers were shown to also inhibit AlkB activity by binding through an allosteric mechanism [54]. Next, the Krylov group was interested in obtaining aptamers to the human repair protein homologue, AlkB homologue 2 (ABH2). Original attempts failed due to challenges with the target (including instability and high positive charge). However, in 2014, NECEEM was coupled to emulsion PCR to efficiently amplify potential sequences. As a result, aptamers were isolated within three rounds that bound specifically to ABH2 with K_d values as low as 85 nM [32].

4. Aptamers That Recognize Mutated Gene Products

If DNA adducts are not repaired, mutations accumulate in the genome. When these mutations occur in oncogenes or tumor suppressor genes, it is possible that the mutations confer a growth advantage (driver mutations), thus resulting in the promotion of cancer [55]. Hot spot mutations may arise from either potentially damage-prone or repair-inaccessible locations in the genome. Ongoing research aims to determine the exact mechanism behind this selection preference [56]. Such mutations occur in tumor samples more frequently than the background [57], and have been identified in several genes. As a result, these mutated gene products provide some of the best-studied targets for chemotherapy [57], which explains why there have been some efforts towards isolating aptamers for these cancer driver gene products.

Codon 12 of the *KRAS* gene is the most frequently mutated codon in human cancers. As a result, many aptamers have been generated to mutant KRAS proteins and peptides [58,59]. In the most recent example, an RNA aptamer was generated that specifically bound to a mutant KRAS protein with a point mutation in codon 12 (KRAS^{V12}). Excitingly, binding to the wild-type KRAS was more than 50 fold lower than the mutant [33]. A second example is the *p53* gene which is considered the "guardian of the genome". *p53* is lost or mutated in about half of human cancer cases [60,61]. The single amino acid substitution p53R175H is one mutation which abolishes p53 function. In 2015, Chen et al. were able to isolate an RNA aptamer that binds to the p53 mutant p53R175H. Remarkably, this RNA aptamer (p53R175H-APT) also displayed a significantly stronger affinity to p53R175H than to the wild-type p53 in both in vitro and in vivo assays [62].

5. Selection Challenges and Considerations

The SELEX process involves iterative rounds of in vitro binding, partitioning and amplification (Figure 3) [63–65]. Despite the simplicity, a major advantage of the process is the flexibility in the enrichment strategy, binding conditions and nucleic acids design and type [66,67]. Due to this flexibility, aptamers have been selected to a wide range of targets, including whole cells, viruses, proteins and small molecules [68]. For reviews on the many modifications and improvements to the SELEX procedure over the past 25 years, see [69–71]. Here, conditions specific to DNA damage and repair targets are highlighted.

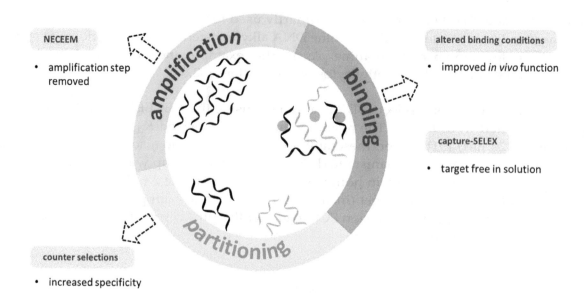

Figure 3. Conceptual representation of classic Systematic Evolution of Ligands by EXponential enrichment (SELEX) and important modifications. Classic SELEX consists of iterative rounds of binding, partitioning and PCR amplification. Single-stranded DNA or RNA libraries are incubated with the target-of-interest (blue circles). A partitioning step removes non-specific sequences (light grey strands). PCR amplification is then used to make multiple copies of the selected sequences (dark grey). Modifications to the classic SELEX process to isolate aptamers for DNA damage and repair targets include: the use of "capture-SELEX" for small molecules allowing them to be selected without immobilization; altered binding conditions to improve binding to strand breaks and improving activity in vivo; rigorous counter selection to ensure binding specificity; and the use of NECEEM for difficult protein targets.

5.1. DNA Adducts

The nucleobases of DNA have molecular weights ranging from approximately 110–150 g/mol. Nucleosides range from 240–285 g/mol, and nucleotides are around 500 g/mol. As a result, the selection targets for DNA damage aptamer libraries are very small, and therefore pose some of the same challenges as small molecule SELEX. Several reviews and methods highlight the conceptual and technical challenges in isolating aptamers to targets of less than 1000 g/mol [11,72]. This explains, in part, the relatively small number of different DNA adduct aptamers as compared to repair proteins, and is consistent with the general trend of fewer small molecule aptamers as compared to aptamers to large targets such as proteins and even cells [68].

The biggest potential break-through in addressing the challenges associated with small molecule aptamer selection was the development of Capture-SELEX, which yields structure-switching aptamers [73]. This method circumvents the needs to immobilize small molecules on a solid-support and further introduces a selection pressure for the selected aptamers to undergo a large structural rearrangement upon binding to the target. This feature is often desired in development detection applications with aptamers [74]. As a result, future aptamer selection efforts to damaged nucleobases and nucleosides should make further use of the Capture-SELEX strategy.

5.2. Strand Breaks

Overcoming electrostatic repulsion of the sugar–phosphate backbone of DNA typically requires the alteration of binding conditions in the selection of aptamers. This may include high concentrations of cations (to shield the charge), lowering the selection pH or adding triplex-stabilizing agents. However, a relatively unexplored strategy is the incorporation of non-natural nucleotides. One potential option would be the use of peptide nucleic acids (PNA), where the negatively charged phosphate backbone is replaced by a neutral amide backbone [75]. As a result, PNA can selectively bind and invade the DNA duplex [76]. There are several examples of DNA aptamers being synthesized as PNA or developing aptamer-PNA conjugates [77]. Furthermore, the Liu lab has described an in vitro selection and amplification system for peptide nucleic acids [78]. Therefore, future efforts should evaluate the use of PNA to improve the binding affinity and specificity of detecting strand breaks.

5.3. Proteins

There are several methods available for selecting aptamers to protein targets; each possessing their own unique advantages and difficulties. Regardless of the method employed, the most important consideration for aptamers in applications involving DNA damage and repair is ensuring specificity. This is particularly critical for targeting both repair proteins and mutated gene products. In the selections described for mutated genes (Section 4), counter selection rounds were imperative to ensure that the resulting aptamers did not bind to the non-mutated, wild-type proteins. Without including these counter measures, it is possible that the aptamers would also recognize and bind to the wild-type proteins. For example, in the polβ selection (Section 3.2), counter selections were not performed. As a result, the aptamers were able to additionally bind and inhibit polκ, a polymerase from a different family. Therefore, highly stringent counter selection steps should be performed to ensure the specificity of the isolated aptamers.

6. Promising Applications

There is a disproportionately large number of publications and patents describing aptamer applications compared to the number of publications describing new aptamers [79]. This trend is consistent across most areas of aptamer research including food safety [80], neuroscience [81], medicine [82] and gene control [83]. In contrast, the opposite appears to be true in the field of DNA damage and repair; there are fewer examples of aptamer applications compared to the number of selections (Table 1). Here, some of the promising future applications are described (Figure 4). The challenges that must be addressed to make these potential applications a reality are summarized.

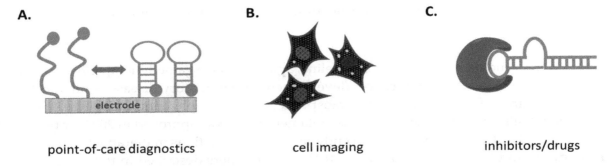

A.

electrode

point-of-care diagnostics

B.

cell imaging

C.

inhibitors/drugs

Figure 4. Conceptual figure highlighting potential applications of aptamers for DNA damage and repair. **(A)** Aptamers (green strands) could be incorporated into several platforms to create rapid analysis or point-of-care diagnostic kits to measured DNA lesion levels; **(B)** Aptamers combined with RNA tools such as "Spinach" could replace antibodies in cell imaging (allowing fluorescent imaging (bright dots) of damage/ repair proteins inside cells); **(C)** Highly specific aptamers (green strand) that inhibit repair proteins and polymerases (brown shape) could be used in cancer treatment and gene therapy.

6.1. Diagnostics

The detection of DNA adducts poses a major analytical challenge due to their very low abundance in genomic samples. As a result, methods must be both very sensitive and highly specific [84]. DNA lesion investigations from the 1980s were typically accomplished using the ^{32}P-postlabeling methodology. This method was capable of detecting lesions at frequencies as low as one lesion in 10^{10} nucleotides [85]. More recently, mass spectrometry (MS) coupled with liquid chromatography–electrospray ionization spectrometry (ESI-LC-MS) has been employed [86], and can currently be used to detect one lesion per 10^8–10^9 nucleotides. Regardless, these methods require significant quantities of purified DNA for quantifications. As a result, there have been several recent efforts toward specifically amplifying DNA adducts or mapping adducts within a genome. As an example, the Sturla group has developed a strategy to PCR-amplify DNA adducts of interest using mutated polymerases capable of specifically incorporating non-natural nucleosides [87]. As another example, the Burrows group has applied nanopore sequencing, and developed a biotin-labelling strategy to enrich and sequence oxidative damage [88,89]. These strategies are powerful, yet time-consuming and costly. Therefore, there is still an unmet need for simple, inexpensive methods that could be used as point-of-care diagnostic tools for decision-making about further DNA damage testing. Aptamers have been readily incorporated into nanoparticle- [90] and electrochemical-based [91] platforms to create point-of-care diagnostics. Therefore, the aptamers described in Table 1, could be easily used to rapidly screen or test for target DNA adducts.

6.2. Cellular Imaging

Immunofluorescent staining of DNA damage and the damage response has led to several important insights into DNA repair processes [92]. However, antibody staining must be performed after fixing of tissue or cell samples [93], and therefore cannot capture real-time or dynamic repair information. To image real-time processes inside cells, Jaffrey and co-workers developed "Spinach", another breakthrough aptamer technology [94]. Spinach, and several newer variants, is an RNA aptamer capable of specifically binding to a small molecule dye. Only upon interaction with the RNA aptamer does the dye fluoresce, creating a "light-up aptamer". These RNA aptamers have been encoded in the genomes of bacteria, yeast and mammalian cells to quantify, image and track specific RNA molecules in real-time [95,96]. More importantly, the Spinach aptamer can be directly coupled to a second aptamer, creating a system where fluorescence is observed only in the presence of the aptamer's target [97]. As a result, metabolites have been imaged and quantified in live cells (see review [98]). Therefore, it is feasible that the same strategy could be applied to imaging and quantification of DNA adducts and repair proteins in live cells and whole animals.

6.3. Therapeutic Targets

With several candidates in the clinical pipeline, aptamers have gained therapeutic visibility [99]. Aptamers have been used to impair cancer development, inflammatory disease, viral infection and cardiovascular illness [99]. However, the most successful example continues to be Pegaptanib, a vascular endothelial growth factor antagonist aptamer which was approved in 2004 for treatment of age-related macular degeneration [100]. Recently, aptamer applications in therapeutics have focused on cancer drugs and targeted drug delivery [101]. The aptamers described in this review could be leveraged for these applications.

In particular, alkylating agents are commonly used in chemotherapy due to their ability to cause DNA damage-induced apoptosis [102]. However, the efficiency of chemotherapeutic agents is strongly reduced by DNA repair systems. Aptamers that bind to the active site of repair proteins or polymerases involved in repair could be used to specifically inhibit their activities in cells [54]. This is a particularly exciting application for aptamers, compared to antibodies, given that aptamers are sufficiently small to fit inside tight binding pockets [27]. As one example, the aptamer selected by the Krylov group, not only bound to the NER repair protein AlkB, but also was capable of efficiently inhibiting catalysis at nanomolar concentrations [31].

As another possible direction, aptamers could be used to target specific cells or "turn on" drug activity in the presence of altered repair processes. It is known that repair processes are altered in cancer cells and that DNA damage and defects in DNA repair can both cause cancer [103]. Certainly, aptamers that recognize cancer cell surface proteins may be useful in delivering DNA damaging chemotherapeutics or DNA repair inhibitors. Alternatively, typical mutated proteins arising from DNA damage (or lack of repair) may be useful for ensuring chemotherapeutics are successfully distributed and/or activated once inside their target cells (in this case, cells that indicate increased DNA damage or altered repair). For example, the aptamer that binds to the mutated p53 protein was only able to interact with the mutated protein in cancer cells in vitro as well as in tumor xenografts [62], and had no impact on the wild-type p53 protein. Therefore, it is feasible that aptamers that bind to mutated gene products could be loaded with drugs that target repair processes, providing a means to target cells that are either cancerous or that may lead to further damage and mutation.

6.4. Application Roadblocks

Despite the exciting potential, it is clear that there have been very limited demonstrations of using aptamers in DNA damage and repair applications. This lack of utility is likely due to several technical and conceptual challenges. For diagnostic applications, perhaps the biggest roadblock is that there is still much to learn concerning the quantitative importance of DNA damage. For example, the linear dose response for genotoxicant-induced gene mutations and chromosomal damage has been challenged [104]. Furthermore, there is a gap in our understanding of the link between sequence-specific DNA adducts and mutational patterns [56,105,106]. Until these questions are addressed, it is not clear if there is a need for rapid point-of-care diagnostics or testing DNA adducts as a biomarkers in disease.

In contrast, cellular imaging with aptamers is a much-needed application, particularly in the field of DNA damage and repair. However, this application has only been very recently demonstrated. It is likely "only a matter of time" before in vivo or cellular imaging of DNA damage and repair processes with aptamers are demonstrated. One challenge is the internalization of the cellular imaging probes. However, researchers have developed many clever transport and delivery strategies to overcome this hurdle (see review [98]), and thus it is no longer a critical issue. However, a bottleneck that has not been tested with this set of aptamers is the in vivo activity. Unfortunately, many of the available aptamers were selected in conditions that are not physiologically relevant. In particular, many aptamers were selected using high concentrations of magnesium. This can be a limitation when genetically-encoding or delivering aptamers into the cellular environment, where free magnesium is much less abundant [107]. Therefore, it is possible that additional aptamer selections, under physiologically relevant conditions, may be needed to enable the successful use of aptamers in cellular imaging of DNA damage and repair.

The majority of the aptamers selected for DNA damage and repair proteins used proof-of-concept targets, including prokaryotic versions of repair proteins. These aptamers would not be useful for therapeutic applications (as they likely do not bind to the human orthologues), but demonstrate the potential for future aptamer selections targeting human repair proteins. In general, the main challenge in applying aptamers that recognize DNA damage and repair targets to therapeutic applications are the same obstacles for all aptamer therapeutic targets. These challenges include efficient internalization of the aptamers into the cell (and nucleus), off-target binding, half-life and stability in vivo and renal clearance. These difficulties, in combination with a general reluctance to divert from conventional antibody-based approaches, account for the long delay in the clinical translation and distribution of therapeutic aptamers [99]. As the number of aptamer modifications increases, these challenges will no longer hinder the therapeutic use of aptamers in general. Subsequently, it is expected that the therapeutic applications for aptamers targeting DNA damage and repair processes will also improve.

7. Conclusions

There are few examples of aptamers that bind to DNA adducts or modified nucleic acids, damage repair proteins and mutated gene products involved in carcinogenesis. However, there are many targets for which future aptamer selection would be beneficial. For example, there are no aptamers that bind to damage resulting from deamination or crosslinking. Furthermore, aptamers could be selected for a long list of repair proteins, including those involved in the NER and HR pathways. Despite the potential, direct evidence of the utility of aptamers in the field of DNA damage and repair is currently limited. Therefore, demonstration of these exciting potential applications should be performed. It is expected that several ongoing advances in the field of aptamers, particularly in the context of targeted therapeutics, will help address current challenges limiting the applications of aptamers in DNA damage and repair. Ultimately, this may motivate the researchers to undertake additional selections to increase the list of aptamers that bind to DNA adducts and repair proteins.

Acknowledgments: The author thanks Shana Sturla for support and insight. This work was supported by the Swiss National Science Foundation.

Abbreviations

ABH2	AlkB homologue 2
AGT	O^6-alkylguanine DNA alkyltransferase
BER	Base excision repair
BIQ	Benzoindoloquinoline
bp	Base pair
DSB	Double-strand break
dsDNA	Double-strand DNA
EMSA	Electrophoretic mobility shift assay
Fpg	Formamidopyrimidine DNA glycosylase
GA	Glycidamide
HR	Homologous recombination
K_d	Dissociation constant
LTR	Long terminal repeat
m^7-GTP	7-methylguanosine 5′-triphosphate
NECEEM	Non-equilibrium capillary electrophoresis of equilibrium mixtures
NER	Nucleotide excision repair
NHEJ	Non non-homologous end joining
N7-meG	N^7-methylguanine
8-oxoG	8-oxoguanine
PNA	Peptide nucleic acid
SELEX	Systematic evolution of ligands by EXPonential enrichment
Sp	Spiroiminodihydantoin

References

1.   Lindahl, T.; Andersson, A. Rate of chain breakage at apurinic sites in double-stranded deoxyribonucleic acid. *Biochemistry* **1972**, *11*, 3618–3623. [CrossRef] [PubMed]

2.   Lindahl, T.; Nyberg, B. Rate of depurination of native deoxyribonucleic acid. *Biochemistry* **1972**, *11*, 3610–3618. [CrossRef] [PubMed]

3.   Lindahl, T.; Nyberg, B. Heat-induced deamination of cytosine residues in deoxyribonucleic acid. *Biochemistry* **1974**, *13*, 3405–3410. [CrossRef] [PubMed]

4.   Knoch, J.; Kamenisch, Y.; Kubisch, C.; Berneburg, M. Rare hereditary diseases with defects in DNA-repair. *Eur. J. Dermatol.* **2012**, *22*, 443–455. [PubMed]

5.   Leone, G.; Pagano, L.; Ben-Yehuda, D.; Voso, M.T. Therapy-related leukemia and myelodysplasia: Susceptibility and incidence. *Haematologica* **2007**, *92*, 1389–1398. [CrossRef] [PubMed]

6.   Wood, M.E.; Vogel, V.; Ng, A.; Foxhall, L.; Goodwin, P.; Travis, L.B. Second malignant neoplasms: Assessment and strategies for risk reduction. *J. Clin. Oncol.* **2012**, *30*, 3734–3745. [CrossRef] [PubMed]

7. Santella, R.M.; Yang, X.Y.; Hsieh, L.L.; Young, T.L. Immunologic methods for the detection of carcinogen adducts in humans. *Prog. Clin. Biol. Res.* **1990**, *340C*, 247–257. [PubMed]

8. Shia, J.; Ellis, N.A.; Klimstra, D.S. The utility of immunohistochemical detection of DNA mismatch repair gene proteins. *Virchows Arch.* **2004**, *445*, 431–441. [CrossRef] [PubMed]

9. Chen, A.; Yang, S. Replacing antibodies with aptamers in lateral flow immunoassay. *Biosens. Bioelectron.* **2015**, *71*, 230–242. [CrossRef] [PubMed]

10. Toh, S.Y.; Citartan, M.; Gopinath, S.C.; Tang, T.H. Aptamers as a replacement for antibodies in enzyme-linked immunosorbent assay. *Biosens. Bioelectron.* **2015**, *64*, 392–403. [CrossRef] [PubMed]

11. McKeague, M.; Derosa, M.C. Challenges and opportunities for small molecule aptamer development. *J. Nucleic Acids* **2012**, *2012*, 748913. [CrossRef] [PubMed]

12. Voskuil, J.L. The challenges with the validation of research antibodies. *F1000Research* **2017**, *6*, 161. [CrossRef] [PubMed]

13. Suzuki, T.; Kamiya, H. Mutations induced by 8-hydroxyguanine (8-oxo-7,8-dihydroguanine), a representative oxidized base, in mammalian cells. *Genes Environ.* **2017**, *39*, 2. [CrossRef] [PubMed]

14. Fleming, A.M.; Ding, Y.; Burrows, C.J. Sequencing DNA for the oxidatively modified base 8-oxo-7,8-dihydroguanine. *Methods Enzymol.* **2017**, *591*, 187–210. [PubMed]

15. Alenko, A.; Fleming, A.M.; Burrows, C.J. Reverse transcription past products of guanine oxidation in rna leads to insertion of a and c opposite 8-oxo-7,8-dihydroguanine and a and g opposite 5-guanidinohydantoin and spiroiminodihydantoin diastereomers. *Biochemistry* **2017**, *56*, 5053–5064. [CrossRef] [PubMed]

16. Rink, S.M.; Shen, J.C.; Loeb, L.A. Creation of rna molecules that recognize the oxidative lesion 7,8-dihydro-8-hydroxy-2′-deoxyguanosine (8-oxodg) in DNA. *Proc. Natl. Acad. Sci. USA* **1998**, *95*, 11619–11624. [CrossRef] [PubMed]

17. Miyachi, Y.; Shimizu, N.; Ogino, C.; Fukuda, H.; Kondo, A. Selection of a DNA aptamer that binds 8-ohdg using gmp-agarose. *Bioorg. Med. Chem. Lett.* **2009**, *19*, 3619–3622. [CrossRef] [PubMed]

18. Zhang, Q.; Wang, Y.; Meng, X.; Dhar, R.; Huang, H. Triple-stranded DNA containing 8-oxo-7,8-dihydro-2′-deoxyguanosine: Implication in the design of selective aptamer sensors for 8-oxo-7,8-dihydroguanine. *Anal. Chem.* **2013**, *85*, 201–207. [CrossRef] [PubMed]

19. Roy, J.; Chirania, P.; Ganguly, S.; Huang, H. A DNA aptamer sensor for 8-oxo-7,8-dihydroguanine. *Bioorg. Med. Chem. Lett.* **2012**, *22*, 863–867. [CrossRef] [PubMed]

20. Stoltenburg, R.; Nikolaus, N.; Strehlitz, B. Capture-selex: Selection of DNA aptamers for aminoglycoside antibiotics. *J. Anal. Methods Chem.* **2012**, *2012*, 415697. [CrossRef] [PubMed]

21. Ghude, P.; Burrows, C.J. Structure-Switching Selex for Selection of Aptamers of Damaged Nucleosides and Nucleobases 8-Oxoguanine and Spiroiminodihydantoin. Ph.D. Thesis, The University of Utah, Salt Lake City, UT, USA, 2015.

22. Haller, A.A.; Sarnow, P. In vitro selection of a 7-methyl-guanosine binding rna that inhibits translation of capped mrna molecules. *Proc. Natl. Acad. Sci. USA* **1997**, *94*, 8521–8526. [CrossRef] [PubMed]

23. Xu, J.; Carrocci, T.J.; Hoskins, A.A. Evolution and characterization of a benzylguanine-binding rna aptamer. *Chem. Commun.* **2016**, *52*, 549–552. [CrossRef] [PubMed]

24. Soukup, G.A.; Ellington, A.D.; Maher, L.J., III. Selection of rnas that bind to duplex DNA at neutral ph. *J. Mol. Biol.* **1996**, *259*, 216–228. [CrossRef] [PubMed]

25. Ayel, E.; Escude, C. In vitro selection of oligonucleotides that bind double-stranded DNA in the presence of triplex-stabilizing agents. *Nucleic Acids Res.* **2010**, *38*, e31. [CrossRef] [PubMed]

26. Srisawat, C.; Engelke, D.R. Selection of RNA aptamers that bind hiv-1 ltr DNA duplexes: Strand invaders. *Nucleic Acids Res.* **2010**, *38*, 8306–8315. [CrossRef] [PubMed]

27. Yoo, S.; Dynan, W.S. Characterization of the RNA binding properties of ku protein. *Biochemistry* **1998**, *37*, 1336–1343. [CrossRef] [PubMed]

28. Vuyisich, M.; Beal, P.A. Controlling protein activity with ligand-regulated RNA aptamers. *Chem. Biol.* **2002**, *9*, 907–913. [CrossRef]

29. Gening, L.V.; Klincheva, S.A.; Reshetnjak, A.; Grollman, A.P.; Miller, H. Rna aptamers selected against DNA polymerase beta inhibit the polymerase activities of DNA polymerases beta and kappa. *Nucleic Acids Res.* **2006**, *34*, 2579–2586. [CrossRef] [PubMed]

30. Drabovich, A.P.; Berezovski, M.; Okhonin, V.; Krylov, S.N. Selection of smart aptamers by methods of kinetic capillary electrophoresis. *Anal. Chem.* **2006**, *78*, 3171–3178. [CrossRef] [PubMed]

31. Krylova, S.M.; Karkhanina, A.A.; Musheev, M.U.; Bagg, E.A.; Schofield, C.J.; Krylov, S.N. DNA aptamers for as analytical tools for the quantitative analysis of DNA-dealkylating enzymes. *Anal. Biochem.* **2011**, *414*, 261–265. [CrossRef] [PubMed]

32. Yufa, R.; Krylova, S.M.; Bruce, C.; Bagg, E.A.; Schofield, C.J.; Krylov, S.N. Emulsion pcr significantly improves nonequilibrium capillary electrophoresis of equilibrium mixtures-based aptamer selection: Allowing for efficient and rapid selection of aptamer to unmodified ABH2 protein. *Anal. Chem.* **2015**, *87*, 1411–1419. [CrossRef] [PubMed]

33. Jeong, S.; Han, S.R.; Lee, Y.J.; Kim, J.H.; Lee, S.W. Identification of rna aptamer specific to mutant kras protein. *Oligonucleotides* **2010**, *20*, 155–161. [CrossRef] [PubMed]

34. Fu, D.; Calvo, J.A.; Samson, L.D. Balancing repair and tolerance of DNA damage caused by alkylating agents. *Nat. Rev. Cancer* **2012**, *12*, 104–120. [CrossRef] [PubMed]

35. Kiga, D.; Futamura, Y.; Sakamoto, K.; Yokoyama, S. An RNA aptamer to the xanthine/guanine base with a distinctive mode of purine recognition. *Nucleic Acids Res.* **1998**, *26*, 1755–1760. [CrossRef] [PubMed]

36. Larguinho, M.; Santos, S.; Almeida, J.; Baptista, P.V. DNA adduct identification using gold-aptamer nanoprobes. *IET Nanobiotechnol.* **2015**, *9*, 95–101. [CrossRef] [PubMed]

37. Wyss, L.A.; Nilforoushan, A.; Eichenseher, F.; Suter, U.; Blatter, N.; Marx, A.; Sturla, S.J. Specific incorporation of an artificial nucleotide opposite a mutagenic DNA adduct by a DNA polymerase. *J. Am. Chem. Soc.* **2015**, *137*, 30–33. [CrossRef] [PubMed]

38. Scharer, O.D. Chemistry and biology of DNA repair. *Angew. Chem. Int. Ed. Engl.* **2003**, *42*, 2946–2974. [CrossRef] [PubMed]

39. Dexheimer, T.S. DNA repair pathways and mechanisms. In *DNA Repair of Cancer Stem Cells*; Mathews, L.A., Cabarcas, S.M., Hurt, E.M., Eds.; Springer: Dordrecht, The Netherlands, 2013; pp. 19–32.

40. Chatterjee, N.; Walker, G.C. Mechanisms of DNA damage, repair, and mutagenesis. *Environ. Mol. Mutagen.* **2017**, *58*, 235–263. [CrossRef] [PubMed]

41. Pei, D.S.; Strauss, P.R. Zebrafish as a model system to study DNA damage and repair. *Mutat. Res.* **2013**, *743–744*, 151–159. [CrossRef] [PubMed]

42. Mondragon, E.; Maher, L.J., III. Anti-transcription factor RNA aptamers as potential therapeutics. *Nucleic Acid Ther.* **2016**, *26*, 29–43. [CrossRef] [PubMed]

43. Fattah, F.; Lee, E.H.; Weisensel, N.; Wang, Y.; Lichter, N.; Hendrickson, E.A. Ku regulates the non-homologous end joining pathway choice of DNA double-strand break repair in human somatic cells. *PLoS Genet.* **2010**, *6*, e1000855. [CrossRef] [PubMed]

44. Wang, H.; Xu, X. Microhomology-mediated end joining: New players join the team. *Cell Biosci.* **2017**, *7*, 6. [CrossRef] [PubMed]

45. Durdikova, K.; Chovanec, M. Regulation of non-homologous end joining via post-translational modifications of components of the ligation step. *Curr. Genet.* **2016**, *63*, 591–605. [CrossRef] [PubMed]

46. Beard, W.A.; Prasad, R.; Wilson, S.H. Activities and mechanism of DNA polymerase beta. *Methods Enzymol.* **2006**, *408*, 91–107. [PubMed]

47. Idriss, H.T.; Al-Assar, O.; Wilson, S.H. DNA polymerase beta. *Int. J. Biochem. Cell Biol.* **2002**, *34*, 321–324. [CrossRef]

48. Fishel, R. Mismatch repair. *J. Biol. Chem.* **2015**, *290*, 26395–26403. [CrossRef] [PubMed]

49. Li, G.M. Mechanisms and functions of DNA mismatch repair. *Cell Res.* **2008**, *18*, 85–98. [CrossRef] [PubMed]

50. Musheev, M.U.; Krylov, S.N. Selection of aptamers by systematic evolution of ligands by exponential enrichment: Addressing the polymerase chain reaction issue. *Anal. Chim. Acta* **2006**, *564*, 91–96. [CrossRef] [PubMed]

51. Yi, C.; He, C. DNA repair by reversal of DNA damage. *Cold Spring Harb. Perspect. Biol.* **2013**, *5*, a012575. [CrossRef] [PubMed]

52. Johannessen, T.C.; Prestegarden, L.; Grudic, A.; Hegi, M.E.; Tysnes, B.B.; Bjerkvig, R. The DNA repair protein alkbh2 mediates temozolomide resistance in human glioblastoma cells. *Neuro Oncol.* **2013**, *15*, 269–278. [CrossRef] [PubMed]

53. Yi, C.; Chen, B.; Qi, B.; Zhang, W.; Jia, G.; Zhang, L.; Li, C.J.; Dinner, A.R.; Yang, C.G.; He, C. Duplex interrogation by a direct DNA repair protein in search of base damage. *Nat. Struct. Mol. Biol.* **2012**, *19*, 671–676. [CrossRef] [PubMed]

54. Krylova, S.M.; Koshkin, V.; Bagg, E.; Schofield, C.J.; Krylov, S.N. Mechanistic studies on the application of DNA aptamers as inhibitors of 2-oxoglutarate-dependent oxygenases. *J. Med. Chem.* **2012**, *55*, 3546–3552. [CrossRef] [PubMed]

55. Tokheim, C.J.; Papadopoulos, N.; Kinzler, K.W.; Vogelstein, B.; Karchin, R. Evaluating the evaluation of cancer driver genes. *Proc. Natl. Acad. Sci. USA* **2016**, *113*, 14330–14335. [CrossRef] [PubMed]

56. Adar, S.; Hu, J.; Lieb, J.D.; Sancar, A. Genome-wide kinetics of DNA excision repair in relation to chromatin state and mutagenesis. *Proc. Natl. Acad. Sci. USA* **2016**, *113*, E2124–E2133. [CrossRef] [PubMed]

57. Chang, M.T.; Asthana, S.; Gao, S.P.; Lee, B.H.; Chapman, J.S.; Kandoth, C.; Gao, J.; Socci, N.D.; Solit, D.B.; Olshen, A.B.; et al. Identifying recurrent mutations in cancer reveals widespread lineage diversity and mutational specificity. *Nat. Biotechnol.* **2016**, *34*, 155–163. [CrossRef] [PubMed]

58. Gilbert, B.A.; Sha, M.; Wathen, S.T.; Rando, R.R. RNA aptamers that specifically bind to a k ras-derived farnesylated peptide. *Bioorg. Med. Chem.* **1997**, *5*, 1115–1122. [CrossRef]

59. Tanaka, Y.; Akagi, K.; Nakamura, Y.; Kozu, T. RNA aptamers targeting the carboxyl terminus of kras oncoprotein generated by an improved selex with isothermal rna amplification. *Oligonucleotides* **2007**, *17*, 12–21. [CrossRef] [PubMed]

60. Pfeifer, G.P.; Denissenko, M.F.; Olivier, M.; Tretyakova, N.; Hecht, S.S.; Hainaut, P. Tobacco smoke carcinogens, DNA damage and p53 mutations in smoking-associated cancers. *Oncogene* **2002**, *21*, 7435–7451. [CrossRef] [PubMed]

61. Hang, B. Formation and repair of tobacco carcinogen-derived bulky DNA adducts. *J. Nucleic Acids* **2010**, *2010*, 709521. [CrossRef] [PubMed]

62. Chen, L.; Rashid, F.; Shah, A.; Awan, H.M.; Wu, M.; Liu, A.; Wang, J.; Zhu, T.; Luo, Z.; Shan, G. The isolation of an rna aptamer targeting to p53 protein with single amino acid mutation. *Proc. Natl. Acad. Sci. USA* **2015**, *112*, 10002–10007. [CrossRef] [PubMed]

63. Tuerk, C.; Gold, L. Systematic evolution of ligands by exponential enrichment: RNA ligands to bacteriophage t4 DNA polymerase. *Science* **1990**, *249*, 505–510. [CrossRef] [PubMed]

64. Ellington, A.D.; Szostak, J.W. In vitro selection of rna molecules that bind specific ligands. *Nature* **1990**, *346*, 818–822. [CrossRef] [PubMed]

65. Robertson, D.L.; Joyce, G.F. Selection in vitro of an RNA enzyme that specifically cleaves single-stranded DNA. *Nature* **1990**, *344*, 467–468. [CrossRef] [PubMed]

66. Jijakli, K.; Khraiwesh, B.; Fu, W.; Luo, L.; Alzahmi, A.; Koussa, J.; Chaiboonchoe, A.; Kirmizialtin, S.; Yen, L.; Salehi-Ashtiani, K. The in vitro selection world. *Methods* **2016**, *106*, 3–13. [CrossRef] [PubMed]

67. Sun, H.; Zu, Y. A highlight of recent advances in aptamer technology and its application. *Molecules* **2015**, *20*, 11959–11980. [CrossRef] [PubMed]

68. McKeague, M.; McConnell, E.M.; Cruz-Toledo, J.; Bernard, E.D.; Pach, A.; Mastronardi, E.; Zhang, X.; Beking, M.; Francis, T.; Giamberardino, A.; et al. Analysis of in vitro aptamer selection parameters. *J. Mol. Evol.* **2015**, *81*, 150–161. [CrossRef] [PubMed]

69. Djordjevic, M. Selex experiments: New prospects, applications and data analysis in inferring regulatory pathways. *Biomol. Eng.* **2007**, *24*, 179–189. [CrossRef] [PubMed]

70. Xi, Z.; Huang, R.; Deng, Y.; He, N. Progress in selection and biomedical applications of aptamers. *J. Biomed. Nanotechnol.* **2014**, *10*, 3043–3062. [CrossRef] [PubMed]

71. Catuogno, S.; Esposito, C.L. Aptamer cell-based selection: Overview and advances. *Biomedicines* **2017**, *5*, 49. [CrossRef] [PubMed]

72. Ruscito, A.; McConnell, E.M.; Koudrian, A.; Velu, R.; Mattic, C.; Hung, V.; McKeague, M.; DeRosa, M.C. In vitro selection and characterization of DNA aptamers to a small molecule target. *Curr. Protoc. Chem. Biol.* **2017**, *9*, 1–36.

73. Yang, K.A.; Pei, R.; Stojanovic, M.N. In vitro selection and amplification protocols for isolation of aptameric sensors for small molecules. *Methods* **2016**, *106*, 58–65. [CrossRef] [PubMed]

74. Nutiu, R.; Li, Y. Structure-switching signaling aptamers. *J. Am. Chem. Soc.* **2003**, *125*, 4771–4778. [CrossRef] [PubMed]

75. Nielsen, P.E.; Egholm, M.; Berg, R.H.; Buchardt, O. Sequence-selective recognition of DNA by strand displacement with a thymine-substituted polyamide. *Science* **1991**, *254*, 1497–1500. [CrossRef] [PubMed]

76. Peffer, N.J.; Hanvey, J.C.; Bisi, J.E.; Thomson, S.A.; Hassman, C.F.; Noble, S.A.; Babiss, L.E. Strand-invasion of duplex DNA by peptide nucleic acid oligomers. *Proc. Natl. Acad. Sci. USA* **1993**, *90*, 10648–10652. [CrossRef] [PubMed]

77. Scheibe, C.; Wedepohl, S.; Riese, S.B.; Dernedde, J.; Seitz, O. Carbohydrate-pna and aptamer-pna conjugates for the spatial screening of lectins and lectin assemblies. *Chembiochem* **2013**, *14*, 236–250. [CrossRef] [PubMed]

78. Brudno, Y.; Birnbaum, M.E.; Kleiner, R.E.; Liu, D.R. An in vitro translation, selection and amplification system for peptide nucleic acids. *Nat. Chem. Biol.* **2010**, *6*, 148–155. [CrossRef] [PubMed]

79. McKeague, M.; Derosa, M.C. Aptamers and selex: Tools for the development of transformative molecular recognition technology. *Aptamers Synth. Antib.* **2014**, *1*, 12–16.

80. McKeague, M.; Velu, R.; De Girolamo, A.; Valenzano, S.; Pascale, M.; Smith, M.; DeRosa, M.C. Comparison of in-solution biorecognition properties of aptamers against ochratoxin A. *Toxins* **2016**, *8*, 336. [CrossRef] [PubMed]

81. McConnell, E.M.; Holahan, M.R.; DeRosa, M.C. Aptamers as promising molecular recognition elements for diagnostics and therapeutics in the central nervous system. *Nucleic Acid Ther.* **2014**, *24*, 388–404. [CrossRef] [PubMed]

82. Kruspe, S.; Giangrande, P.H. Aptamer-sirna chimeras: Discovery, progress, and future prospects. *Biomedicines* **2017**, *5*, 45. [CrossRef] [PubMed]

83. Schneider, C.; Suess, B. Identification of rna aptamers with riboswitching properties. *Methods* **2016**, *97*, 44–50. [CrossRef] [PubMed]

84. Klaene, J.J.; Sharma, V.K.; Glick, J.; Vouros, P. The analysis of DNA adducts: The transition from (32)p-postlabeling to mass spectrometry. *Cancer Lett.* **2013**, *334*, 10–19. [CrossRef] [PubMed]

85. Phillips, D.H.; Arlt, V.M. Genotoxicity: Damage to DNA and its consequences. *EXS* **2009**, *99*, 87–110. [PubMed]

86. Guo, J.; Yun, B.H.; Upadhyaya, P.; Yao, L.; Krishnamachari, S.; Rosenquist, T.A.; Grollman, A.P.; Turesky, R.J. Multiclass carcinogenic DNA adduct quantification in formalin-fixed paraffin-embedded tissues by ultraperformance liquid chromatography-tandem mass spectrometry. *Anal. Chem.* **2016**, *88*, 4780–4787. [CrossRef] [PubMed]

87. Wyss, L.A.; Nilforoushan, A.; Williams, D.M.; Marx, A.; Sturla, S.J. The use of an artificial nucleotide for polymerase-based recognition of carcinogenic o^6-alkylguanine DNA adducts. *Nucleic Acids Res.* **2016**, *44*, 6564–6573. [CrossRef] [PubMed]

88. Riedl, J.; Ding, Y.; Fleming, A.M.; Burrows, C.J. Identification of DNA lesions using a third base pair for amplification and nanopore sequencing. *Nat. Commun.* **2015**, *6*, 8807. [CrossRef] [PubMed]

89. Ding, Y.; Fleming, A.M.; Burrows, C.J. Sequencing the mouse genome for the oxidatively modified base 8-oxo-7,8-dihydroguanine by og-seq. *J. Am. Chem. Soc.* **2017**, *139*, 2569–2572. [CrossRef] [PubMed]

90. McKeague, M.; Foster, A.; Miguel, Y.; Giamberardino, A.; Verdin, C.; Chan, J.Y.S.; DeRosa, M.C. Development of a DNA aptamer for direct and selective homocysteine detection in human serum. *RSC Adv.* **2013**, *3*, 24415–24422. [CrossRef]

91. Schoukroun-Barnes, L.R.; Glaser, E.P.; White, R.J. Heterogeneous electrochemical aptamer-based sensor surfaces for controlled sensor response. *Langmuir* **2015**, *31*, 6563–6569. [CrossRef] [PubMed]

92. Bennett, B.T.; Bewersdorf, J.; Knight, K.L. Immunofluorescence imaging of DNA damage response proteins: Optimizing protocols for super-resolution microscopy. *Methods* **2009**, *48*, 63–71. [CrossRef] [PubMed]

93. Luise, C.; Nuciforo, P. Immunohistochemistry protocol for γH2AX detection (formalin-fixed paraffin-embedded sections). *Protoc. Exch.* **2006**. [CrossRef]

94. Paige, J.S.; Wu, K.Y.; Jaffrey, S.R. RNA mimics of green fluorescent protein. *Science* **2011**, *333*, 642–646. [CrossRef] [PubMed]

95. Nilaratanakul, V.; Hauer, D.A.; Griffin, D.E. Development and characterization of sindbis virus with encoded fluorescent rna aptamer spinach2 for imaging of replication and immune-mediated changes in intracellular viral rna. *J. Gen. Virol.* **2017**, *98*, 992–1003. [CrossRef] [PubMed]

96. Huang, K.; Doyle, F.; Wurz, Z.E.; Tenenbaum, S.A.; Hammond, R.K.; Caplan, J.L.; Meyers, B.C. Fastmir: An RNA-based sensor for in vitro quantification and live-cell localization of small rnas. *Nucleic Acids Res.* **2017**, *45*, e130. [CrossRef] [PubMed]

97. McKeague, M.; Wong, R.S.; Smolke, C.D. Opportunities in the design and application of rna for gene expression control. *Nucleic Acids Res.* **2016**, *44*, 2987–2999. [CrossRef] [PubMed]

98. Alsaafin, A.; McKeague, M. Functional nucleic acids as in vivo metabolite and ion biosensors. *Biosens. Bioelectron.* **2017**, *94*, 94–106. [CrossRef] [PubMed]

99. Zhou, J.; Rossi, J. Aptamers as targeted therapeutics: Current potential and challenges. *Nat. Rev. Drug Discov.* **2017**, *16*, 440. [CrossRef] [PubMed]

100. Ng, E.W.; Shima, D.T.; Calias, P.; Cunningham, E.T., Jr.; Guyer, D.R.; Adamis, A.P. Pegaptanib, a targeted anti-vegf aptamer for ocular vascular disease. *Nat. Rev. Drug Discov.* **2006**, *5*, 123–132. [CrossRef] [PubMed]

101. Poolsup, S.; Kim, C.Y. Therapeutic applications of synthetic nucleic acid aptamers. *Curr. Opin. Biotechnol.* **2017**, *48*, 180–186. [CrossRef] [PubMed]

102. Soll, J.M.; Sobol, R.W.; Mosammaparast, N. Regulation of DNA alkylation damage repair: Lessons and therapeutic opportunities. *Trends Biochem. Sci.* **2017**, *42*, 206–218. [CrossRef] [PubMed]

103. Kelley, M.R.; Logsdon, D.; Fishel, M.L. Targeting DNA repair pathways for cancer treatment: What's new? *Futur. Oncol.* **2014**, *10*, 1215–1237. [CrossRef] [PubMed]

104. Thomas, A.D.; Fahrer, J.; Johnson, G.E.; Kaina, B. Theoretical considerations for thresholds in chemical carcinogenesis. *Mutat. Res. Rev. Mutat. Res.* **2015**, *765*, 56–67. [CrossRef] [PubMed]

105. Li, W.; Hu, J.; Adebali, O.; Adar, S.; Yang, Y.; Chiou, Y.Y.; Sancar, A. Human genome-wide repair map of DNA damage caused by the cigarette smoke carcinogen benzo[a]pyrene. *Proc. Natl. Acad. Sci. USA* **2017**, *114*, 6752–6757. [CrossRef] [PubMed]

106. Hu, J.; Selby, C.P.; Adar, S.; Adebali, O.; Sancar, A. Molecular mechanisms and genomic maps of DNA excision repair in escherichia coli and humans. *J. Biol. Chem.* **2017**, *292*, 15588–15597. [CrossRef] [PubMed]

107. McKeague, M.; Wang, Y.H.; Smolke, C.D. In vitro screening and in silico modeling of rna-based gene expression control. *ACS Chem. Biol.* **2015**, *10*, 2463–2467. [CrossRef] [PubMed]

12

Computational Methods for Modeling Aptamers and Designing Riboswitches

Sha Gong [1], **Yanli Wang** [2], **Zhen Wang** [2] and **Wenbing Zhang** [2,*]

[1]   Hubei Key Laboratory of Economic Forest Germplasm Improvement and Resources Comprehensive Utilization, Hubei Collaborative Innovation Center for the Characteristic Resources Exploitation of Dabie Mountains, Huanggang Normal University, Huanggang 438000, China; shagong@hgnu.edu.cn

[2]   Department of Physics, Wuhan University, Wuhan 430072, China; 13212746736@163.com (Y.W.), zhenwang@whu.edu.cn (Z.W.)

*   Correspondence: wbzhang@whu.edu.cn

Abstract: Riboswitches, which are located within certain noncoding RNA region perform functions as genetic "switches", regulating when and where genes are expressed in response to certain ligands. Understanding the numerous functions of riboswitches requires computation models to predict structures and structural changes of the aptamer domains. Although aptamers often form a complex structure, computational approaches, such as RNAComposer and Rosetta, have already been applied to model the tertiary (three-dimensional (3D)) structure for several aptamers. As structural changes in aptamers must be achieved within the certain time window for effective regulation, kinetics is another key point for understanding aptamer function in riboswitch-mediated gene regulation. The coarse-grained self-organized polymer (SOP) model using Langevin dynamics simulation has been successfully developed to investigate folding kinetics of aptamers, while their co-transcriptional folding kinetics can be modeled by the helix-based computational method and BarMap approach. Based on the known aptamers, the web server Riboswitch Calculator and other theoretical methods provide a new tool to design synthetic riboswitches. This review will represent an overview of these computational methods for modeling structure and kinetics of riboswitch aptamers and for designing riboswitches.

Keywords: riboswitch; aptamer; mRNA structure; gene regulation

---

1. Introduction

Many noncoding RNAs (ncRNAs) have been found to bear important and diverse biological functions in all domains of life, including catalysis, protection of genomes, and regulation of cell activities [1–3]. This diversity in biological functions is attributed to the remarkable structure variety accommodated by RNAs. Riboswitches, which present a fundamental example of ncRNAs, are involved in cellular regulation through vast structural rearrangement in response to the intracellular physical signals, such as metabolites [1,4,5] and ions [6–10]. Among previously validated riboswitches, metabolite-specific riboswitches are the most widespread and the nature of their ligands is well defined in most cases, except for several "riboswitch-like", presumably *cis*-acting, RNA structures for which no ligand has been found yet [11,12]. In order to function, most riboswitches usually consist of two domains: a conserved aptamer domain that is responsible for ligand binding, and an expression platform that converts changes in the aptamer domain into changes in gene expression. In contrast to these riboswitches, the S_{MK} (SAM-III) riboswitch is the one that can use single domain for both ligand binding and gene regulation [13–15]. Aptamer domains, which are typically 35~200 nucleotides in length [16–18], often form a ligand-binding pocket (the aptamer structure) to bind the ligands with high specificity. To date, according to the aptamer structures (Figure 1), more than 30 riboswitch classes have been found in all three kingdoms of life [19–21]. Due to their specificity and function as genetic regulators, riboswitches represent a novel class of molecular target for developing antibiotics and chemical tools [22]. Thus, a comprehensive understanding of riboswitches is important to facilitate the design for riboswitch-targeted drug, molecular robotics, and new molecular sensors.

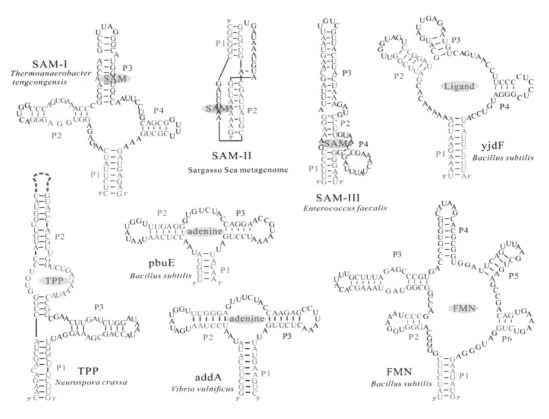

Figure 1.   Structural models of aptamers from several extensively studied riboswitch candidates. The structures can bind their nature ligands, such as *S*-adenosylmethionine (SAM), thiamine pyrophosphate (TPP), adenine and flavin mononucleotide (FMN), to form a ligand bound conformation. Except the SAM-II riboswitch from the Sargasso Sea metagenome [23,24] and TPP riboswitch aptamer from *Neurospora crassa* [25], other aptamers are from bacteria. Nucleotides within helices P1, P2, P3, P4, P5, and P6 found within the bound aptamers, are colored differently. The dash line denotes the long helix region in the structure of TPP riboswitch aptamer.

The signal-dependent conformational shifts of riboswitches, usually between two distinct functional states, i.e., ligand bound state and unbound state, regulate the downstream gene expression (Figure 2). One of the alternative states serves as the genetic off state (OFF state) by forming an intrinsic terminator hairpin or a repression stem to repress gene expression [26–28]. The other state acts as the genetic on state (ON state), which induces gene expression by preventing the formation of these regulatory elements. Riboswitches can also regulate RNA splicing by controlling the structural flexibility near the relevant splice site [25]. During the regulatory activities, one of the two structures is adopted by riboswitches, depending on whether the aptamer domain can form the pocket and bind its ligand on time or not. Therefore, to investigate the regulation mechanism of these functional ncRNAs, one of the major challenges is the information of the aptamer structure (Figure 2). In contrast to proteins, a much smaller number of RNA structures have been solved by using the traditional experimental methods [29], such as nuclear magnetic resonance (NMR) spectroscopy and X-ray crystallography [15,30]. Sensitive X-ray crystallography must take special care to avoid RNA aggregation and misfolding prior to crystal formation, and NMR experiments easily suffer from poor long-range correlations for RNA. These create a great demand to obtain information of RNA structures by using theoretical approaches. Up to now, many packages and methods have been developed to predict RNA secondary structure (two-dimensional (2D)) [31–33], as well as the three-dimensional (3D) structure for small RNAs [32–38]. These methods can quickly produce structure for a given RNA based on the input data. Some computational methods, such as RNAComposer and Rosetta [34,35,39], have been successfully applied to modeling 3D structure for several complex aptamers. As classical experimental methods may be limited in applicability to RNA [40], these theoretical approaches are likely to circumvent the bottleneck from experimental methods.

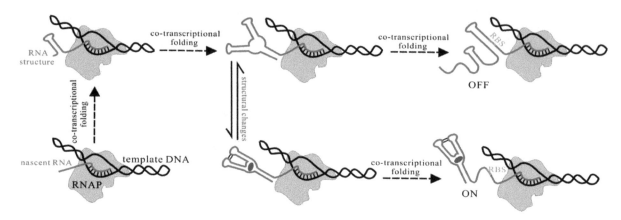

Figure 2. Schematic representation of riboswitch behaviors in cells. Nascent chain of the riboswitch and the ligand are colored red and blue respectively. The green line denotes the ribosome binding site (RBS) and RNA polymerase (RNAP) is denoted by yellow-green. The structure, structural changes, and co-transcriptional folding should be investigated for fully understanding the regulation mechanisms of riboswitches.

Another scientific challenge for the elucidation of the riboswitch function is to model aptamer kinetics, which is intrinsic to folding and conformational switching. Traditional molecular dynamics (MD) simulations [41,42], are able to provide a direct link between structure and dynamics. Nonetheless, due to the extreme complexity of force fields, a large number of atoms and a high number of degrees of freedom in RNA molecules, the detailed all of the atom simulations are difficult to produce trajectories in the time frame relevant for function of aptamers, which usually need to undergo large structural changes for purposes. Coarse-grained structural model with a less exhaustive representation is therefore particularly efficient to deal with these problems [43,44]. For example, the folding kinetics of $pbuE$ [45], $addA$, and S_{MK} riboswitch aptamers [43,46], have already been studied by using the coarse-grained SOP model. Another efficient alternative based on coarse-grained model is to investigate the kinetics of RNA 2D structure [47,48], which can also capture enough detail to understand the functions of aptamers.

Like protein, which can fold as soon as the N-terminal part emerges from the ribosome (co-translational folding) [49], nascent RNA also folds spontaneously as the nucleotides are synthesized by RNA polymerase in living cells (co-transcriptional folding) [50]. As riboswitches fold co-transcriptionally, folding patterns of the aptamer domain can direct the folding of the downstream expression platform. In fact, since many riboswitches regulate transcription, ligand binding can only occur during transcription for these to control gene expression [26,51–53]. Hence, the co-transcriptional folding kinetics is crucial for understanding the intracellular function of aptamers. Besides the optical-trapping assay and other experimental approaches [28,50,54], the kinetic Monte Carlo (MC) method was used to study the co-transcriptional folding of riboswitches [55], but it only considered the base pairing interactions closed in the native structures. By combining the master equation and the free energy landscape, BarMap and the helix-based computational methods have been applied to modeling the co-transcriptional folding behaviors for several riboswitches [56–60]. The results suggest that the aptamer domain folds into the pocket structure as soon as the nucleotides are transcribed [57,58], while the riboswitch without a separate aptamer domain is more likely to form an alternative structure instead of the pocket during the transcription [56]. These computation models, which can predict stable and metastable structures, kinetics, and transition states, bridge the gap in understanding the relationship between the structure and biological function of aptamers. Furthermore, computational RNA design has also made a great progress to construct synthetic riboswitches by using different strategies [61,62]. Here, we will provide a collection of these computational methods for modeling the structure and kinetics of aptamers, and for designing riboswitches.

2. Computational Method for Predicting Aptamer Structures

Since the structure of RNA determines its biological function, a complete understanding of the aptamer structure is the necessary prerequisite to understand the riboswitch-mediated regulation processes in the cell. RNAs fold into complex structures; the linear ribonucleotide sequence is the determinant of base-pairing interactions (2D structure), which in turn, determines the spatial shape (3D structure). Since most computational methods use the input of 2D structure to produce the RNA 3D structure [29,32,63], the precise prediction of 2D structure becomes more and more important. Early computational approaches for predicting RNA 2D structure (e.g., RNAfold [64]), only find the structure with the lowest free energy for a given RNA. As the functional structure may not be the one with the lowest energy, methods such as *mfold* and RNAsubopt [31,65] are developed to predict a set of low-energy structures. Other prediction methods, such as PPfold and RNAalifold [66,67], are based on evolutionary considerations. For a given aptamer, these methods can quickly produce 2D structure and the free energy. The methods for the modeling structure do not consist of a process that assumes co-transcriptional folding. Recently, RNA 2D structure prediction has been reinforced by incorporating the constraints from the experiments [65].

2.1. RNAComposer

RNA 2D structure is a crucial step in the functional characterization, but a thorough understanding of aptamer functions depends critically on the 3D structure, which is the key determinant of their interactions with ions and other molecules in cell. Based on the 2D structure tree graph representation and homology of structural elements, the RNAComposer method (Table 1) was developed to automatically predict the 3D structure for large RNAs [29]. As a knowledge-based method, it uses RNA sequence and 2D structure topology in dot-bracket notation as an input for 3D structure prediction (Figure 3). In this notation [64], unpaired nucleotides and the nucleotides that are involved in base pairs are represented by dots and brackets, respectively; square brackets and curly brackets refer to first-order pseudoknots and higher order structures, respectively. Although the input RNA 2D structure can be obtained by using the methods that are incorporated within the RNAComposer system: RNAfold, RNAstructure, and Contrafold, experimentally adjusted 2D structure is able to largely improve quality of the prediction [39].

The input RNA 2D structure first is divided into stems, loops, and single strands in the program. 3D structure elements corresponding to these fragmentations are searched within the structure elements dictionary, which is tailored from the RNA FRABASE database and consists of a 3D structural element with good structural properties [68]. After the searching process, the 3D elements whose heavy-atom root mean square deviation (RMSD) is lower than 1.0 Å relative to the parent PDB structure, are selected according to the 2D structure topology, sequence similarity, and so on. By merging the selected 3D structure elements, the initial RNA 3D structure is obtained, and then refined by the energy minimization in the torsion angle space and Cartesian atom coordinate space to get the final 3D structure. RNAComposer has been used to accurately build the 3D structure of several complex riboswitch families [39], such as FMN riboswitch aptamer, TPP, and purine riboswitch aptamers. It is offered at two sites: http://rnacomposer.ibch.poznan.pl and http://rnacomposer.cs.put.poznan.pl. By typing the 2D structure of THI riboswitch aptamer in Figure 3 on the website [69], the related 3D structure will be released in PDB formatted file to users within 10 s. RNAComposer is automated, efficient, especially suited for RNA 3D structure prediction of large RNAs, but it highly depends on the 3D structural elemental dictionary, and the applicability of the method is limited to RNAs with a few complex kink turn motifs.

Sequence
5'GGAGAGUAGAUGAUUCGCGUUAAGCGUGUG
UGAAUGGGAUGUCGUCACACAACGAAGCGA
GAGCGCGGUGAAUCAUUGCAUCCGCUCCA3'

Input
5' ((((....(((((((((((......(((((((...[[[....)))))))...
..((....)))))).)))))))))..]]]].)))). 3'

Output

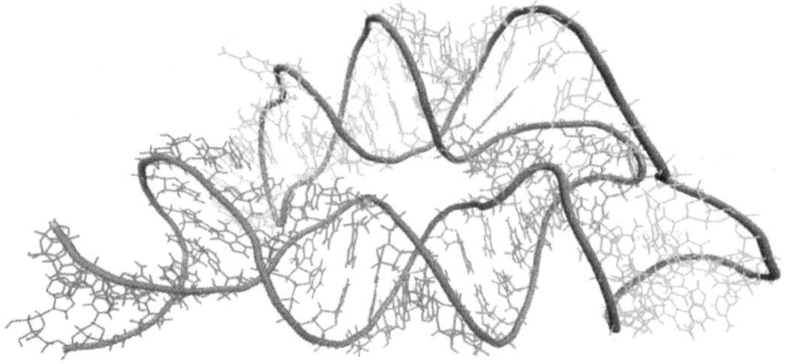

Figure 3. Three-dimensional (3D) structure for THF riboswitch aptamer (PDB code: 3SD3) predicted by using RNAComposer program online.

2.2. Rosetta and Discrete Molecular Dynamics

RNAComposer is motif library-based, while Rosetta is a fragment-based method that is available online [35]. In the Rosetta approach, according to the RNA 2D structure and the experimental proximity mapping data, low-resolution models are generated by using Fragment Assembly of RNA (FARNA) [34], where models are assembled using RNA fragments from a crystallographic database via a MC algorithm. The Rosetta all-atom energy function is then used to minimize a small number of the low-resolution models with the lowest Rosetta energy scores. To select a representative set of 3D models, the largest and lowest-energy subsets of models that fall within a certain RMSD threshold of each other are collected to reflect the native fold of the RNA. Taking the aptamer from *Fusobacterium nucleatum* double glycine riboswitch as an example, the Rosetta 3D modeling can give a similar prediction as RMdetect and JAR3D [70–73].

Table 1. Methods which have been used for modeling 3D structure for aptamers.

Methods	Description	Availability	Reference
RNAComposer	A motif library-based method that uses the dictionary tailored from RNA FRABASED database to build initial 3D structure.	Web server	[39]
Rosetta	A fragment-based method that uses FARFAR optimizes RNA conformations in the context of a physically realistic energy function.	Local installation	[72]
RMdetect	A bioinformatics tool for identifying known 3D structural modules on genomic sequences.	Local installation	[73]
JAR3D	Scoring sequences to motif groups based on sequences' ability to form the same pattern of interactions in motif.	Web server	[70]
RAGTOP	Predicting RNA topologies by a coarse-grained sampling of 3D graphs guided by statistical knowledge-based potentials.	Not available online	[74]
iFoldRNA	Incorporating SHAPE into discrete molecular dynamics to predict RNA structure.	Web server	[75]

Also, starting from 2D structures, a hierarchical computational approach (RAGTOP) was modified to predict 10 representative riboswitch aptamers with diverse structural features [74], by combining the coarse-grained graph sampling approach [76]. Through integration of computational and experimental methods, a three-bead coarse-grained model of RNA for discrete molecular dynamics simulation have gotten precise 3D structure predictions for the M-box riboswitch and TPP riboswitch aptamers [77], and for RNA in the range of a few hundred nucleotides [40]. In this coarse-grained model, each RNA nucleotide is represented by three beads corresponding to the base, sugar, and phosphate groups. The potential terms includes bonded, non-bonded interactions, and additional potential terms based on the experimental hydroxyl radical probing data. This method uses RNA sequences and base pairing as inputs to generate structures, and then applies replica exchange simulations with the potential to find a representative structure. Based on this method, the web server iFoldRNA was created for prediction of RNA 3D structure [75].

The 3D structures of aptamers modeled by these approaches suggest that the aptamer domains often form a compact structure involving many complicated tertiary interactions. Since the entire prediction of these approaches depends on the input data, the correct 2D structure is critical for the accurate 3D structure modeling. Experimental data provides powerful constraints to reinforce 2D structure prediction, but these methods currently can only achieve subhelix-resolution accuracy or near-atomic accuracy for RNAs.

3. Computation Model to Characterize Structural Changes in Aptamers

A key event in the biological function of riboswitches is the structural change within the aptamer domain. This change can lead to a change of the folding pattern within the expression platform, thereby directly modulating the gene expression. For effective flipping of riboswitches, the structural change of aptamers must be achieved within the certain time window. Therefore, characterizing the structural changes of aptamers is also important for fully understanding their function in the cell. The conventional all atom MD simulation has been widely used to describe the time-dependent motions of biological molecules [78]. However, even though modern parallelization of MD simulation is able to model trajectories on the order of milliseconds, it still fails to address the majority of biological processes, including folding or unfolding of aptamers that occur on much longer timescales (in seconds). In order to model time-dependent structural changes of aptamers, an effective solution is to use a coarse-grained structural model.

3.1. The Master Equation Approach

The master equation approach or kinetic MC method based on coarse-grained system [79–81], namely RNA 2D structure, is advantageous for accessing behaviors at long time scales, even minutes or hours [80]. In the master equation approach, structural changes are usually modeled on the RNA energy landscape, which specifies the conformation space and the transition rates between conformations. For a given RNA, the conformation space is sampled by all of the possible 2D structures that are constructed by using the formation or disruption of an entire helix as the elementary step to allow for large structural changes. Their free energies can be calculated using the Turner energy parameters [82,83]. The transition rates between states are obtained based on the free energy landscape analysis [48,80,81]: (1) formation; (2) disruption of a helix; and, (3) helix formation with concomitant partial melting of a competing helix. With these key concepts, the folding process can be described by the master equation:

$$dp_i(t)/dt = \sum_{i \neq i} [k_{j \to i} p_j(t) - k_{i \to j} p_i(t)] \qquad (1)$$

where $p_i(t)$ is the population of state i at time t and $k_{i \to j}$ is the transition rate from state i to state j. This computational method which integrates RNA 2D structure with the static energy landscape, provides a basic idea to predict the folding kinetics of the Hepatitis delta virus ribozyme and S_{MK} riboswitch [56,80].

3.2. Coarse-Grained SOP Model

Besides these approaches, the coarse-grained SOP model using Langevin dynamics simulation can also be used to study the kinetics of RNA molecules by characterizing the folding landscape [43,45,46]. Actually, many complex biological processes, such as unfolding and refolding of various RNA and proteins [44,84,85], are described with great success by using the SOP model. In this coarse-grained model, each nucleotide, as well as the ligand, is represented as a single site. The total potential energy of the bound aptamer is

$$V_T = V_{APT} + V_{APT-L} \tag{2}$$

where V_{APT} is the energy function of the aptamer; and, V_{APT-L} is the interaction between the ligand and the aptamer. The dynamics of the system can be simulated by using Brownian dynamics or the Langevin equation in the overdamped limit. During the force ramp simulation, the 5′-end of RNA is attached to a spring pulled with a constant speed, while the 3′-end is fixed. The free energy profiles are obtained using

$$G_z = -k_B T \ln P(z) \tag{3}$$

where k_B and T are the Boltzmann constant and temperature, respectively; and, the probability $P(z)$ of the extension between z and $z + dz$, is calculated from the folding trajectories. Based on the theory of mean first-passage times [84], the transition rate between two folding states can be calculated from the time traces of the extension.

The kinetics of the SAM-III and *addA* riboswitch aptamers have been successfully studied with this approach [43,46]. As the crystal structure of *pbuE* riboswitch aptamer is not available, its atomic structure is produced via conformational sampling with MD by substituting the sequence of *pbuE* aptamer into the crystal structure of *addA* aptamer (PDB code: 1Y26), when considering the structural similarity between the two aptamers. But despite this structural similarity, their folding behaviors are different in the pulling simulation [43,45]. In *addA* riboswitch aptamer (Figure 1), the unfolding occurs in the order of F → P2|P3 → P3 → U, while the unfolding order of *pbuE* aptamer is F → P2|P3 → P2 → U, where F and U denote the fully folded state and unfolded state, respectively; Pi is the hairpin structure with helix Pi; P2|P3 denotes the state with helices P2 and P3. The different unfolding order of P2 and P3 in the two aptamers suggests that the riboswitches carry out different regulatory activities in bacteria, even though they belong to the same class. Helix P3 in *pbuE* riboswitch is unfolded ahead of helix P2, because of its instability. This can explain why helix P3 is disrupted in OFF state of *pbuE* riboswitch but keeps folded in that of *addA* riboswitch [28,86]. Due to this greater conformational change in *pbuE* riboswitch, its OFF state can hardly transit to the aptamer structure, implying an irreversible kinetic riboswitch [57]. On the contrary, *addA* riboswitch is able to quickly reach equilibrium between the OFF state and the aptamer structure, which is consistent with a reversible thermodynamic switch [28]. The different unfolding kinetics of the aptamers under force thus can provide the information of their function.

4. Methods to Predict the Structure Transitions during Transcription

RNA folding occurs in two different ways [87]: (i) folding after synthesis of the entire RNA molecule (refolding); and, (ii) sequential folding occurs during transcription, namely co-transcriptional folding. In vivo, most RNAs fold co-transcriptionally, due to the sequential nature of RNA synthesis. The sequential folding during transcription is crucial for riboswitches, especially the kinetic riboswitches, to exert their regulation. Since kinetic switches are trapped in one state depending on whether the trigger is present at the time of folding, the mechanism of their function can only be understood in the context of co-transcriptional folding. For example, the full length *pbuE* riboswitch quickly folds to OFF state, which hinders adenine binding, while the aptamer structure that is responsible for ligand binding is not observed [53,57]. However, as nucleotides of the aptamer domain are transcribed first, the sequential folding may allow for the aptamer to form the pocket and bind to the ligand before formation of OFF state during transcription. Thus, the co-transcriptional folding kinetics of aptamers during transcription plays an important role for understanding their function in living cells.

In the case of co-transcriptional folding, as the RNA chain grows, the whole transcription process can be divided into a series of transcription steps [48,57,59], with each corresponding to adding one nucleotide. RNA folding kinetics within each step is modeled on the energy landscape as a certain RNA chain. But a link between two consecutive steps should be constructed due to the sequential folding in the transcription context. Based on this idea, the BarMap approach and helix-based computational approach have been developed to investigate the folding behaviors of several riboswitches under different transcription conditions [56–58,88].

4.1. BarMap

The BarMap approach integrates RNA 2D structure with the dynamic energy landscape to explore co-transcriptional folding kinetics [59]. The main idea of this approach is to model RNA kinetics on individual landscapes, where external triggers are considered as discrete changes. For successive kinetic simulations, a map between states of adjacent landscapes is computed to define the transfer of population densities.

In the case of transcription, the folding is treated as a process on a time-varying landscape in this approach. For each RNA elongation step, an energy landscape is first computed using the barriers program, which can simulate RNA folding kinetics [89,90]. Then, BarMap constructs a map between the landscapes of two successive steps, as shown in Figure 4a. According to this map, the final population at the previous step can be mapped to the initial population on the landscape at the current step. Finally, starting with the first landscape with more than one state, RNA folding kinetics can be simulated by the treekin program [59], which integrates the master equation for arbitrary times through calculating the matrix exponential. The amount of time t on a particular landscape corresponds to the elongation time of the polymerase. Co-transcriptional folding behaviors of RNAs can be obtained from the relationship between the population density and the transcription step that is given by the approach.

The BarMap approach has been used to study the RNA thermometer, RNA refolding during pore translocation, and co-transcriptional folding [59,88]. The effects of transcriptional pause sites, transcription rates and ligand concentrations, can easily be included in the approach by specifying the amount of time t and changing the binding energy added to all of the ligand-competent states, respectively. Using the example of a recently designed theophylline-dependent RS10 riboswitch, the BarMap approach predicts the folding behaviors in good agreement with experimental observations [88].

4.2. The Helix-Based Computational Method

In helix-based computation model [60], the folding time window for each RNA elongation step also depends on the time required for RNAP to transcribe the relevant nucleotide. At each step, the population kinetics is calculated in the same manner as the refolding kinetics that are calculated in the master equation approach [48,60]: first, the conformation space is generated and the transition rates are calculated; then, the population relaxation within the folding time window is described with the master equation, where the initial population at the current step is determined by the folding results of the previous step. Based on possible structural changes as the RNA chain is elongated by one nucleotide, a link between the initial population distribution at the current step and the final population distribution at the previous step is constructed in Figure 4b. Like the BarMap approach, the effects of the transcription speed, transcriptional pause, and ligand concentration can be mimicked by modifying the folding time window for each step and the binding energy in this method.

The helix-based computational method has been used to explore the regulation mechanisms of several well-studied riboswitches, such as *pbuE* [57], S_{MK} [56], *metF*, and *yitJ* riboswitches [58]. The results show an excellent agreement of predicted trajectories with that from experiments and other methods [50,55,91]. The folding behavior of *addA* riboswitch aptamer, as shown in Figure 5, suggests that as the chain grows, the nascent RNA chain folds through a series of discrete intermediate states (from S0 to S5). When the first 25th nucleotide is transcribed, the open RNA chain S0 begins to form structure S1, which is replaced by S2 from step on 32. As the 49th nucleotide is free to form structures, S3 is formed and occupies most of the population till the 59th step. From step on 59, S3 begins to transit to S4. When helix P1 can be nucleated, the chain folds into S5.

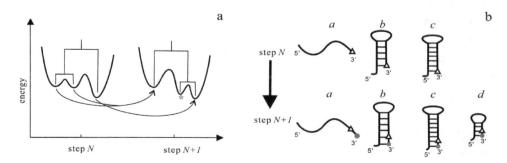

Figure 4. Relationship of landscapes in the BarMap approach (**a**) and structures in the helix-based computational method (**b**) between two adjacent steps. In (**a**), there are three types of events in landscapes: (i) A one-to-one correspondence between two minima (right); (ii) A two-to-one correspondence (left); and, (iii) a new local minima (marked by *) appears in the landscape at step $N + 1$. In (**b**), the triangles and red dots denote the newly transcribed nucleotide at step N and $N + 1$, respectively. For structures belonging to the first three types, their initial population at step $N + 1$ "directly inherit" from step N, while for type d, the initial population of the structure at step $N + 1$ is zero.

This aptamer domain can fold into the pocket S5 as soon as the relevant nucleotides are transcribed (Figure 5a,d), which has also been found in *pbuE* [57], *yitJ*, and *metF* riboswitches [58]. During the refolding process (Figure 5b,e), the entire molecule folds into S5 mainly through structure I and S3. Although the refolding pathway is different from the co-transcriptional folding pathway, the aptamer domain also can form S5 within 0.1 s, implying that the aptamer domain is highly evolved. In contrast to these riboswitch aptamers, the SAM-III riboswitch, which utilizes a single domain to exert functions, quickly folds to ON state instead of the pocket (OFF) structure (as shown in Figure 1) under both the transcription context and the refolding condition (Figure 5) [56]. This thermodynamic switch is not sensitive to co-transcriptional folding kinetics, so it can be understood by the equilibrium properties. The co-transcriptional folding pathway of *addA* aptamer is similar to that of the *pbuE* aptamer [57], possibly because of the conservation within the aptamer domains. For many riboswitches, since the expression platforms are required to form a terminator or a repression stem [1,4], their sequences could be helpful to decipher the different folding intermediates as well.

Previous studies suggest that transcriptional pause plays a key role for riboswitches and other RNAs to exert function [92–95]. The major transcriptional pause sites found within several riboswitches are located immediately after the aptamer domains [51,86,96]. As the time window that is allowed for ligands binding is limited during transcription, the pauses in these regions can give the aptamers extra time to bind to the ligands but prevent the unbound functional states from being formed. Their effects have been assessed by the helix-based computational method and experimental approaches [51,57,58,93]. The results from the helix-based computational method suggest that removing the pause sites leads to a demand for an even higher ligand concentration to trigger the switch [57,58].

The good agreement of the results from these theoretical approaches with the experiments and other methods implies that they provide a reliable tool to understand the function of aptamers in the riboswitch-mediated gene regulation. However, all of these approaches based on 2D structure prediction ignore tertiary interactions and the effects of ions. Although the ligand binding can be mimicked by modifying the binding energy from experiments, the 2D structure model still cannot precisely predict the ligand–RNA interactions. Hence, there is a significant requirement in incorporating these factors to fully understand the function of riboswitches.

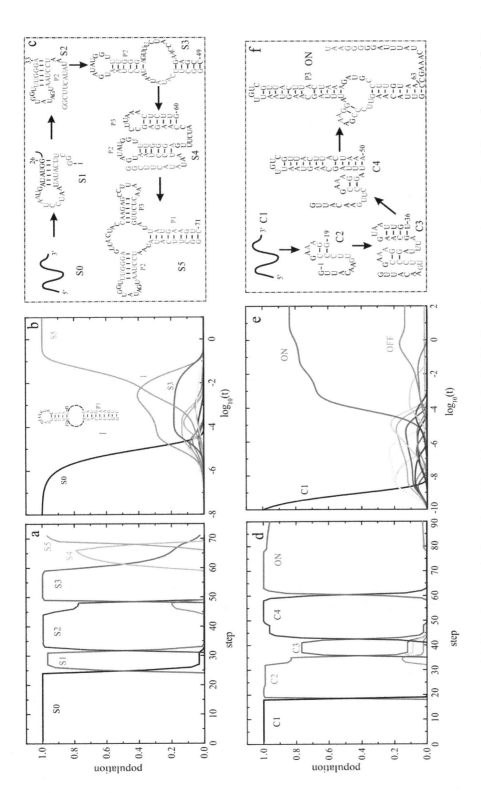

Figure 5. The co-transcriptional folding kinetics (**a,d**) and refolding kinetics (**b,e**) of *addA* riboswitch aptamer (up panel) and S_{MK} riboswitch (bottom panel). The transcription rates for *addA* aptamer and S_{MK} riboswitch are 25 nt/s and 50 nt/s, respectively, and the main intermediate states formed during the transcription are shown in (**c,f**). The dash dot lines in (**b**) for *addA* riboswitch aptamer denote the unpaired nucleotides within the large inter-loop. OFF state of S_{MK} riboswitch is the structure that is responsible for SAM binding shown in Figure 1.

5. Computation Design for Synthetic Riboswitches

Artificial riboswitches established recently demonstrate that they can be used as a new tool for the drug-regulated expression of viral genes [97]. To design artificial riboswitches or other RNA devices for industrial and medical applications, different computation strategies have been developed over the years [62,98–102]. Based on a biophysical model of translation initiation, a web server called Riboswitch Calculator can be used to design synthetic riboswitches form various RNA aptamers [62]. In this model, a riboswitch is considered as a long mRNA molecule. The interactions of RNA–RNA, RNA–ribosome and ligand–RNA, control the translation initiation rate, which can be calculated as:

$$r = \exp(-\beta \Delta G_{total}) \tag{4}$$

where β, ΔG_{total} are the apparent Boltzmann coefficient and total energy change between mRNA and initiating 30S ribosomal subunit. The energies of RNA folding and the ribosome binding to mRNA are calculated using the ViennaRNA suite and RBS Calculator [64,103], respectively.

Riboswitch Calculator provides a useful tool to design translation regulating riboswitches. The four inputs are necessary for the design: the sequence and structural constraint of an aptamer, the ligand binding energy, and the protein coding section. These inputs are used by the optimization algorithm to generate an initial set of riboswitch candidates. Rounds of random mutations, evaluation, selection, and recombination are performed using these candidates. The biophysical model calculations are employed for the fitness evaluation to select the candidates that meet the objective function requirements, such as reaching targeted translation rates in OFF and ON states. The selected candidates will be sent back for the next round to find the riboswitch candidates with the highest fitness.

Transcription regulating riboswitches can also be designed by using the computational method [61]. The design algorithm used the aptamer with the known sequence and secondary structure. When considering the terminators always with a minimal size and the U stretch, sequences with lengths between 6 and 20 nucleotides are randomly generated to create a spacer database. The following step is to generate the sequences that are complementary to the subsequences of the 3′ part of the given aptamer. With these terms, a riboswitch candidate can be created by concatenating the aptamer, a spacer, a complementary sequence, and the U stretch. Using the RNAfold program [64], these candidates are evaluated by the folding simulations to select the elite members. A theophylline-specific riboswitch that is generated by this approach can activate transcription on binding of the target ligand [61].

These computational methods can be applied to constructing synthetic riboswitches, but the evaluation of synthetic riboswitch candidates is based on thermodynamic calculations in both of the approaches. Besides the above approaches, other methods that use different strategies are developed to design RNA devices recently [99–101,104]. It is known that co-transcriptional folding kinetics is fundamental to the action of functional RNAs in cell. However, it is not well incorporated into these methods, possibly because of the complexity of co-transcriptional folding. Although the transcription context may pose a serious challenge in the quest for designing RNA devices, these methods have made a great progress in the development of computational methods for designing RNA devices.

6. Conclusions

Since functions of biological molecules are determined by the formed structures, precise structure precondition is crucial to a complete understanding of many RNA-mediated processes, such as the regulation activities of riboswitches. Two domains of riboswitches, namely, the aptamer domain and the expression platform, often share nucleotides. The aptamer domain usually needs to form a unique ligand-binding pocket for specifically binding its ligand, which in turn, locks the conformation of the aptamer domain and directs the folding of the downstream expression platform. However, if the aptamer fails to bind its ligand, part of the nucleotides within the aptamer domain will pair with the nucleotides within the expression platform to form an alternative structure. For the riboswitches with one single domain, conformational change induced by the ligand binding would result in the formation of the structure that is different from the unbound functional state.

As structural changes within the aptamer domain produce a substantial change in the gene expression, it therefore demands to characterize structure and kinetics of aptamers for a thorough understanding of the riboswitch-mediated regulation. Given the fact that most aptamers have been extensively studied experimentally, yet they are not well characterized in some aspects. In the last several decades, many computational methods have been developed to investigate aptamers. One primary concern for aptamers is to precisely model the structure that is responsible for ligand recognizing and binding. Effective approaches, such as the motif library-based and fragment-based methods, have made significant progress in 3D structure prediction by integrating knowledge-based algorithm with experimental data or physics-based model. However, 3D modeling for aptamers and other large RNAs has not achieved a consistent atomic accuracy and the data may not enable statistical meta-analysis currently. Most of them are not suitable for predicting the 3D structure for highly complex RNAs. Furthermore, RNA chaperones may play a role in RNA folding [87,105,106], which are not considered in the model. High quality 3D structure prediction may require more experimental mapping data and significant human insight to build accurate models. Further advancements may hold great promise for making progress toward the goal.

Another important issue for aptamers comes from describing their great structural changes. Modern parallelization of MD simulation largely improves the computation efficiency, but it still cannot address the structural rearrangements of aptamers that occur on a seconds' timescale. The coarse-grained SOP model and other computational methods have been applied to characterizing the kinetics of several riboswitch aptamers with great success. These demonstrate that the coarse-grained structural model, which enhances the conformation sampling, can be used to study the large scale conformational fluctuations of RNAs.

Also based on the coarse-grained structural model, a number of computational methods, such as RNAkinetics [107], Kinefold [108], and COFOLD [47] are developed to simulate co-transcriptional folding pathways of mRNAs from the early 1980s when key experiments show that structure formation happens co-transcriptionally [109]. In order to investigate the riboswitch-mediated regulation mechanisms, ligand binding should be taken into account. By incorporating the effect of ligand binding, the BarMap package and helix-based computational method have been successfully used to predict the co-transcriptional folding for several riboswitches. The application of these approaches in riboswitches implies that the 2D structure model can capture enough details of their behaviors in living cells.

Since conformational changes in aptamers can be induced by ligands, some other theoretical approaches focus on the prediction about interactions between ligands and aptamers or other features [110–112]. In recent years, computational methods for designing riboswitches have also made excellent progress in synthetic biology. As discussed above, several notable limitations of these computational methods still exist. Continuing developments in computational and hybrid methods are expected to overcome these limitations.

Acknowledgments: This work was partly supported by the National Natural Science Foundation of China under Grants No. 31600592 (to Sha Gong), 31270761 and 11574234 (to Wenbing Zhang). It was supported in part from Hubei Key Laboratory of Economic Forest Germplasm Improvement and Resources Comprehensive Utilization, Hubei Collaborative Innovation Center for the Characteristic Resources Exploitation of Dabie Mountains [2017BX08].

Author Contributions: Sha Gong, Yanli Wang, Zhen Wang and Wenbing Zhang designed the paper. Sha Gong and Wenbing Zhang wrote the paper. All authors have contributed in the improving of the manuscript, and read and approved its final version.

Abbreviations

ncRNA	noncoding RNA
SOP	Self-Organized Polymer
MC	Monte Carlo
FMN	Flavin Mononucleotide
SAM	S-adenosylmethionine
TPP	Thiamine Pyrophosphate
NMR	Nuclear Magnetic Resonance
MD	Molecular Dynamics
RNAP	RNA polymerase
RMSD	Root Mean Square Deviation
FARNA	Fragment Assembly of RNA
RAGTOP	RNA-As-Graph-Topologies

References

1. Breaker, R.R. Riboswitches and the RNA world. *Cold Spring Harb. Perspect. Biol.* **2012**, *4*, 1–15. [CrossRef] [PubMed]
2. Winkler, W.C. Riboswitches and the role of noncoding RNAs in bacterial metabolic control. *Curr. Opin. Chem. Biol.* **2005**, *9*, 594–602. [CrossRef] [PubMed]
3. Cech, T.R.; Steitz, J.A. The noncoding RNA revolution—Trashing old rules to forge new ones. *Cell* **2014**, *157*, 77–94. [CrossRef] [PubMed]
4. Chen, J.; Gottesman, S. Riboswitch regulates RNA. *Science* **2014**, *345*, 876–877. [CrossRef] [PubMed]
5. Tucker, B.J.; Breaker, R.R. Riboswitches as versatile gene control elements. *Curr. Opin. Struct. Biol.* **2005**, *15*, 342–348. [CrossRef] [PubMed]
6. Baker, J.L.; Sudarsan, N.; Weinberg, Z.; Roth, A.; Stockbridge, R.B.; Breaker, R.R. Widespread Genetic Switches and Toxicity Resistance Proteins for Fluoride. *Science* **2012**, *335*, 233–235. [CrossRef] [PubMed]
7. Dambach, M.; Sandoval, M.; Updegrove, T.B.; Anantharaman, V.; Aravind, L.; Waters, L.S.; Storz, G. The Ubiquitous yybP-ykoY Riboswitch Is a Manganese-Responsive Regulatory Element. *Mol. Cell* **2015**, *57*, 1099–1109. [CrossRef] [PubMed]
8. Dann, C.E.; Wakeman, C.A.; Sieling, C.L.; Baker, S.C.; Irnov, I.; Winkler, W.C. Structure and Mechanism of a Metal-Sensing Regulatory RNA. *Cell* **2007**, *130*, 878–892. [CrossRef] [PubMed]
9. Price, I.R.; Gaballa, A.; Ding, F.; Helmann, J.D.; Ke, A. Mn^{2+}-Sensing Mechanisms of yybP-ykoY Orphan Riboswitches. *Mol. Cell* **2015**, *57*, 1110–1123. [CrossRef] [PubMed]
10. Cromie, M.J.; Shi, Y.; Latifi, T.; Groisman, E.A. An RNA Sensor for Intracellular Mg^{2+}. *Cell* **2006**, *125*, 71–84. [CrossRef] [PubMed]
11. Li, S.; Breaker, R.R. Identification of 15 candidate structured noncoding RNA motifs in fungi by comparative genomics. *BMC Genom.* **2017**, *18*, 785. [CrossRef] [PubMed]
12. Weinberg, Z.; Lünse, C.E.; Corbino, K.A.; Ames, T.D.; Nelson, J.W.; Roth, A.; Perkins, K.R.; Sherlock, M.E.; Breaker, R.R. Detection of 224 candidate structured RNAs by comparative analysis of specific subsets of intergenic regions. *Nucleic Acids Res.* **2017**, *45*, 10811–10823. [CrossRef] [PubMed]
13. Fuchs, R.T.; Grundy, F.J.; Henkin, T.M. S-adenosylmethionine directly inhibits binding of 30S ribosomal subunits to the SMK box translational riboswitch RNA. *Proc. Natl. Acad. Sci. USA* **2007**, *104*, 4876–4880. [CrossRef] [PubMed]
14. Lu, C.; Smith, A.M.; Ding, F.; Chowdhury, A.; Henkin, T.M.; Ke, A. Variable sequences outside the SAM-binding core critically influence the conformational dynamics of the SAM-III/SMK box riboswitch. *J. Mol. Biol.* **2011**, *409*, 786–799. [CrossRef] [PubMed]
15. Wilson, R.C.; Smith, A.M.; Fuchs, R.T.; Kleckner, I.R.; Henkin, T.M.; Foster, M.P. Tuning riboswitch regulation through conformational selection. *J. Mol. Biol.* **2011**, *405*, 926–938. [CrossRef] [PubMed]
16. DebRoy, S.; Gebbie, M.; Ramesh, A.; Goodson, J.R.; Cruz, M.R.; van Hoof, A.; Winkler, W.C.; Garsin, D.A. A riboswitch-containing sRNA controls gene expression by sequestration of a response regulator. *Science* **2014**, *345*, 937–940. [CrossRef] [PubMed]

17. Rinaldi, A.J.; Lund, P.E.; Blanco, M.R.; Walter, N.G. The Shine-Dalgarno sequence of riboswitch-regulated single mRNAs shows ligand-dependent accessibility bursts. *Nat. Commun.* **2016**, *7*, 8976. [CrossRef] [PubMed]

18. Barrick, J.E.; Breaker, R.R. The distributions, mechanisms, and structures of metabolite-binding riboswitches. *Genome Biol.* **2007**, *8*, R239. [CrossRef] [PubMed]

19. Breaker, R.R. Prospects for Riboswitch Discovery and Analysis. *Mol. Cell* **2011**, *43*, 867–879. [CrossRef] [PubMed]

20. Li, S.; Hwang, X.Y.; Stav, S.; Breaker, R.R. The yjdF riboswitch candidate regulates gene expression by binding diverse azaaromatic compounds. *RNA* **2016**, *22*, 530–541. [CrossRef] [PubMed]

21. Serganov, A.; Nudler, E. A Decade of Riboswitches. *Cell* **2013**, *152*, 17–24. [CrossRef] [PubMed]

22. Deigan, K.E.; FerrÉ-D'AmarÉ, A.R. Riboswitches: Discovery of Drugs That Target Bacterial Gene-Regulatory RNAs. *Acc. Chem. Res.* **2011**, *44*, 1329–1338. [CrossRef] [PubMed]

23. Liberman, J.A.; Wedekind, J.E. Riboswitch structure in the ligand-free state. *Wiley Interdiscip. Rev. RNA* **2012**, *3*, 369–384. [CrossRef] [PubMed]

24. Gilbert, S.D.; Rambo, R.P.; Van Tyne, D.; Batey, R.T. Structure of the SAM-II riboswitch bound to S-adenosylmethionine. *Nat. Struct. Mol. Biol.* **2008**, *15*, 177–182. [CrossRef] [PubMed]

25. Cheah, M.T.; Wachter, A.; Sudarsan, N.; Breaker, R.R. Control of alternative RNA splicing and gene expression by eukaryotic riboswitches. *Nature* **2007**, *447*, 497–500. [CrossRef] [PubMed]

26. Mandal, M. A Glycine-Dependent Riboswitch That Uses Cooperative Binding to Control Gene Expression. *Science* **2004**, *306*, 275–279. [CrossRef] [PubMed]

27. Diegelman-Parente, A.; Bevilacqua, P.C. A mechanistic framework for Co-transcriptional folding of the HDV genomic ribozyme in the presence of downstream sequence. *J. Mol. Biol.* **2002**, *324*, 1–16. [CrossRef]

28. Lemay, J.-F.; Desnoyers, G.; Blouin, S.; Heppell, B.; Bastet, L.; St-Pierre, P.; Massé, E.; Lafontaine, D.A. Comparative Study between Transcriptionally- and Translationally-Acting Adenine Riboswitches Reveals Key Differences in Riboswitch Regulatory Mechanisms. *PLoS Genet.* **2011**, *7*, e1001278. [CrossRef] [PubMed]

29. Popenda, M.; Szachniuk, M.; Antczak, M.; Purzycka, K.J.; Lukasiak, P.; Bartol, N.; Blazewicz, J.; Adamiak, R.W. Automated 3D structure composition for large RNAs. *Nucleic Acids Res.* **2012**, *40*, 1–12. [CrossRef] [PubMed]

30. Chen, B.; Zuo, X.; Wang, Y.-X.; Dayie, T.K. Multiple conformations of SAM-II riboswitch detected with SAXS and NMR spectroscopy. *Nucleic Acids Res.* **2012**, *40*, 3117–3130. [CrossRef] [PubMed]

31. Zuker, M. Mfold web server for nucleic acid folding and hybridization prediction. *Nucleic Acids Res.* **2003**, *31*, 3406–3415. [CrossRef] [PubMed]

32. Xu, X.; Zhao, P.; Chen, S.J. Vfold: A web server for RNA structure and folding thermodynamics prediction. *PLoS ONE* **2014**, *9*. [CrossRef] [PubMed]

33. Reuter, J.S.; Mathews, D.H. RNAstructure: Software for RNA secondary structure prediction and analysis. *BMC Bioinform.* **2010**, *11*, 129. [CrossRef] [PubMed]

34. Das, R.; Baker, D. Automated de novo prediction of native-like RNA tertiary structures. *Proc. Natl. Acad. Sci. USA* **2007**, *104*, 14664–14669. [CrossRef] [PubMed]

35. Das, R.; Karanicolas, J.; Baker, D. Atomic accuracy in predicting and designing noncanonical RNA structure. *Nat. Methods* **2010**, *7*, 291–294. [CrossRef] [PubMed]

36. Jonikas, M.A.; Radmer, R.J.; Altman, R.B. Knowledge-based instantiation of full atomic detail into coarse-grain RNA 3D structural models. *Bioinformatics* **2009**, *25*, 3259–3266. [CrossRef] [PubMed]

37. Parisien, M.; Major, F. The MC-Fold and MC-Sym pipeline infers RNA structure from sequence data. *Nature* **2008**, *452*, 51–55. [CrossRef] [PubMed]

38. Shi, Y.-Z.; Wang, F.-H.; Wu, Y.-Y.; Tan, Z.-J. A coarse-grained model with implicit salt for RNAs: Predicting 3D structure, stability and salt effect. *J. Chem. Phys.* **2014**, *141*, 105102. [CrossRef] [PubMed]

39. Purzycka, K.J.; Popenda, M.; Szachniuk, M.; Antczak, M.; Lukasiak, P.; Blazewicz, J.; Adamiak, R.W. Automated 3D RNA Structure Prediction Using the RNAComposer Method for Riboswitches. In *Methods in Enzymology*; Elsevier Inc.: Waltham, MA, USA, 2015; Volume 553, pp. 3–34, ISBN 0076-6879.

40. Ding, F.; Lavender, C.A.; Weeks, K.M.; Dokholyan, N.V. Three-dimensional RNA structure refinement by hydroxyl radical probing. *Nat. Methods* **2012**, *9*, 603–608. [CrossRef] [PubMed]

41. Higgs, P.G. RNA secondary structure: Physical and computational aspects. *Q. Rev. Biophys.* **2000**, *33*, 199–253. [CrossRef] [PubMed]

228

Aptamers in Biotechnology

42. Sharma, M.; Bulusu, G.; Mitra, A. MD simulations of ligand-bound and ligand-free aptamer: Molecular level insights into the binding and switching mechanism of the add A-riboswitch. *RNA* **2009**, *15*, 1673–1692. [CrossRef] [PubMed]

43. Lin, J.-C.; Thirumalai, D. Relative stability of helices determines the folding landscape of adenine riboswitch aptamers. *J. Am. Chem. Soc.* **2008**, *130*, 14080–14081. [CrossRef] [PubMed]

44. Hyeon, C.; Thirumalai, D. Mechanical Unfolding of RNA: From Hairpins to Structures with Internal Multiloops. *Biophys. J.* **2007**, *92*, 731–743. [CrossRef] [PubMed]

45. Lin, J.-C.; Hyeon, C.; Thirumalai, D. Sequence-dependent folding landscapes of adenine riboswitch aptamers. *Phys. Chem. Chem. Phys.* **2014**, *16*, 6376–6382. [CrossRef] [PubMed]

46. Lin, J.-C.C.; Thirumalai, D. Kinetics of allosteric transitions in *S*-adenosylmethionine riboswitch are accurately predicted from the folding landscape. *J. Am. Chem. Soc.* **2013**, *135*, 16641–16650. [CrossRef] [PubMed]

47. Proctor, J.R.; Meyer, I.M. CoFold: An RNA secondary structure prediction method that takes co-transcriptional folding into account. *Nucleic Acids Res.* **2013**, *41*, e102. [CrossRef] [PubMed]

48. Zhao, P.; Zhang, W.-B.; Chen, S.-J. Predicting Secondary Structural Folding Kinetics for Nucleic Acids. *Biophys. J.* **2010**, *98*, 1617–1625. [CrossRef] [PubMed]

49. Holtkamp, W.; Kokic, G.; Jäger, M.; Mittelstaet, J.; Komar, A.A.; Rodnina, M.V. Cotranslational protein folding on the ribosome monitored in real time. *Science* **2015**, *350*, 1104–1107. [CrossRef] [PubMed]

50. Frieda, K.L.; Block, S.M. Direct Observation of Cotranscriptional Folding in an Adenine Riboswitch. *Science* **2012**, *338*, 397–400. [CrossRef] [PubMed]

51. Wickiser, J.K.; Winkler, W.C.; Breaker, R.R.; Crothers, D.M. The Speed of RNA Transcription and Metabolite Binding Kinetics Operate an FMN Riboswitch. *Mol. Cell* **2005**, *18*, 49–60. [CrossRef] [PubMed]

52. Hennelly, S.P.; Novikova, I.V.; Sanbonmatsu, K.Y. The expression platform and the aptamer: Cooperativity between Mg^{2+} and ligand in the SAM-I riboswitch. *Nucleic Acids Res.* **2013**, *41*, 1922–1935. [CrossRef] [PubMed]

53. Lemay, J.-F.; Penedo, J.C.; Tremblay, R.; Lilley, D.M.J.; Lafontaine, D.A. Folding of the Adenine Riboswitch. *Chem. Biol.* **2006**, *13*, 857–868. [CrossRef] [PubMed]

54. Watters, K.E.; Strobel, E.J.; Yu, A.M.; Lis, J.T.; Lucks, J.B. Cotranscriptional folding of a riboswitch at nucleotide resolution. *Nat. Struct. Mol. Biol.* **2016**, *23*, 1124–1131. [CrossRef] [PubMed]

55. Lutz, B.; Faber, M.; Verma, A.; Klumpp, S.; Schug, A. Differences between cotranscriptional and free riboswitch folding. *Nucleic Acids Res.* **2014**, *42*, 2687–2696. [CrossRef] [PubMed]

56. Gong, S.; Wang, Y.; Wang, Z.; Wang, Y.; Zhang, W. Reversible-Switch Mechanism of the SAM-III Riboswitch. *J. Phys. Chem. B* **2016**, *120*, 12305–12311. [CrossRef] [PubMed]

57. Gong, S.; Wang, Y.; Zhang, W. Kinetic regulation mechanism of pbuE riboswitch. *J. Chem. Phys.* **2015**, *142*, 15103. [CrossRef] [PubMed]

58. Gong, S.; Wang, Y.; Zhang, W. The regulation mechanism of yitJ and metF riboswitches. *J. Chem. Phys.* **2015**, *143*, 45103. [CrossRef] [PubMed]

59. Hofacker, I.L.; Flamm, C.; Heine, C.; Wolfinger, M.T.; Scheuermann, G.; Stadler, P.F. BarMap: RNA folding on dynamic energy landscapes. *RNA* **2010**, *16*, 1308–1316. [CrossRef] [PubMed]

60. Zhao, P.; Zhang, W.; Chen, S.-J. Cotranscriptional folding kinetics of ribonucleic acid secondary structures. *J. Chem. Phys.* **2011**, *135*, 245101. [CrossRef] [PubMed]

61. Wachsmuth, M.; Findeiß, S.; Weissheimer, N.; Stadler, P.F.; Mörl, M. De novo design of a synthetic riboswitch that regulates transcription termination. *Nucleic Acids Res.* **2013**, *41*, 2541–2551. [CrossRef] [PubMed]

62. Espah Borujeni, A.; Mishler, D.M.; Wang, J.; Huso, W.; Salis, H.M. Automated physics-based design of synthetic riboswitches from diverse RNA aptamers. *Nucleic Acids Res.* **2016**, *44*, 1–13. [CrossRef] [PubMed]

63. Zhao, Y.; Huang, Y.; Gong, Z.; Wang, Y.; Man, J.; Xiao, Y. Automated and fast building of three-dimensional RNA structures. *Sci. Rep.* **2012**, *2*, 734. [CrossRef] [PubMed]

64. Hofacker, I.L. RNA Secondary Structure Analysis Using the Vienna RNA Package. In *Current Protocols in Bioinformatics*; John Wiley & Sons, Inc.: Hoboken, NJ, USA, 2009; pp. 1–16, ISBN 0471250953.

65. Puton, T.; Kozlowski, L.P.; Rother, K.M.; Bujnicki, J.M. CompaRNA: A server for continuous benchmarking of automated methods for RNA secondary structure prediction. *Nucleic Acids Res.* **2013**, *41*, 4307–4323. [CrossRef] [PubMed]

66. Bernhart, S.H.; Hofacker, I.L.; Will, S.; Gruber, A.R.; Stadler, P.F. RNAalifold: Improved consensus structure prediction for RNA alignments. *BMC Bioinform.* **2008**, *9*, 474. [CrossRef] [PubMed]

67. Sükösd, Z.; Knudsen, B.; Vaerum, M.; Kjems, J.; Andersen, E.S. Multithreaded comparative RNA secondary structure prediction using stochastic context-free grammars. *BMC Bioinform.* **2011**, *12*, 103. [CrossRef] [PubMed]

68. Popenda, M.; Szachniuk, M.; Blazewicz, M.; Wasik, S.; Burke, E.K.; Blazewicz, J.; Adamiak, R.W. RNA FRABASE 2.0: An advanced web-accessible database with the capacity to search the three-dimensional fragments within RNA structures. *BMC Bioinform.* **2010**, *11*, 231. [CrossRef] [PubMed]

69. Miranda-Rios, J.; Navarro, M.; Soberon, M. A conserved RNA structure (thi box) is involved in regulation of thiamin biosynthetic gene expression in bacteria. *Proc. Natl. Acad. Sci. USA* **2001**, *98*, 9736–9741. [CrossRef] [PubMed]

70. Zirbel, C.L.; Roll, J.; Sweeney, B.A.; Petrov, A.I.; Pirrung, M.; Leontis, N.B. Identifying novel sequence variants of RNA 3D motifs. *Nucleic Acids Res.* **2015**, *43*, 7504–7520. [CrossRef] [PubMed]

71. Rahrig, R.R.; Leontis, N.B.; Zirbel, C.L. R3D align: Global pairwise alignment of RNA 3D structures using local superpositions. *Bioinformatics* **2010**, *26*, 2689–2697. [CrossRef] [PubMed]

72. Kladwang, W.; Chou, F.C.; Das, R. Automated RNA structure prediction uncovers a kink-turn linker in double glycine riboswitches. *J. Am. Chem. Soc.* **2012**, *134*, 1404–1407. [CrossRef] [PubMed]

73. Cruz, J.A.; Westhof, E. Sequence-based identification of 3D structural modules in RNA with RMDetect. *Nat. Methods* **2011**, *8*, 513–519. [CrossRef] [PubMed]

74. Kim, N.; Zahran, M.; Schlick, T. Computational Prediction of Riboswitch Tertiary Structures Including Pseudoknots by RAGTOP. In *Methods in Enzymology*; Elsevier Inc.: Waltham, MA, USA, 2015; Volume 553, pp. 115–135, ISBN 1557-7988 (Electronic) 0076-6879.

75. Krokhotin, A.; Houlihan, K.; Dokholyan, N.V. iFoldRNA v2: Folding RNA with constraints. *Bioinformatics* **2015**, *31*, 2891–2893. [CrossRef] [PubMed]

76. Kim, N.; Laing, C.; Elmetwaly, S.; Jung, S.; Curuksu, J.; Schlick, T. Graph-based sampling for approximating global helical topologies of RNA. *Proc. Natl. Acad. Sci. USA* **2014**, *111*, 4079–4084. [CrossRef] [PubMed]

77. Krokhotin, A.; Dokholyan, N. V Computational methods toward accurate RNA structure prediction using coarse-grained and all-atom models. *Methods Enzymol.* **2015**, *553*, 65–89. [CrossRef] [PubMed]

78. Wang, Y.; Gong, S.; Wang, Z.; Zhang, W. The thermodynamics and kinetics of a nucleotide base pair. *J. Chem. Phys.* **2016**, *144*, 115101. [CrossRef] [PubMed]

79. Gusarov, I.; Nudler, E. Control of intrinsic transcription termination by N and NusA: The basic mechanisms. *Cell* **2001**, *107*, 437–449. [CrossRef]

80. Chen, J.; Gong, S.; Wang, Y.; Zhang, W. Kinetic partitioning mechanism of HDV ribozyme folding. *J. Chem. Phys.* **2014**, *140*, 25102. [CrossRef] [PubMed]

81. Chen, J.; Zhang, W. Kinetic analysis of the effects of target structure on siRNA efficiency. *J. Chem. Phys.* **2012**, *137*, 225102. [CrossRef] [PubMed]

82. Xia, T.; SantaLucia, J.; Burkard, M.E.; Kierzek, R.; Schroeder, S.J.; Jiao, X.; Cox, C.; Turner, D.H. Thermodynamic Parameters for an Expanded Nearest-Neighbor Model for Formation of RNA Duplexes with Watson−Crick Base Pairs. *Biochemistry* **1998**, *37*, 14719–14735. [CrossRef] [PubMed]

83. Mathews, D.H.; Sabina, J.; Zuker, M.; Turner, D.H. Expanded sequence dependence of thermodynamic parameters improves prediction of RNA secondary structure. *J. Mol. Biol.* **1999**, *288*, 911–940. [CrossRef] [PubMed]

84. Hyeon, C.; Morrison, G.; Thirumalai, D. Force-dependent hopping rates of RNA hairpins can be estimated from accurate measurement of the folding landscapes. *Proc. Natl. Acad. Sci. USA* **2008**, *105*, 9604–9609. [CrossRef] [PubMed]

85. Hyeon, C.; Dima, R.I.; Thirumalai, D. Pathways and Kinetic Barriers in Mechanical Unfolding and Refolding of RNA and Proteins. *Structure* **2006**, *14*, 1633–1645. [CrossRef] [PubMed]

86. Rieder, R.; Lang, K.; Graber, D.; Micura, R. Ligand-Induced Folding of the Adenosine Deaminase A-Riboswitch and Implications on Riboswitch Translational Control. *ChemBioChem* **2007**, *8*, 896–902. [CrossRef] [PubMed]

87. Schroeder, R.; Grossberger, R.; Pichler, A.; Waldsich, C. RNA folding in vivo. *Curr. Opin. Struct. Biol.* **2002**, *12*, 296–300. [CrossRef]

88. Badelt, S.; Hammer, S.; Flamm, C.; Hofacker, I.L. Thermodynamic and kinetic folding of riboswitches. *Methods Enzymol.* **2015**, *553*, 193–213. [CrossRef] [PubMed]

89. Wolfinger, M.T.; Svrcek-Seiler, W.A.; Flamm, C.; Hofacker, I.L.; Stadler, P.F. Efficient computation of RNA folding dynamics. *J. Phys. A. Math. Gen.* **2004**, *37*, 4731–4741. [CrossRef]

90. Flamm, C.; Hofacker, I.L.; Stadler, P.F.; Wolfinger, M.T. Barrier Trees of Degenerate Landscapes. *Z. Phys. Chem.* **2002**, *216*, 155. [CrossRef]

91. Geis, M.; Flamm, C.; Wolfinger, M.T.; Tanzer, A.; Hofacker, I.L.; Middendorf, M.; Mandl, C.; Stadler, P.F.; Thurner, C. Folding kinetics of large RNAs. *J. Mol. Biol.* **2008**, *379*, 160–173. [CrossRef] [PubMed]

92. Huang, W.; Kim, J.; Jha, S.; Aboul-ela, F. A mechanism for S-adenosyl methionine assisted formation of a riboswitch conformation: A small molecule with a strong arm. *Nucleic Acids Res.* **2009**, *37*, 6528–6539. [CrossRef] [PubMed]

93. Chauvier, A.; Picard-Jean, F.; Berger-Dancause, J.-C.; Bastet, L.; Naghdi, M.R.; Dubé, A.; Turcotte, P.; Perreault, J.; Lafontaine, D.A. Transcriptional pausing at the translation start site operates as a critical checkpoint for riboswitch regulation. *Nat. Commun.* **2017**, *8*, 13892. [CrossRef] [PubMed]

94. Perdrizet II, G.A.; Artsimovitch, I.; Furman, R.; Sosnick, T.R.; Pan, T. Transcriptional pausing coordinates folding of the aptamer domain and the expression platform of a riboswitch. *Proc. Natl. Acad. Sci. USA* **2012**, *109*, 3323–3328. [CrossRef] [PubMed]

95. Hollands, K.; Sevostiyanova, A.; Groisman, E.A. Unusually long-lived pause required for regulation of a Rho-dependent transcription terminator. *Proc. Natl. Acad. Sci. USA* **2014**, *111*, E1999–E2007. [CrossRef] [PubMed]

96. Wong, T.N.; Pan, T. RNA Folding During Transcription: Protocols and Studies. In *Methods in Enzymology*; Elsevier Inc.: San Diego, CA, USA, 2009; Volume 468, pp. 167–193, ISBN 1557-7988 (Electronic)r0076-6879.

97. Ketzer, P.; Kaufmann, J.K.; Engelhardt, S.; Bossow, S.; von Kalle, C.; Hartig, J.S.; Ungerechts, G.; Nettelbeck, D.M. Artificial riboswitches for gene expression and replication control of DNA and RNA viruses. *Proc. Natl. Acad. Sci. USA* **2014**, *111*, E554–E562. [CrossRef] [PubMed]

98. Zhang, F.; Carothers, J.M.; Keasling, J.D. Design of a dynamic sensor-regulator system for production of chemicals and fuels derived from fatty acids. *Nat. Biotechnol.* **2012**, *30*, 354–359. [CrossRef] [PubMed]

99. Carothers, J.M.; Goler, J.A.; Juminaga, D.; Keasling, J.D. Model-Driven Engineering of RNA Devices to Quantitatively Program Gene Expression. *Science* **2011**, *334*, 1716–1719. [CrossRef] [PubMed]

100. Townshend, B.; Kennedy, A.B.; Xiang, J.S.; Smolke, C.D. High-throughput cellular RNA device engineering. *Nat. Methods* **2015**, *12*, 989–994. [CrossRef] [PubMed]

101. Endoh, T.; Sugimoto, N. Rational Design and Tuning of Functional RNA Switch to Control an Allosteric Intermolecular Interaction. *Anal. Chem.* **2015**, *87*, 7628–7635. [CrossRef] [PubMed]

102. Wei, K.Y.; Smolke, C.D. Engineering dynamic cell cycle control with synthetic small molecule-responsive RNA devices. *J. Biol. Eng.* **2015**, *9*, 21. [CrossRef] [PubMed]

103. Espah Borujeni, A.; Channarasappa, A.S.; Salis, H.M. Translation rate is controlled by coupled trade-offs between site accessibility, selective RNA unfolding and sliding at upstream standby sites. *Nucleic Acids Res.* **2014**, *42*, 2646–2659. [CrossRef] [PubMed]

104. McKeague, M.; Wong, R.S.; Smolke, C.D. Opportunities in the design and application of RNA for gene expression control. *Nucleic Acids Res.* **2016**, *44*, 2987–2999. [CrossRef] [PubMed]

105. Zemora, G.; Waldsich, C. RNA folding in living cells. *RNA Biol.* **2010**, *7*, 634–641. [CrossRef] [PubMed]

106. Herschlag, D. RNA Chaperones and the RNA Folding Problem. *J. Biol. Chem.* **1995**, *270*, 20871–20874. [CrossRef] [PubMed]

107. Danilova, L.V.; Pervouchine, D.D.; Favorov, A.V.; Mironov, A.A. RNAKinetics: A web server that models secondary structure kinetics of an elongating RNA. *J. Bioinform. Comput. Biol.* **2006**, *4*, 589–596. [CrossRef] [PubMed]

108. Xayaphoummine, A.; Bucher, T.; Isambert, H. Kinefold web server for RNA/DNA folding path and structure prediction including pseudoknots and knots. *Nucleic Acids Res.* **2005**, *33*, 605–610. [CrossRef] [PubMed]

109. Kramer, F.R.; Mills, D.R. Secondary structure formation during RNA synthesis. *Nucleic Acids Res.* **1981**, *9*, 5109–5124. [CrossRef] [PubMed]

110. Gong, Z.; Zhao, Y.; Chen, C.; Xiao, Y. Role of Ligand Binding in Structural Organization of Add A-riboswitch Aptamer: A Molecular Dynamics Simulation. *J. Biomol. Struct. Dyn.* **2011**, *29*, 403–416. [CrossRef] [PubMed]

111. Philips, A.; Milanowska, K.; Lach, G.; Boniecki, M.; Rother, K.; Bujnicki, J.M. MetalionRNA: Computational predictor of metal-binding sites in RNA structures. *Bioinformatics* **2012**, *28*, 198–205. [CrossRef] [PubMed]
112. Philips, A.; Milanowska, K.; Ach, G.L.; Bujnicki, J.M. LigandRNA: Computational predictor of RNA—Ligand interactions. *RNA* **2013**, 1605–1616. [CrossRef] [PubMed]

Permissions

The contributors of this book come from diverse backgrounds, making this book a truly international effort. This book will bring forth new frontiers with its revolutionizing research information and detailed analysis of the nascent developments around the world.

We would like to thank all the contributing authors for lending their expertise to make the book truly unique. They have played a crucial role in the development of this book. Without their invaluable contributions this book wouldn't have been possible. They have made vital efforts to compile up to date information on the varied aspects of this subject to make this book a valuable addition to the collection of many professionals and students.

This book was conceptualized with the vision of imparting up-to-date information and advanced data in this field. To ensure the same, a matchless editorial board was set up. Every individual on the board went through rigorous rounds of assessment to prove their worth. After which they invested a large part of their time researching and compiling the most relevant data for our readers.

The editorial board has been involved in producing this book since its inception. They have spent rigorous hours researching and exploring the diverse topics which have resulted in the successful publishing of this book. They have passed on their knowledge of decades through this book. To expedite this challenging task, the publisher supported the team at every step. A small team of assistant editors was also appointed to further simplify the editing procedure and attain best results for the readers.

Apart from the editorial board, the designing team has also invested a significant amount of their time in understanding the subject and creating the most relevant covers. They scrutinized every image to scout for the most suitable representation of the subject and create an appropriate cover for the book.

The publishing team has been an ardent support to the editorial, designing and production team. Their endless efforts to recruit the best for this project, has resulted in the accomplishment of this book. They are a veteran in the field of academics and their pool of knowledge is as vast as their experience in printing. Their expertise and guidance has proved useful at every step. Their uncompromising quality standards have made this book an exceptional effort. Their encouragement from time to time has been an inspiration for everyone.

The publisher and the editorial board hope that this book will prove to be a valuable piece of knowledge for researchers, students, practitioners and scholars across the globe.

List of Contributors

Takahito Mukai
Department of Life Science, College of Science, Rikkyo University, 3-34-1 Nishi-Ikebukuro, Toshima-ku, Tokyo 171-8501, Japan

Andrey A. Buglak
A. N. Bach Institute of Biochemistry, Research Center of Biotechnology, Russian Academy of Sciences, Leninsky prospect 33, 119071 Moscow, Russia
Physical Faculty, St. Petersburg State University, 7/9 Universitetskaya naberezhnaya, 199034 St. Petersburg, Russia

Alexey V. Samokhvalov, Anatoly V. Zherdev and Boris B. Dzantiev
A. N. Bach Institute of Biochemistry, Research Center of Biotechnology, Russian Academy of Sciences, Leninsky prospect 33, 119071 Moscow, Russia

Lujun Hu and Linlin Wang
State Key Laboratory of Food Science and Technology, School of Food Science and Technology, Jiangnan University, Wuxi 214122, China

Wenwei Lu, Jianxin Zhao and Hao Zhang
State Key Laboratory of Food Science and Technology, School of Food Science and Technology, Jiangnan University, Wuxi 214122, China
International Joint Research Center for Probiotics & Gut Health, Jiangnan University, Wuxi 214122, China

Wei Chen
State Key Laboratory of Food Science and Technology, School of Food Science and Technology, Jiangnan University, Wuxi 214122, China
International Joint Research Center for Probiotics & Gut Health, Jiangnan University, Wuxi 214122, China
Beijing Innovation Centre of Food Nutrition and Human Health, Beijing Technology and Business University, Beijing 100048, China

Ulrich Hahn
Chemistry Department, Institute for Biochemistry and Molecular Biology, MIN-Faculty, Universität Hamburg, Martin-Luther-King-Platz 6, D-20146 Hamburg, Germany

Bruno Macedo and Yraima Cordeiro
Faculty of Pharmacy, Federal University of Rio de Janeiro (UFRJ), Av. Carlos Chagas Filho 373, Bloco B, Subsolo, Sala 17, Rio de Janeiro, RJ 21941-902, Brazil

Wesley O. Tucker, Andrew B. Kinghorn, Lewis A. Fraser, Yee-Wai Cheung and Julian A. Tanner
School of Biomedical Sciences, Li Ka Shing Faculty of Medicine, The University of Hong Kong, 21 Sassoon Road, Hong Kong, China

Maria A. Vorobyeva, Anna S. Davydova and Alya G. Venyaminova
Institute of Chemical Biology and Fundamental Medicine, Siberian Division of Russian Academy of Sciences, Lavrentiev Ave., 8, 630090 Novosibirsk, Russia

Pavel E. Vorobjev and Dmitrii V. Pyshnyi
Institute of Chemical Biology and Fundamental Medicine, Siberian Division of Russian Academy of Sciences, Lavrentiev Ave., 8, 630090 Novosibirsk, Russia
Department of Natural Sciences, Novosibirsk State University, Pirogova St., 2, 630090 Novosibirsk, Russia

Pascal Röthlisberger and Marcel Hollenstein
Institut Pasteur, Department of Structural Biology and Chemistry, Laboratory for Bioorganic Chemistry of Nucleic Acids, CNRS UMR3523, 28, rue du Docteur Roux, 75724 Paris CEDEX 15, France

Cécile Gasse
Institute of Systems & Synthetic Biology, Xenome Team, 5 rue Henri Desbruères Genopole Campus 1, University of Evry, F-91030 Evry, France

Peggy Reich, Dieter Frense and Dieter Beckmann
Institut für Bioprozess- und Analysenmesstechnik e.V., 37308 Heilbad Heiligenstadt, Germany

Regina Stoltenburg
UFZ – Helmholtz Centre for Environmental Research, 06120 Halle, Germany

Beate Strehlitz
UFZ – Helmholtz Centre for Environmental Research, 04318 Leipzig, Germany

Emma M. Hays, Wei Duan and Sarah Shigdar
Centre for Molecular and Medical Research, School of Medicine, Deakin University, 75 Pigdons Road, Waurn Ponds, Victoria 3216, Australia

Maureen McKeague
Department of Health Sciences and Technology, ETH Zürich, Schmelzbergstrasse 9, 8092 Zurich, Switzerland

Sha Gong
Hubei Key Laboratory of Economic Forest Germplasm Improvement and Resources Comprehensive Utilization, Hubei Collaborative Innovation Center for the Characteristic Resources Exploitation of Dabie Mountains, Huanggang Normal University, Huanggang 438000, China

Yanli Wang, Zhen Wang and Wenbing Zhang
Department of Physics, Wuhan University, Wuhan 430072, China

Index

Printed in the USA
CPSIA information can be obtained
at www.ICGtesting.com
JSHW051402091023
49903JS00006B/241